Geophysical Monograph Series

Geophysical Monograph Series

Geophysical Monograph 211

Subduction Dynamics

From Mantle Flow to Mega Disasters

Gabriele Morra
David A. Yuen
Scott D. King
Sang-Mook Lee
Seth Stein

Editors

This Work is a co-publication between the American Geophysical Union and John Wiley and Sons, Inc.

WILEY

Published under the aegis of the AGU Publications Committee

Brooks Hanson, Director of Publications

© 2016 by the American Geophysical Union, 2000 Florida Avenue, N.W., Washington, D.C. 20009
For details about the American Geophysical Union, see www.agu.org.

Published by John Wiley & Sons, Inc., Hoboken, New Jersey
Published simultaneously in Canada

For general information on our other products and services or for technical support, please contact our Customer Care Department within the United States at (800) 762-2974, outside the United States at (317) 572-3993 or fax (317) 572-4002.

Wiley also publishes its books in a variety of electronic formats. Some content that appears in print may not be available in electronic formats. For more information about Wiley products, visit our web site at www.wiley.com.

Library of Congress Cataloging-in-Publication Data is available.

ISBN: 978-1-118-88885-8

Cover images: Part of the famous map of Marie Tharp oceanic floor (global map), showing continental features, underwater ranges, trenches and other plate tectonic subduction zone features; Models of subduction zone topped by four atmospheric layers, ionosphere 1, 2, 3 and 4 placed one above the other, as they represent the atmosphere response at different heights.

Printed in the United States of America

10 9 8 7 6 5 4 3 2 1

CONTENTS

CONTRIBUTORS

Lingmin Cao
South China Sea Institute of Oceanology
Chinese Academy of Sciences
Guangzhou, China

Chuanxu Chen
Sanya Institute of Deep-sea Science and Engineering
Chinese Academy of Sciences
Sanya, China

Robert J. Geller
Department of Earth and Planetary Science
Graduate School of Science
University of Tokyo
Hongo, Bunkyo-ku, Tokyo, Japan

M. A. Jadamec
Department of Earth, Environmental, and
Planetary Sciences
Brown University
Providence, Rhode Island, USA
and
Department of Earth and Atmospheric Sciences
University of Houston
Houston, Texas, USA

Scott D. King
Department of Geosciences
Virginia Tech University
Blacksburg, Virginia, USA

Changyeol Lee
Faculty of Earth and Environmental Sciences
Chonnam National University
Gwangju, Republic of Korea

Sang-Mook Lee
School of Earth and Environmental Sciences
Seoul National University
Gwanak-Gu, South Korea

Juan Li
Key Laboratory of Earth and Planetary Physics
Institute of Geology and Geophysics
Chinese Academy of Sciences
Beijing, China

Petra Maierová
Center for Lithospheric Research
Czech Geological Survey
Prague, Czech Republic

Gabriele Morra
Department of Physics and
School of Geosciences
University of Louisiana at Lafayette
Lafayette, Louisiana, USA

Francesco Mulargia
Department of Physics and Astronomy
University of Bologna
Bologna, Italy

Giovanni Occhipinti
Institut de Physique du Globe de Paris
Sorbonne Paris Cité
Université Paris Diderot, France

Chang Whan Oh
Department of Earth and Environmental Sciences
and Earth and Environmental Science System
Research Center
Chonbuk National University
Jeonju, South Korea

Jin Qian
Key Laboratory of Marine Geology and Environment
Institute of Oceanology
Chinese Academy of Sciences
Qingdo, China

In-Chang Ryu
Department of Geology
Kyungpook National University
Daegu, Republic of Korea

Byung-Dal So
Research Institute of Natural Sciences
Chungnam National University
Daejeon, South Korea

Philip B. Stark
Department of Statistics
University of California, Berkeley
Berkeley, California, USA

Seth Stein
Department of Earth and Planetary Sciences
Northwestern University
Evanston, Illinois, USA

Nicola Tosi
Department of Astronomy and Astrophysics
Technische Universität Berlin
Berlin, Germany
and
Department of Planetary Physics
German Aerospace Center (DLR)
Berlin, Germany

Shiguo Wu
Sanya Institute of Deep-Sea Science and Engineering,
Chinese Academy of Sciences
Sanya, China

David A. Yuen
School of Environment Studies,
China University of Geosciences
Wuhan, China
and
Minnesota Supercomputing Institute
and Department of Earth Sciences
University of Minnesota
Minneapolis, Minnesota, USA

Changlei Zhao
Research Institute of Petroleum Explorations
and Development
China National Petroleum Corporation
Hangzhou, China

Siqi Zhang
Geophysics and Geodynamics Group
Macquarie University
Sydney, Australia

INTRODUCTION

The Impact of Subduction Dynamics on Mantle Flow, Continental Tectonics, and Seismic Hazard

Gabriele Morra[1], David A. Yuen[2], Scott D. King[3], Sang-Mook Lee[4], and Seth Stein[5]

Subduction of tectonic plates provides the main driving force for mantle convection, but also causes the greatest natural hazards such as megathrust earthquakes, often with accompanying tsunami waves. This book aims at expanding the traditional view in which subduction dynamics is usually presented, showing the wide range of natural phenomena related to the subduction process. Chapters in this book treat diverse topics ranging from the response of the ionosphere to earthquake and tsunamis, to the origin of midcontinental volcanism thousands kilometers distant from the subduction zone; from the puzzling deep earthquakes triggered in the interior of the descending slabs, to the detailed pattern of accretionary wedges in convergent zones; from the induced mantle flow in the deep mantle, to the nature of the paradigms of earthquake occurrence, showing how they relate to the subduction process (Figure I.1).

Megathrust earthquakes are the greatest destructive earthquakes and are caused by the sudden 10–20 meter slip of hundreds of kilometers along convergent plate boundaries. Until the present, the seismic hazard relating convergent margins, where plates subduct, has been based mostly on the limited historical data derived from earthquake catalogues [*Stein et al.,* 2012]. Unfortunately, as shown by the unexpected enormous damages caused by Sumatra and Tohoku earthquakes, catalogues have proved to be too limited in scope to predict the maximum magnitude of an earthquake, or to predict the size of the attendant tsunami (see Chapter 10 by Robert Geller and colleagues). While past attempts to relate the size of the largest earthquakes to the dynamics of the slab (dip, convergence, age, presence of heterogeneities) have failed due to their excessive simplicity, the combination of new data and improved numerical models based on thermal-mechanical principles [e.g., *van Dinther,* 2013] promises a new generation of time-dependent hazard maps that have more physics builtin.

This book is the result of a meeting that was held August 19 to 22, 2012, on Jeju Island, South Korea, with about 50 researchers from East Asia, North America, and Europe [*Morra et al.,* 2013]. On the first day of the meeting, most presentations focused on the causes of the Sumatra (2004) and Tohoku (2011) earthquakes. The opening talk was a review of the ionospheric response to subduction-induced thrust earthquakes and the associated tsunamis, which has been extended into Chapter 9 by Giovanni Occhipinti. From the presentations that followed, it emerged that the detailed analysis of the tsunami height along the coast of Tohoku is incompatible with the simple model based on the rebound of an overriding plate. Alternative possibilities were proposed from which two hypotheses emerged, one involving the release of gravitational potential energy at the accretionary wedge [*McKenzie and Jackson,* 2012], and the other the slip of a spline fault in crustal wedge, providing boost in elastic energy in driving the tsunami. While the discussion continues within the scientific community [e.g., *Tappin et al.,* 2014], this controversy shows how the seismic hazard estimation at subduction zones

[1]*Department of Physics and School of Geosciences, University of Louisiana at Lafayette, Louisiana, USA*

[2]*Minnesota Supercomputing Institute and Department of Earth Sciences, University of Minnesota, Minneapolis, Minnesota, USA and School of Environmental Studies, China University of Geosciences, Wuhan, China*

[3]*Department of Geosciences, Virginia Tech University, Blacksburg, Virginia, USA*

[4]*School of Earth and Environmental Sciences, Seoul National University, Gwanak-Gu, South Korea*

[5]*Department of Earth and Planetary Sciences, Northwestern University, Evanston, Illinois, USA*

Subduction Dynamics: From Mantle Flow to Mega Disasters, Geophysical Monograph 211, First Edition.
Edited by Gabriele Morra, David A. Yuen, Scott D. King, Sang-Mook Lee, and Seth Stein.
© 2016 American Geophysical Union. Published 2016 by John Wiley & Sons, Inc.

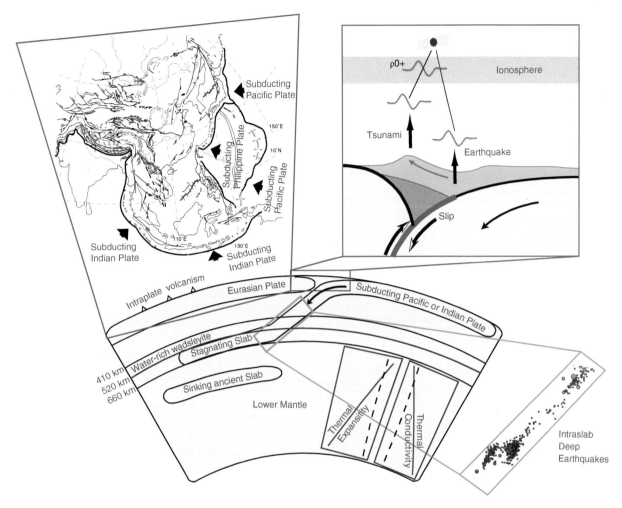

Figure I.1 Sketch of some of the dynamics investigated in this book. The history of the effects of the subduction process are largely observed in the tectonic structures of the overriding plate, as shown by the chapters focusing on the Eurasian continent (see Figure I.2). The most dramatic events caused by Subducting plates are the large thrust earthquakes (see Chapter 10 by *Geller et al.*). Megathrusts and tsunamis also cause waves in the iono-sphere, as illustrated in the panel on the top right. Rayleigh waves propagate due to both the quake and to the tsunami (see Chapter 9 by Occhipinti). The down-going slab induces a complex mantle flow, as shown by M. A. Jadamec in Chapter 7. Inside the slab, earthquakes are detected down to the transition between upper and lower mantle, but their mechanism is still debated (see So and Yuen in Chapter 8, e.g. about the hypocenter distribution within the Tonga slab on the bottom right panel). Thermal expansivity and conductivity gradients estimates are shown in the bottom right panels. Their role in mantle dynamics is analized in *Tosi et al.*, Chapter 6.

should be rethought to take into account the bimodal dynamics of the crustal wedge.

In order to improve our ability to estimate the seismic hazard due to the motion of the solid Earth, we must understand the details of subduction dynamics. The present paradigms emerge mainly from reconstructing and numeri-cally modeling plate tectonics in the past. Speed of comput-ing increased from gigaflops in 1990 to petaflops today, a factor of a million. Thus, investigation of the origin of plate tectonics in the next 20 years is expected to rely heavily on new numerical algorithms and the ability to run numeri-cal models on the fastest and largest supercomputers.

The pull of the slab in the upper mantle is the main driver of the motion of the plates on the surface [*King*, 2001]. The detection of plate kinematics through GPS data and their reconstruction based on the magnetic isochrones in the seafloor constrain to 10–20 Myrs the time necessary for a plate to sink through the upper mantle. At longer timescales, the dynamics behind Earth evolution are less understood [*Torsvik et al.*, 2014]. From hundreds of Myrs to Gyrs, the Wilson cycle of the forma-tion and breakup of supercontinents, global reorgani-zation of plate motion [*King et al.*, 2002], and cycles controlling plate size and morphology [*Morra et al.*, 2013]

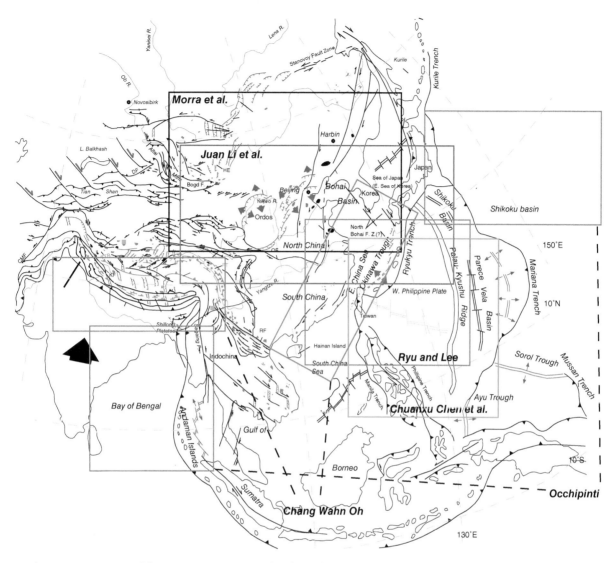

Figure I.2 Summary of the Asian regions covered in this volume overlaying a overview of the Asian tectonics. The map is modified after [*Yin*, 2010]. Chapter 2 from Chang Wahn Oh and Chapter 9 by Giovanni Occhipinti cover two separate regions.

are only some of the global geodynamic processes that control the global Earth evolution [*Funiciello et al.*, 2008]. Five chapters of this volume focus on this long-term tectonic evolution, all of them related to the subduction process at the eastern and southern Eurasian margins. (Detailed map in Figure I.2.)

In Chapter 4, Chuanxu Chen and coworkers analyze detailed data on accretionary wedges essential to understanding the development of thrust earthquakes. Focusing on the Manila trench, they find that the segmentation of the incoming plate correlates well with the distribution of seismicity. They suggest that the seafloor morphology might be carved by the heterogeneities created during the rifting process and control seismicity at the precollision zone.

An investigation of the causes of the very deep earthquakes within the subducting slab is presented in Chapter 8 by Byung-Dal So and David A. Yuen. Here the authors look for a relationship between the deviatoric stress regime and the distribution of deep earthquakes in the subducted lithosphere. The authors suggest that a new mechanism based on shear heating around the 410-km phase transition and consequent reduction in elastic modulus might account for part of this seismicity.

Many talks at the Jeju meeting focused on Asian tectonics and this is reflected in this volume. In Chapter 2, Chang Whan Oh reviews the geological evidence that the Dabie-Hongseong collision belt in eastern Asia might have been caused by a westward-migration slab break-off migration from 245 Ma to 240 Ma. Using the same approach, he relates the Himalayan collision to an eastward migrating slab break-off from ~50 Ma to ~25 Ma. In Chapter 3, In-Chang Ryu and Changyeol Lee approach another mystery of Asian tectonics: the origin of the adakitic arc magmatism, which extends from South China to eastern Japan, and appeared during the Cretaceous, 140–60 Ma. They notice that the traditional hypothesis that links the volcanism to a migration of the Pacific-Izanagi ridge is incompatible with most recent plate reconstructions, and propose instead the opposite migration of an intracontinental mantle plume.

Focusing on the evolution of Asia during the Cenozoic, Chapter 5 by Gabriele Morra and coworkers investigates the causes of the volcanism in central and eastern Asia, proposing that it emerges from a volatile-rich layer at the top of the transition zone, fed by the underlying stagnating Pacific slab. They propose that the volatiles slowly rise from the subducted crust and form fluid-rich microregions at the top of the transition zone from which they rise as clusters of diapirs.

Juan Li and coworkers, in Chapter 1, constrain the thickness of the Pacific stagnating slab beneath northeast Asia. Through the method of caustic waveform, they model both P and S waves and estimate an extra thickening of 50–60 km, supporting the hypothesis that the subducted slab may be piling up under northeast Asia, therefore partially explaining the missing slab from Pacific subduction.

Finally two chapters investigate subduction-induced mantle flow. The role of thickening of the subducted slab is addressed in Chapter 6 by Nicola Tosi and coworkers who show that the variations of thermal expansivity and conductivity with temperature and pressure induce a severe change of the strength in the downgoing slab. In particular, they find that the decrease of expansivity with depth enhances buckling and slab thickening in the lower mantle.

The response of the upper-mantle flow to subduction is addressed by M. A. Jadamec who shows in Chapter 7 how a rheological formulation closer to laboratory data, implementing a fully nonlinear mantle rheology, impacts significantly on the mantle response. She shows that the dynamical support of the mantle to the downgoing slab is reduced by the nonlinearity of the mantle-rocks creep, impacting the slab dip, the stress distribution in the slabs, and the stress state of the surface plates.

REFERENCES

Funiciello, F., C. Faccenna, A. Heuret, S. Lallemand, E. Di Giuseppe, and T. W. Becker (2008), Trench migration, net rotation and slab-mantle coupling, *Earth Planet. Sci. Lett.*, *271*(1–4), 233–240.

King, S. D. (2001), Subduction zones: Observations and geodynamic models, *Phys. Earth Planet. Inter.*, *127*(1–4), 9–24.

King, S. D., J. P. Lowman, and C. W. Gable (2002), Episodic tectonic plate reorganizations driven by mantle convection, *Earth Planet. Sci. Lett.*, *203*, 83–91.

McKenzie, D., and J. Jackson (2012), Tsunami earthquake generation by the release of gravitational potential energy, *Earth Planet. Sci. Lett.*, *345–348*(0), 1–8.

Morra, G., M. Seton, L. Quevedo, and R. D. Müller (2013a), Organization of the tectonic plates in the last 200 Myr, *Earth Planet. Sci. Lett.*, *373*, 93–101.

Morra, G., R. J. Geller, S. T. Grilli, S. Karato, S. King, S.-M. Lee, P. Tackley, and D. Yuen (2013b), Growing understanding of subduction dynamics indicates need to rethink seismic hazards, *Eos, Transactions American Geophysical Union*, *94*(13), 125–126 .

Stein, S., R. J. Geller, and M. Liu (2012), Why earthquake hazard maps often fail and what to do about it, *Tectonophysics*, *562*, 1–25.

Tappin, D. R., S. T. Grilli, J. C. Harris, R. J. Geller, T. Masterlark, J. T. Kirby, F. Shi, G. Ma, K. K. S. Thingbaijam, and P. M. Mai (2014), Did a submarine landslide contribute to the 2011 Tohoku tsunami? *Marine Geology*, *357*, 344–361.

Torsvik, T. H., R. van der Voo, P. V. Doubrovine, K. Burke, B. Steinberger, L. D. Ashwal, and A. L. Bull (2014), Deep mantle structure as a reference frame for movements in and on the Earth, *Proceedings of the National Academy of Sciences*, *111*(24), 8735–8740.

Van Dinther, Y., T. V. Gerya, L. A. Dalguer, P. M. Mai, G. Morra, and D. Giardini, (2013), The seismic cycle at subduction thrusts: Insights from seismo-thermo-mechanical models, *J. Geophys. Res.: Sol Ea*, *118*, 6183–6202, doi:10.1002/2013JB010380.

Yin, A. (2010), Cenozoic tectonic evolution of Asia: A preliminary synthesis, *Tectonophysics*, *488*, 293–325, doi:10.1016/j.tecto.2009.06.002.

1

Evidence from Caustic Waveform Modeling for Long Slab Thickening above the 660-km Discontinuity under Northeast Asia: Dynamic Implications

Juan Li[1], Nicola Tosi[2], Petra Maierová[3], and David A. Yuen[4]

ABSTRACT

Knowledge of the thickness and age of slabs is of great importance to further our understanding of the deformation of the subducting lithosphere and, in general, of regional mantle rheology. The technique of seismic tomography has been widely used to identify thermally induced mantle heterogeneities. However, because of the contamination of complicated caustic waveforms, its resolving power is relatively poor when it comes to image slabs in the mantle transition zone. In this study, we aim at constraining the thickness of the slab in the mantle transition zone beneath northeast Asia through the method of caustic waveform modeling for both P and S waves. We detect a high-velocity layer associated with a thickened slab lying in the transition zone and estimate its thickness to be ~140±20 km, corresponding to a thickening of 50–60 km at the long-lying segment. Numerical simulations indicate that downgoing material may be piling up under northeast Asia. The part of the Pacific plate subducting at the Japan trench thus can be interpreted to be in a buckling/folding retreating mode. We argue that thickening of the slab due to buckling instabilities could partly explain the enigma of the missing Pacific slab from the Cenozoic era. In addition, the slab might be close to the threshold of instability. Based on numerical simulations, mantle circulation may become temporarily layered. Cold and dense lithospheric material, after accumulating at the 660-km discontinuity, can suddenly sink into the lower mantle, resulting in an avalanche that, perhaps, might occur in the next tens of millions of years. This will be accompanied by a resurgence of volcanic activity, precipitated by upwellings emanating from the lower mantle close to the site of the avalanche.

[1]Key Laboratory of Earth and Planetary Physics, Institute of Geology and Geophysics, Chinese Academy of Sciences, Beijing, China

[2]Department of Astronomy and Astrophysics, Technische Universität Berlin, Berlin, Germany; Department of Planetary Physics, German Aerospace Center (DLR), Berlin, Germany

[3]Center for Lithospheric Research, Czech Geological Survey, Prague, Czech Republic

[4]School of Environment Studies, China University of Geosciences, Wuhan, China; Minnesota Supercomputing Institute and Department of Earth Sciences, University of Minnesota, Minneapolis, Minnesota, USA

1.1. INTRODUCTION

Subducted oceanic slabs represent the major source of buoyancy driving mantle flow [e.g., *Ricard et al.*, 1993]. Characterizing the process of subduction is crucial to better understand the thermal and chemical evolution of our planet. In the past decades, seismic-wave tomography, based mainly on travel-time analysis, has proved able to resolve subduction-related thermal anomalies [e.g., *van der Hilst*, 1991; *Fukao et al.*, 2001; *Huang and Zhao*, 2006; *Li et al.*, 2008; *Fukao and Obayashi*, 2013]. However, intrinsic problems of damping-associated inversion techniques used in tomography as well as the poor resolution

Subduction Dynamics: From Mantle Flow to Mega Disasters, Geophysical Monograph 211, First Edition.
Edited by Gabriele Morra, David A. Yuen, Scott D. King, Sang-Mook Lee, and Seth Stein.
© 2016 American Geophysical Union. Published 2016 by John Wiley & Sons, Inc.

at depth inhibit the possibility to place robust constraints of the detailed structure of subducting slabs. A recent study devoted to quantifying the uncertainty in travel-time tomography reveals that the root mean square velocity perturbations for many acceptable models compatible with the same travel-time dataset can vary from 0.3% to 1.3% in the upper mantle [*de Wit et al.*, 2012]. Furthermore, in the transition zone, the limited ray coverage of the first arrivals caused by waveform triplication results in an even more ambiguous image.

Constraining the thickness of subducted slabs is of great importance in understanding problems related to mantle rheology and evolution of the lithosphere-mantle system. Slabs can undergo large deformation: they can bend, stretch, and thicken [*Gurnis and Hager*, 1988; *King*, 2001]. In the upper- to midlower mantle, thick blobs of seismically fast anomalies have been identified by different authors in subduction regions beneath Tonga, Marianas, and Kuril [e.g., *Creager and Jordan*, 1986; *van der Hilst*, 1995; *Fukao and Obayashi*, 2013]. Scaling laws derived from the theory of buckling of viscous sheets have been successfully employed to explain the apparent thickening of the subducted lithosphere beneath the transition zone as revealed by a few seismic studies [*Ribe et al.*, 2007]. Fully numerical simulations of viscous flow with composite rheology have been performed to further explore the possible mechanisms leading to the emergence of long-wavelength thickening of slabs in the lower mantle [*Běhounková and Čížková*, 2008]. Furthermore, 3D numerical models have shown that a relatively weak slab, with a viscosity not more than 100 times larger than that of the ambient mantle, can achieve a broad variety of shapes [*Loiselet et al.*, 2010]. In spite of this focus on slab behavior in the middle to lower mantle, different styles of slab deformation have been identified in the upper mantle transition zone. The thickness of the subducted lithosphere in the upper part of the Bullen region [*Bullen*, 1963], which was proposed to be the transition region between the upper and lower mantle, however, is not well revolved. A slab with an initial thickness of ~80–90 km [*Zhao et al.*, 1994] has been imaged beneath the Japan trench, behind which a vast amount of material lies subhorizontally over the 660-km discontinuity [e.g., *Huang and Zhao*, 2006; *Li and van der Hilst*, 2010, *Fukao and Obayashi*, 2013] (hereafter referred to as the 660). The resolved high-velocity layer in the stagnant slab, however, seems to occupy the whole mantle transition zone with an uncertainty of around ±90 km or even more.

In this paper, we discuss our finding of a thickened slab in the upper mantle transition zone beneath northeast Asia and the possible mechanism of its formation, and we speculate on the dynamic implication associated with its fate. The constraint on the thickness of the slab is based on P and S wave models of *Li et al.* [2013] obtained from the method of caustic waveform modeling. We first show the sensitivity of the method to the structure around the 660-km discontinuity, and then briefly introduce the data sources and main features of the resolved P and S velocity models. New waveform datasets are added to verify the correctness of the obtained model, and a series of uncertainty tests based on waveform matching are performed, which provide a robust constraint on the thickness of the stagnant segment of the slab. 2D finite-volume numerical simulations were also performed in order to understand the dynamic implications of a thickened stagnant slab in the transition zone. We will argue that thickening of the slab due to buckling instability may partly explain the enigma of the missing Pacific slab. The slab may be close to be unstable, and the continuous piling up of lithospheric material may trigger an event of mantle avalanche beneath northeast Asia.

1.2. CAUSTIC WAVEFORM MODELING AND DATA SOURCES

Caustic waveforms are caused by seismic triplication [*Aki and Richards*, 2009], occurring in the presence of a first-order discontinuity with positive velocity jump or a sharp increase in velocity gradient (Figure 1.1). Because the pathways of triplicate phases lie very close to each other in the lithosphere and crust, differences in relative time and amplitude between different seismic branches, and especially the location of the caustic points (Figure 1.1b), turn out to be quite sensitive to the velocity structure around the discontinuity. For example, the location of cusp-B is sensitive to the velocity gradient in the lower-mantle transition zone right above the 660; the location of cusp-C is sensitive to the velocity below the 660; and the travel-time difference between AB and CD branches is sensitive to velocity jump across the 660.

The triplication method has been applied to constrain the upper mantle discontinuities and core-mantle boundary (CMB) structure [e.g., *Lay and Helmberger*, 1983; *Grand and Helmberger*, 1984; *Wang and Yao*, 1991; *Tajima and Grand*, 1995; *Tajima and Grand*, 1998; *Brudzinski and Chen*, 2000; *Brudzinski and Chen*, 2003; *Song et al.*, 2004]. However, due to the sparse distribution of seismic instrumentations, previous studies were largely based on individual or limited seismogram analysis within a large aperture seismological array with which it is difficult to resolve the trade-off between the depth and sharpness of the discontinuity and velocity variation [e.g., *Tajima and Grand*, 1998; *Wang et al.*, 2006]. To illustrate how a velocity variation around the 660 can influence triplication waveforms, we show synthetic transversal displacements (Figure 1.2a) calculated from three different velocity

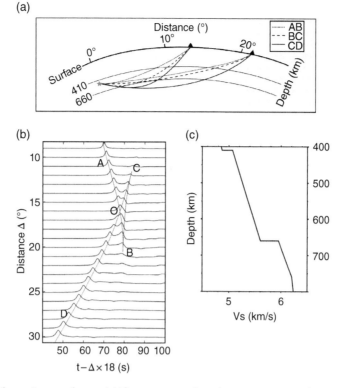

Figure 1.1 Diagram of caustic waveforms. (a) Three consecutive phases are expected to arrive at the receiver side. These are the direct AB phase, a wave diving above the discontinuity; the reflected BC phase, a wave reflected from the discontinuity; and the CD phase diving below the discontinuity (b) Synthetic transversal displacement for an assumed epicentral depth of 520 km. The locations of caustic points are very sensitive to the velocity structure around the discontinuity. (c) Iasp91 model used for the synthetic waveform calculation.

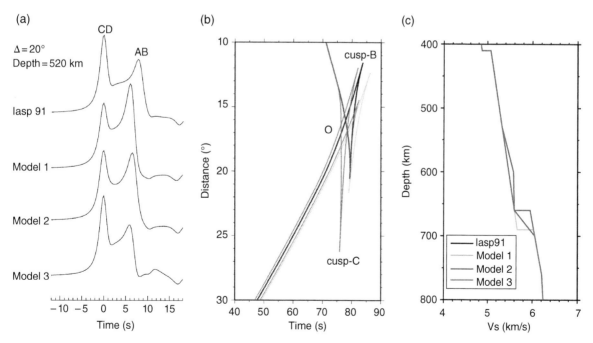

Figure 1.2 (a) Synthetic seismograms at 20° epicentral distance (Δ) for transversal component S wave calculated for iasp91 model and three different models (c). The depth of the event is 520 km. The trade-off between the depth of the interface and velocity variation around the 660-km discontinuity is not well resolved due to the similarity of the waveforms in a single seismogram. (b) Predicted caustic-wave travel-time after aligning seismograms with the epicentral distance. Significant discrepancy appeared in the major features of caustic points, which can be used to discriminate between different models. (c) Different models used for calculation of (a) and (b). Note that the reference model iasp91 (black line) is overlapped with the other three models.

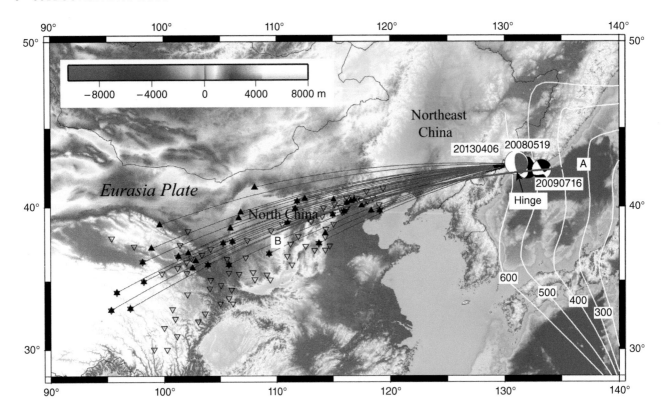

Figure 1.3 Map showing the location of deep earthquakes and regional seismic stations. The beach-ball symbols mark deep earthquake events used for caustic waveform modeling in our paper and previous study [*Li et al.*, 2013], with the red ball indicating the new event. Only events with both P and SH waveforms clearly identified are shown here. The filled triangles represent Chinese regional seismic network stations used for the new event, and the inverted blank triangles represent stations used in our previous study. Brown lines bounded by two sections of black lines highlight portions of the ray traveling below the 660 for event 20130406. The thick blue line AB marks the cross section in Figure 1.8, with "Hinge" indicating the location where the subducting slab encounters the 660 [*Li et al.*, 2008].

models (Figure 1.2c), which are proposed to characterize deep structure in subduction regions [e.g., *Tajima and Grand*, 1998; *Wang et al.*, 2006; *Wang and Niu*, 2010; *Ye et al.*, 2011]. In Model 1, the 660-km discontinuity is depressed to 690 km; in Model 2, the 660-km discontinuity is a layer with a thickness of ~50 km instead of a sharp velocity jump; in Model 3, there is a gentle velocity gradient (dV/dh = 0.0036 s⁻¹) across a layer with a thickness of ~60 km located just above the 660. Over a certain distance, the amplitude of AB and CD phases varies between the models. However, the differences arising from a single seismogram are not clear enough for discriminating the right model (Figure 1.2a). On the other hand, the location of cusp-B, cusp-C, and crossover point O varies significantly (Figure 1.2b). For example, the locations of cusp-B and O are shifted by 0.9° in Model 1 compared with the reference iasp91 model; the AB branch terminates at a much longer distance (~27°) in Model 3.

With the rapidly increasing installation of broadband seismic stations, the record section of seismograms as a function of epicentral distance within a limited azimuth range becomes much more useful and reliable for exploration of the vertical deep-mantle structure [e.g., *Wang and Niu*, 2010; *Ye et al.*, 2011]. In recent years, more than 800 broadband seismic instrumentations have been installed on mainland China, making it feasible to investigate the deep-earth structure in detail through the dense array technique [*Zheng et al.*, 2010; *Niu and Li*, 2011].

The velocity structure of the mantle transition zone described in the next section is based mainly on our new findings obtained through caustic waveform modeling [*Li et al.*, 2013]. Here we want to emphasize three important points: (1) In contrast to the traditional single or limited seismogram analysis, we take advantage of a well-constructed fan-shaped profile within a narrow azimuth range (Figure 1.3), and use the major features of the whole stack of seismic records, e.g., the location of caustic points, the crossover distance of the AB and CD branches, and the relative time between the AB and CD phases as well, with the goal of obtaining an

Table 1.1 Information of earthquakes used in caustic waveform modeling.

| Event ID | Origin time | Location | | Magnitude | Depth (km) | | |
		Lat.(°)	Lon.(°)	Mw	PDE	CMT	Used
20080519[a]	19/05/2008 00:08:36.31	42.50	131.87	5.6	513	522	519
20090716[a]	16 /07/2009 06:29:04.76	42.37	133.00	5.3	477	485	477
20130406	06/04/2013 00:29:55.00	42.73	130.97	5.8	563	575	570[b]

[a] Events are from *Li et al.* [2013]. Only events with both P and SH waveform used are illustrated here and in Figure 1.3.
[b] Depth determined by array stacking method [*Li et al.*, 2008].

accurate seismic image of the upper mantle transition zone beneath northeast Asia. (2) Only events with both P and SH waveforms clearly recorded (Table 1.1) are plotted in Figure 1.3. We used exactly the same station-receiver geometry for both P and SH waves to ensure a close match between the ray paths traversed by the P and S waves. (3) We considered a new dataset associated with one recent deep earthquake (20130406) that occurred near the border of China, Russia, and the Japan Sea (Table 1.1). This dataset, which is characterized by a relatively high signal-to-noise ratio (SNR) for both P and SH waves, was added to verify the overall features of aligned records and the fitness between the observation and theoretical seismograms. We also note that, compared to the dataset of the event 20080519 used in our previous study, the waveform of this new event seems to be a little noisy, and only 35 stations are selected. Therefore, we did not update the velocity models by using all those events simultaneously. The modification to the model is small considering the uncertainty estimates that will be addressed in the following section.

1.3. SLAB IMAGE IN THE MANTLE TRANSITION ZONE

The comparison between the observed and synthetic seismograms calculated from our preferred models for the new event is shown in Figure 1.4. A reflectivity synthetic code [*Wang*, 1999] is applied to generate theoretical seismograms, and a Gaussian wavelet is used to represent the source-time function. The preferred P and SH velocity models obtained from caustic waveform modeling are shown in Figure 1.5.

For the transversal component, the observed AB branch extends to as far as 23°, much farther than ~19° as predicted by iasp91 reference model; the CD branch begins to emerge at a distance of ~12.5°, in contrast to ~10.5° predicted by iasp91. We clearly observe a broadened BOD-zone with significantly delayed AB wave after a distance of 16°, consistent with our previous study. In the aligned vertical waveforms, the AB phase appears to be weaker around a distance of 23°; the CD phase begins to emerge around 13°, much farther than

the value of ~11° calculated from the iasp91 model. Due to the limitations imposed by the station coverage, it is difficult to determine exactly the crossover distance of AB and CD phases for both types of waveform. For the detailed description of the caustic waveform modeling and comparison of seismic waveforms of other previous events, we refer the reader to *Li et al.* [2013].

In Figure 1.4, we can see that locations of caustic points and crossover distance of different branches as well as the relative time difference are matched well overall for both vertical and transversal components. Although there is some mismatch between the synthetic and observed waveforms, we argue that the local shallow structure and lateral velocity variations may contribute to this discrepancy. Waveform modeling for a 3D velocity structure should be applied to account for lateral velocity variations and for the interaction between the slab and deep mantle.

One of the major features of both P and SH models is that a high-velocity layer with a thickness of ~140 km lies just above the 660-km discontinuity (Figure 1.5), which could not be well constrained from seismic tomographic images [e.g., *Huang and Zhao*, 2006]. The slab appears to be significantly thickened in the upper mantle transition zone when compared with its initial thickness of ~80–90 km [*Zhao et al.*, 1994] near the Japan trench. To our knowledge, this is the first time the thickness of the stagnant slab in the mantle transition zone is self-consistently resolved from both P and S waveforms. Thickened and broad fast velocity anomalies in the lower mantle beneath subduction zones have been recognized by several authors [e.g., *Ribe et al.*, 2007; *Ren et al.*, 2007; *Běhounková and Čížková*, 2008]. However, seismic investigation of the thickening of slabs at the bottom of the upper mantle as well as the dynamic implications of this process are quite limited. In the following sections, we will mainly pay attention to this point. We will first present an analysis of the uncertainty for our estimate of the slab thickness; we will then discuss results from numerical simulations in order to explore possible mechanisms and implications of slab thickening in the mantle transition zone.

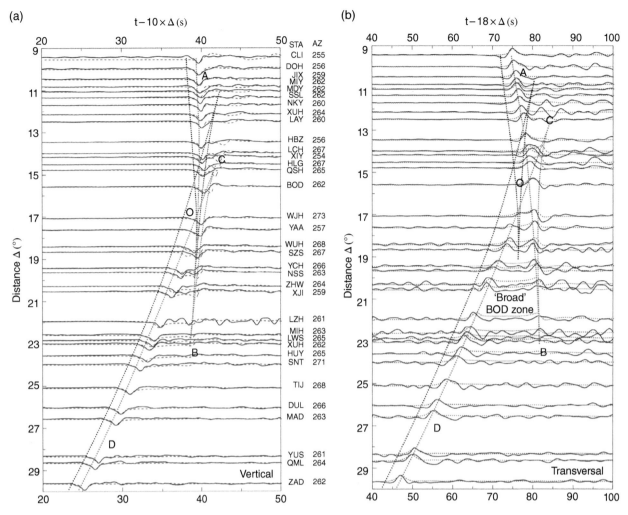

Figure 1.4 Comparison of the observed (black solid line) and synthetic waveforms (blue dashed line) calculated from the preferred P and S velocity models for the new event 20130406. (a) and (b) refer to vertical and transversal waveforms, respectively.

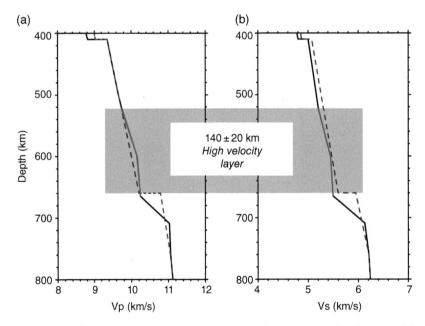

Figure 1.5 Velocity models obtained from caustic waveform modeling: (a) V_P model; (b) V_S model [*Li et al.*, 2013].

1.4. UNCERTAINTY ESTIMATES OF THICKNESS OF SLAB

Because of the great trade-off between depth of discontinuity and velocity variation, it is hard to provide a quantitative and accurate uncertainty estimate of the obtained velocity model. We adopted here a scheme similar to *Brudzinski and Chen* [2000]. We explored the model by perturbing the best fitting models until synthetic seismograms showed large misfits in the locations of cusp-B, cusp-C, and O point, or in relative time between AB and CD phases. We focused on two parameters: the thickness of the higher-velocity layer just above the 660 and the velocity variation above the 660. While perturbing one parameter, we tried to keep the other parts of the structure undisturbed. We rejected perturbations that produced misfits larger than ±0.5 s in relative time between AB and CD phases, or 0.5° shift in either cusp-B, cusp-C, or O location relative to the preferred models. We choose these values because significant discrepancies of the major features will appear in the aligned waveforms. However, different cut-off of these values will definitely affect the uncertainty estimation of the slab thickness to some degree.

A series of uncertainty tests was made based on waveform matching. Figure 1.6 shows how the locations of caustic points (cusp-C and cusp-B), crossover distance (point O), and relative time between the AB and CD branches change upon perturbing the thickness of the 660. If we take the above-mentioned rules to be the standard for acceptable models, we arrive at an estimated thickness of the slab to be 140 ± 20 km, which yields a much better constraint to the thickness of the stagnant slab trapped in the mantle transition zone (Figure 1.5).

1.5. DYNAMIC SIMULATION OF SLAB THICKENING

We carried out a series of 2D Cartesian numerical simulations of subduction that can help obtain a better physical understanding of the mechanisms and implications of a thickened slab. The finite-volume-based code YACC was used for the simulation [e.g., *Tosi et al.*, 2013]. YACC has been benchmarked for non-Boussinesq thermal convection [*King et al.*, 2010]. In Figure 1.7, we show three snapshots of the temperature field obtained from a simulation in which a significant thickening of the slab upon interaction with the 660-km discontinuity is observed. These results are from the reference model described by *Tosi et al.* in Chapter 6 of this volume, to which we refer for a detailed description.

Here we list the essential features of the model. In the framework of the Extended Boussinesq Approximation [e.g., *Christensen and Yuen*, 1985] with variable thermal expansivity and conductivity [*Tosi et al.*, 2013], we employed a purely thermal convection model in a two-dimensional box with a depth of 2890 km and width of 11,560 km. A 20-km wide weak zone inclined by 30°, with a viscosity of 10^{20} Pa·s is prescribed between an oceanic and an overriding plate. At the trench, the oceanic plate has a thickness of 100 km and a temperature distribution corresponding to that obtained from a half-space cooling model for an age of ~120 Ma. The overriding plate, which is assumed to have a uniform thickness of 100 km and an age of 120 Myr, is kept fixed with a no-slip upper-boundary condition throughout the simulation. After initiating the subduction kinematically by prescribing for the subducting plate a surface velocity of 5 cm/a over a time interval of 4 Myr, we changed its boundary condition to free slip, thereby allowing the slab to fall under its

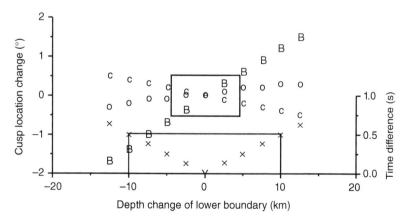

Figure 1.6 Uncertainty estimation of the slab thickness. We explore the model by perturbing the best fitting models until synthetic seismograms show large misfits in the location of cusp-B (B marks), cusp-C (C marks), and O point (O marks) or in the relative time between AB and CD phases (X marks). If the variation of the upper boundary is taken into account, the uncertainty in the thickness of the slab is estimated to be ±20 km.

(a)

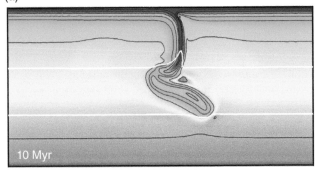

(b)

(c)

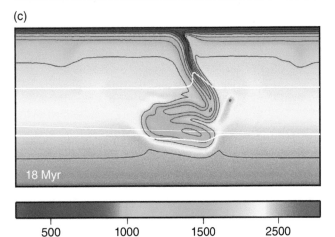

Figure 1.7 Numerical simulation from a model of purely thermal subduction. We applied a mixed rheology: brittle with surface yield stress of 500 MPa, along with Newtonian and power-law with n = 3.5 (with different activation parameters for the upper and lower mantle). (a) The slab goes down until encountering the 660-km discontinuity; (b) Buckling instabilities occur; (c) Upon buckling, the slab material keeps piling up above the 660-km discontinuity, and the thickness of the slab is enhanced eventually. Details of model setting can be found in Chapter 6 by *Tosi et al.*

own weight. We took into account the effects on buoyancy and latent heat due to the exothermic phase transition from olivine to spinel at 410-km depth, and to the endothermic one from spinel to perovksite at 660-km

depth. In addition, we imposed a 10-fold viscosity jump at the 660-km discontinuity and considered pressure-, temperature-, and phase-dependent coefficients of thermal expansion and conduction following the parameterization introduced by *Tosi et al.* [2013].

Returning now to Figure 1.7, the slab descends along the weak zone and quickly starts to sink nearly vertically because of the choice of keeping the trench at a fixed position. The negative thermal buoyancy drives the slab until it encounters a significant resistance near the 660-km discontinuity (Figure 1.7a), which is caused both by the endothermic ringwoodite-perovskite phase transition and by the imposed viscosity jump. As a result, the slab slows down, and buckling instabilities occur (Figure 1.7b). Upon buckling, more and more slab material keeps piling up above the 660, with the consequence that the overall thickness of the subducted lithosphere is significantly enhanced eventually (Figure 1.7c).

1.6. DISCUSSION AND DYNAMIC IMPLICATION

Recent seismic tomography results have revealed a variety of configurations of subducted slabs beneath different subduction regions. Along the western margin of the Pacific plate, the slab beneath the Mariana arc appears to sink into the lower mantle in a nearly vertical fashion [e.g., *van der Hilst et al.*, 1991]; beneath southern Kurile, Japan, and Izu-Bonin arcs, slabs are found to be lying horizontally and stagnate over the 660 [e.g., *Fukao et al.*, 2001; *Huang and Zhao*, 2006; *Li and van der Hilst*, 2010; *Fukao and Obayashi*, 2013]; beneath the Tonga arc, they appear instead to penetrate into the lower mantle after lying horizontally in the mantle transition zone [*van der Hilst*, 1995]. The spatial and temporal variations in slab strength and the history of subduction forces play an important role in the partitioning of mass across the phase transition boundary, thus leading to the different subduction behaviors [*King and Ita*, 1995; *Billen*, 2008].

It is well known that the thickness of the lithosphere does not remain uniform during the subduction process. Slabs tend to thicken while sinking into the lower mantle. This has often been interpreted as evidence for buckling of cold and stiff lithosphere [*Gaherty and Hager*, 1994; *Ribe et al.*, 2007; *Běhounková and Čížková*, 2008; *Lee and King*, 2011; *Tosi et al.*, Chapter 6 of this volume]. Seismic tomography studies have also revealed the feature of apparent broadening of subducted lithosphere in the middle or lower mantle [e.g., *van der Hilst*, 1995]. For example, the thickness of slabs may increase from 50 km to 100 km above the 660 to more than 400 km in the middle mantle beneath Java and Central America. Beneath the Mariana arc and Tonga, the thickness of slabs in the lower mantle is also suggested to increase by a factor of up to five [e.g., *van der Hilst*, 1995]. Pure compression alone due to the increase of viscosity with depth [e.g., *Bunge et al.*, 1996] cannot account for the apparent observed significant thickening of slabs. A periodic buckling mechanism was proposed to account for the broad fast velocity anomalies beneath the mantle transition zone [*Ribe et al.*, 2007]. The characteristic shapes of the observed anomalies agree well with those predicted by the scaling laws for buckling [*Ribe*, 2003].

Because of its poor depth resolution, seismic tomography is not able to resolve well the variations in the slab thickness. In the mantle transition zone, the inverted image from first arrivals is even blurred due to complications caused by caustic waveforms. The thickness of initial part of the slab beneath Japan is constrained to be ~80–90 km [*Zhao et al.*, 1994] from travel-time inversion of local, regional, and teleseismic records. This feature is consistent with a slab age of 120 Ma when assuming a potential temperature of 1180°C [*Deal and Nolet*, 1999]. However, our caustic waveform modeling of both P and S waves revealed a thickened slab trapped in the upper mantle transition zone beneath northeast Asia, which appears to be thickened by 50–60 km at the long-standing segment, indicating ongoing piling or thickening under northeast Asia today.

One of the earliest works on folding of viscous plumes falling vertically onto a density or viscosity interface was that of *Griffiths and Turner* [1988]. They suggested that buckling might occur when subducted oceanic lithosphere interacts with a density and/or viscosity interface around the 660 in the deep mantle. Our simulations deal with a more realistic case in which the dynamic effect of the γ-olivine to perovksite phase transition is included. We also took into account the temperature and pressure dependence of the coefficients of thermal expansion and conduction. As discussed by *Tosi et al.* [Chapter 6 of this volume], the occurrence of buckling instability near the 660 is due to the joint effects of viscosity increase, the endothermic ringwoodite-perovskite phase transition, and decrease of thermal expansivity upon compression (see Figure 1.7a).

We argue that the observed thickening of slab trapped in the mantle transition zone beneath northeast Asia reflects piling of slab. Several factors might contribute to this piling. The first is the viscosity increase across the boundary of the lower and upper mantle. A viscosity contrast of 10 between the lower and upper mantle is applied in our simulation. The compressional stress axes of intermediate-depth and deep earthquakes are all parallel or subparallel to the down-dip motion of the subducting part of slab, indicating a compression-dominated regime in the downgoing slab [*Li et al.*, 2013]. A recent

mid-mantle scattered waves study has constrained the viscosity of the upper part of the lower mantle beneath the same region to be in a range of $1.0 \times 10^{22} - 1.6 \times 10^{23}$ Pa s [*Li and Yuen*, 2014], also suggesting a possible viscosity contrast across the upper-lower mantle boundary.

Second, the accelerated effect of grain-size reduction due to phase transition of ringwoodite in a cold subducting slab [*Karato*, 2003; *Yamazaki et al.*, 2005] results in a nonlinear diffuse creep deformation. In addition, the grain size in the cold portion of the slab is expected to be less than ~100 μm, compared with an ambient mantle grain size of the order of millimeters. Viscosity of the slab subducting through the transition zone caused by the grain-size reduction alone is estimated to be four to six orders of magnitude lower than that of the surrounding mantle, if the temperature effect is not considered [*Yamazaki et al.*, 2005]. Such a softer slab could easily bend, deform, and pile up above the 660-km discontinuity. *Stegman et al.* [2010] indeed showed with 3D numerical models of subduction dynamics that only when using a relatively small viscosity contrast of ~100–300 between subducting plate and upper mantle, sinking plates exhibit recumbent folds atop the lower mantle. Furthermore, weak slabs can be maintained by subduction due to the low growth rate of ringwoodite [*Yamazaki et al.*, 2005].

Third, the pressure-, temperature-, and phase-dependent thermal expansivity and conductivity also contribute to the thickening of the slab. *Tosi et al.*, [Chapter 6 of this volume] has demonstrated that the two parameters exert a dramatic impact on the dynamics of subduction. An increased propensity toward locally layered convection is observed, with slabs trapped in the mantle transition zone when both thermal-dynamic parameters are allowed to vary together [*Tosi et al.*, 2013; *Tosi et al.*, Chapter 6 of this volume].

Although not allowed in our simulation, the trench retreat, or the rolling back of the slab might facilitate the quasi-periodic buckling of the slab in the mantle transition zone. Recently, *Cizkova and Bina* [2013] explored the time-dependent interplay between trench retreat, slab buckling, and slab stagnation. Their 2D simulations show that if the overriding plate can slip freely, an oscillating behavior of the subducting plate can occur, resulting in a subhorizontal distribution of buckle folds above the 660-km discontinuity. *Billen* [2010] proposed that trench retreat prior to slabs entering the mantle transition zone is a possible mechanism for trapping slabs in the mantle transition zone. We argue that the subducting Pacific slab along the Japan trench can be interpreted as being in a "buckling/folding retreating mode," as described, for example, by *Ribe* [2010]. With the continuation of buckling, the overall thickness of the slab above the 660-km discontinuity is enlarged significantly (Figure 1.7c) as inferred from our seismic analysis.

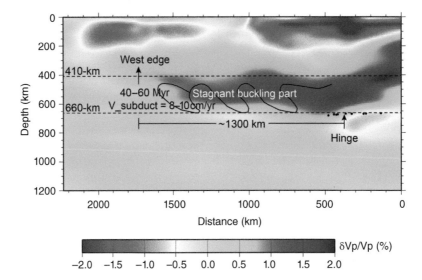

Figure 1.8 Estimation of the length of the stagnant slab trapped in the mantle transition zone. The cross section is along line AB in Figure 1.3. The black dots mark the undulation of the 660-km discontinuity investigated by S-to-P scattered wave [*Li et al.*, 2008]; the "hinge" is the location where the slab encountered the 660, and a rapid change of the depth of the 660 occurred here. The length is estimated to be ~1300 km from the hinge to the west edge of the slab, and the age of the slab is no more than 40–60 Myr even after considering the significant slab piling effect. The depth section of the P wave velocity model [*Obayashi and Fukao*, 2013] is also plotted, with the blue and red colors indicating fast and slow velocity anomaly, respectively.

Piling of subducted material could help explain the enigmatic issue that part of the Pacific plate appears to be missing. It has been estimated that in the past 150 Myr, approximately 13,000 km of lithosphere has descended into the Japan and western North American trenches. The total amount of subducted Pacific plate at the Japan trench over the last 50 Myr has been calculated to be ~4000–5000 km in a reference frame fixed with respect to the overriding plate [*Engebretson et al.*, 1992], which is consistent with the assumption of an average subduction rate ~8–10 cm/yr from the early Cenozoic. Tomographic images of the deep mantle presented in a variety of seismic studies indicate that the western edge of the stagnant slab is located approximately 2000–2500 km west of the Japan trench [e.g., *Huang and Zhao*, 2006; *Li et al.*, 2008; *Fukao and Obayashi*, 2013], resulting in a severely limited amount of material that may have been subducted (Figure 1.8). Assuming the piling of the slab is uniform along the stagnant segment, the actual trapped length of the slab can be estimated by multiplying the imaged length of ~1300 km by a thickening factor of 1.6–1.8. The buckling and piling of slab in the mantle transition zone can account for at least 800–1000 km of "missing" Pacific plate.

We can also estimate that the age of the west edge of the subducting Pacific slab is not older than 40–60 Myr, even after accounting for the thickening effect of the slab due to buckling. This is consistent with the drifting history of the Pacific plate, which changed its direction from NNW to NW at ~50 Ma [*Sun et al.*, 2007]. In recent years, the destruction of North China Craton (NCC) has been an important topic in the study of continental evolution. Different mechanisms including thermal erosion [e.g., *Xu*, 2001, 2007; *Menzies et al.*, 2007], delamination [*Wu et al.*, 2002; *Gao et al.*, 2004], Pacific subduction [*Niu.*, 2005], or even the India-Eurasia collision and mantle plume activities have been proposed by different authors [e.g., *Liu et al.*, 2004; *Wilde et al.*, 2003]. Recent studies suggested that the Pacific subduction [*Zhu et al.*, 2012] and the subduction of the ridge between the Pacific and the Izanagi plates [*Ling et al.*, 2013] are the principal triggers for the destruction of the NCC. We argue that the ongoing piling and stagnation occurred after 50 Ma, and thus the stagnation of the Pacific slab observed so far has no direct relationship to the destruction of the NCC, which took place with a peak age of ~120–125 Ma [e.g., *Zhu et al.*, 2011; *Ling et al.*, 2013]. At least it cannot be the direct cause of the reactivation of the stable North China Craton.

The continuous piling up of slabs may have important implications for the fate of the descending lithosphere under northeast Asia (Figure 1.9). Tomographic images are only a snapshot in time and that slab morphology and mass flux into the lower mantle may be a transient phenomenon [*Griffiths et al.*, 1995]. With the ongoing piling of the trapped cold lithosphere, we argue that the slab might be on the margin of instability. The 3D

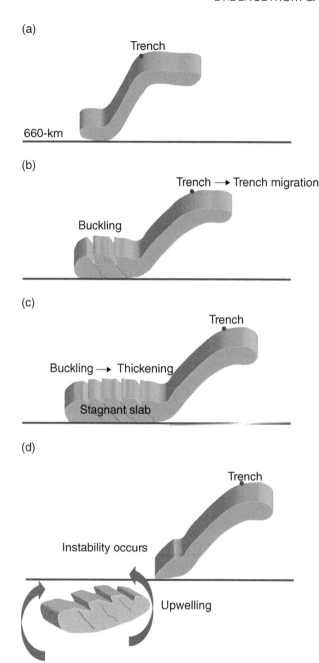

Figure 1.9 Schematic plot of subduction process and the instability. (a) Initial phase: The point indicates initial position of the trench. (b) After the slab encounters the 660, it bends and buckling starts to occur. Trench rollback facilitates slab stagnation. (c) With the buckling and piling of cold materials, the trapped stagnant slab is thickened. This kind of subducting slab can be classified to be in a buckling/folding retreating mode. (d) Slab detachment occurs accompanying a flush of upwellings.

numerical simulations of mantle convection [*Honda et al.*, 1993b] indicate that, when considering the olivine family of phase transitions, cold material is likely to accumulate at the phase-change boundary between the

upper and lower mantle, eventually resulting in a large-scale gravitational instability. To the west of the Japan trench, convection seems to be temporarily hindered by the phase changes. A large volume of cold and dense material has been accumulating at the 660-km discontinuity. This could in turn rapidly sink into the lower mantle, leading to a mantle avalanche. The occurrence of instability, which may be in the future 5–10 Myr, would be accompanied by a flush of volcanic activity [*Honda et al.*, 1993a; *Honda et al.*, 1993b], precipitated by mantle upwellings emanating from the lower mantle (Figure 1.9).

1.7. CONCLUSIONS

The motivation of this paper is to stimulate more multidisciplinary work revolving around the study of the structure and dynamics of the mantle beneath northeast Asia, where significant material appears to be stagnating in the transition zone. We have partly addressed the issue employing caustic waveform modeling to image the slab. We inferred the presence of a long stagnant slab with a thickness around 140 km, possibly indicating ongoing piling of subducted oceanic lithosphere. In addition, we showed numerical simulations indicating that the subducting lithosphere exhibits buckling instabilities due to a combination of viscosity jump across the upper-lower mantle interface, phase transitions, and variable thermodynamic parameters that naturally lead to a significant thickening of the slab. This thickened slab in the mantle transition zone can partly account for the missing volume of the Pacific slab subducted beneath the Japan trench for the last 50 Ma. We also speculate that such a thickened stagnant slab might be close to an impending instability with a rapid flush of cold material possibly occurring in the next few million years.

ACKNOWLEDGMENT

This work is supported by NSFC (J. Li, Grants 41322026 and 41274065). We thank Gabriele Morra, Taras Gerya, Weidong Sun, and Yongtai Yang for helpful discussion and Scott King for comments and suggestions that helped improve the manuscript significantly. An anonymous reviewer provided constructive and critical suggestions. Center of China National Seismic Network at Institute of Geophysics, China Earthquake Administration provided the seismic data. GMT software was used in figure plotting. This research was also supported by CMG and Petrology program of US NSF. N. Tosi acknowledges support from the Helmholtz Gemeinschaft (project VH-NG-1017).

REFERENCES

Aki, K., and P. G. Richards (2002), *Quantitative Seismology* (2nd ed.), University Science Books.

Běhounková, M., and H. Čížková (2008), Long-wavelength character of subducted slabs in the lower mantle, *Earth Planet. Sci. Lett.*, *275*(1–2), 43–53.

Billen, M. I. (2008), Modeling the dynamics of subducting slabs, *Ann. Rev. Earth and Planet. Sci.*, *36*, 325–356, doi:10.1146/annurev.earth.36.031207.124129.

Billen, M. I. (2010), Slab dynamics in the transition zone, *Physics of the Earth and Planetary Interiors*, *183*(1–2), 296–308.

Brudzinski, M. R., and W. P. Chen (2000), Variations in P wave speeds and outboard earthquakes: Evidence for a petroligic anomaly in the mantle transition zone, *J. Geophys. Res.*, *105*(B9), 21661–21682.

Brudzinski, M. R., and W. P. Chen (2003), A petrologic anomaly accompanying outboard earthquakes beneath Fiji-Tonga: Corresponding evidence from broadband P and S waveforms, *J. Geophys. Res.*, *108*(B6), doi:10.1029/2002JB002012.

Bullen, B. (1963), *Introduction to the Theory of Seismology*, Cambridge Univ. Press, New York.

Bunge, H. P., M. A. Richards, and J. R. Baumgardner (1996), Effect of depth-dependent viscosity on the planform of mantle convection, *Nature*, *379*(6564), 436–438.

Christensen, U. R., and D. A. Yuen (1985), Layered convection induced by phase transitions, *J. Geophys. Res.*, *90*, 10291–10300.

Cizkova, H., and C. R. Bina (2013), Effects of mantle and subduction-interface rheologies onslab stagnation and trench rollback, *Earth Planet. Sci. Lett.*, *379*, 95–103.

Creager, K. C., and T. H. Jordan (1986), Slab penetration into the lower mantle beneath the Mariana and other island arcs of the northwest Pacific, *J. Geophys. Res.*, *91*(B3), 3573–3589, doi:10.1029/JB091iB03p03573.

Deal, M. M., and G. Nolet (1999), Slab temperature and thickness from seismic tomography 2. Izu-Bonin, Japan, and Kuril subduction zones, *J. Geophys. Res-Sol Ea*, *104*(B12), 28803–28812, doi: 10.1029/1999jb900254.

de Wit, R. W. L., J. Trampert, and R. D. van der Hilst (2012), Toward quantifying uncertainty in travel time tomography using the null-space shuttle, *J. Geophys. Res.*, *117*(B3), B03301, doi:10.1029/2011jb008754.

Engebretson, D. C., K. P. Kelley, H. J. Cashman, (1992), 180 million years of subduction, *GSA Today*, *2* (5), 93–95

Fukao, Y., and M. Obayashi (2013), Subducted slabs stagnant above, penetrating through, and trapped below the 660 km discontinuity, *Journal of Geophysical Research: Solid Earth*, *118*(11), 2013JB010466, doi:10.1002/2013JB010466.

Fukao, Y., S. Widiyantoro, and M. Obayashi (2001), Stagnant slabs in the upper and lower mantle transition region, *Rev. Geophys.*, *39*(3), 291–323.

Gaherty, J. B., and B. H. Hager (1994), Compositional vs thermal buoyancy and the evolution of subducted lithosphere, *Geophys. Res. Lett.*, *21*(2), 141–144, doi: 10.1029/93gl03466.

Gao, S., R. L. Rudnick, H. L. Yuan, X. M. Liu, Y. S. Liu,, W. L. Xu, W. L. Ling, J. Ayers, X. C. Wang, and Q. H. Wang (2004), Recycling lower continental crust in the North China Craton, *Nature*, *432*, 892–897.

Grand, S. P., and D. V. Helmberger (1984), Upper mantle shearstructure beneath the northwest Atlantic Ocean, *J. Geophys. Res.-Sol Ea*, *89*(B13), 11465–11475, doi:10.1029/JB089iB13p11465.

Griffiths, R. W., and J. S. Turner (1988), Folding of viscous plumes impinging on a density or viscosity interface, *Geophys. J.*, *95*(2), 397–419, doi:10.1111/j.1365-246X.1988.tb00477.x.

Griffiths, R. W., R. I. Hackney, and R. D. Vanderhilst (1995), A laboratory investigation of effects of trench migration on the descent of subducted slabs, *Earth Planet. Sci. Lett.*, *133*(1–2), 1–17, doi:10.1016/0012-821x(95)00027-A.

Gurnis, M., and B. H. Hager (1988), Controls of the structure of subducted slabs, *Nature*, *335*(6188), 317–321.

Honda, S., S. Balachandar, D. A. Yuen, and D. Reuteler (1993a), 3-Dimensional mantle dynamics with an endothermic phase-transition, *Geophys. Res. Lett.*, *20*(3), 221–224, doi: 10.1029/92gl02976.

Honda, S., D. A. Yuen, S. Balachandar, and D. Reuteler (1993b), 3-dimensional instabilities of mantle convection with multiple phase-transitions, *Science*, *259*(5099), 1308–1311, doi:10.1126/science.259.5099.1308.

Huang, J. L., and D. P. Zhao (2006), High-resolution mantle tomography of China and surrounding regions, *J. Geophys. Res.-Sol Ea*, *111*(B9), doi:Artn B09305Doi 10.1029/2005jb004066.

Karato, S.-I. (2003), *The Dynamic Structure of the Deep Earth–An Interdisciplinary Approach*, Princeton University Press, Princeton and Oxford.

King, S. D. (2001), Subduction zones: observations and geodynamic models, *Physics of the Earth and Planetary Interiors*, *127*(1–4), 9–24, doi:10.1016/S0031-9201(01)00218-7.

King, S. D., and J. Ita (1995), Effect of slab rheology on masstransport across a phase-transition boundary, *J. Geophys. Res.-Sol Ea*, *100*(B10), 20211–20222.

King, S. D., C. Lee, P. E. van Keken, W. Leng, S. J. Zhong, E. Tan, N. Tosi, and M. C. Kameyama (2010), A community benchmark for 2-D Cartesian compressible convection in the Earth's mantle, *Geophys. J. International*, *180*(1), 73–87.

Lay, T., and D. V. Helmberger (1983), A lower mantle S-wave triplication and the shear velocity structure of D″, *Geophys. J. Royal Astronom. Soc.*, *75*(3), 799–837, doi:10.1111/j.1365-246X.1983.tb05010.x.

Lee, C., and S. D. King (2011), Dynamic buckling of subducting slabs reconciles geological and geophysical observations, *Earth Planet. Sci. Lett.*, *312*(3–4), 360–370.

Li, C., and R. D. van der Hilst (2010), Structure of the upper mantle and transition zone beneath Southeast Asia from traveltime tomography, *J. Geophys. Res.*, *115*(B7), B07308, doi:10.1029/2009jb006882.

Li, C., R. D. van der Hilst, E. R. Engdahl, and S. Burdick (2008), A new global model for P wave speed variations in Earth's mantle, *Geochem. Geophys. Geosys.*, *9*.

Li, J., X. Wang, X. J. Wang, and D. A. Yuen (2013), P and SH velocity structure in the upper mantle beneath Northeast

China: Evidence for a stagnant slab in hydrous mantle transition zone, *Earth Planet. Sci. Lett.*, *367*, 71–81, doi: 10.1016/j.epsl.2013.02.026.

Li, J., and D. Yuen (2014), Mid-mantle heterogeneities associated with Izanagi plate: implications for regional mantle viscosity, *Earth Planet. Sci. Lett.*, *385*, 137–144, doi:10.1016/j.epsl.2013.10.042.

Ling M. X., Y. Li, X. Ding, F. Z. Teng, X. Y. Yang, W. M. Fan, Y. G. Xu, and W. D. Sun (2013), Destruction of the North China Craton induced by ridge subductions. *J. Geol.*, *121*, 197–213.

Liu, M., X. Cui, and F. Liu (2004), Cenozoic rifting and volcanism in eastern China: a mantle dynamic link to the Indo-Asian collision? *Tectonophysics*, *393*(1–4), 29–42.

Loiselet, C., J. Braun, L. Husson, C. L. de Veslud, C. Thieulot, P. Yamato, and D. Grujic (2010), Subducting slabs: Jellyfishes in the Earth's mantle, *Geochem. Geophys. Geosys.*, *11*, doi:Artn Q08016Doi 10.1029/2010gc003172.

Menzies, M., Y. G. Xu, H. F. Zhang, and W. M. Fan (2007), Integration of geology, geophysics and geochemistry: A key to understanding the North China Craton, *Lithos*, *96*, 1–21.

Niu, F., and J. Li (2011), Component zaimuths of the CEArray stations estimated from P-wave particle motion, *Earthquake Science*, *24*, 3–13, doi:10.1007/s11589-011-0764-8.

Niu, Y. L. (2005), Generation and evolution of basaltic magmas: some basic concepts and a new view on the origin of Mesozoic-Cenozoic basaltic volcanism in eastern China, *Geol. J. China Univ.*, *11*, 9–46.

Ren, Y., E. Stutzmann, R. D. van der Hilst, and J. Besse (2007), Understanding seismic heterogeneities in the lower mantle beneath the Americas from seismic tomography and plate tectonic history, *J. Geophys. Res.-Sol Ea*, *112*(B1), B01302, doi:10.1029/2005JB004154.

Ribe, N. M. (2003), Periodic folding of viscous sheets, *Phys. Rev. E.*, *68*(3).

Ribe, N. M. (2010), Bending mechanics and mode selection in free subduction: a thin-sheet analysis, *Geophys. J. International*, *180*(2), 559–576.

Ribe, N. M., E. Stutzmann, Y. Ren, and R. van der Hilst (2007), Buckling instabilities of subducted lithosphere beneath the transition zone, *Earth Planet. Sci. Lett.*, *254*(1–2), 173–179, doi:10.1016/j.epsl.2006.11.028.

Ricard, Y., M. Richards, C. Lithgow-Bertelloni, and Y. L. Stunff (1993), A geodynamic model of mantle density heterogeneity, *J. Geophys. Res.*, *98*(B12), 21895–21909.

Song, A. T.-R., D. V. Helmberger, and S. P. Grand (2004), Low-velocity zone atop the 410-km seismic discontinuity in the northwestern United States, *Nature*, *427*(6974), 530–533, doi:http://www.nature.com/nature/journal/v427/n6974/suppinfo/nature02231_S1.html.

Stegman, D. R., R. Farrington, F. A. Capitanio, and W. P. Schellart (2010), A regime diagram for subduction styles from 3-D numerical models of free subduction, *Tectonophysics*, *483*(1–2), 29–45.

Sun, W. D., X. Ding, Y. H. Hu, and X. H. Li (2007), The golden transformation of the Cretaceous plate subduction in the west Pacific, *Earth Planet. Sci. Lett.*, *262*, 533–542.

Tajima, F., and S. P. Grand (1995), Evidence of high-velocity anomalies in the transition zone associated with southern Kurile subduction zone, *Geophys Res Lett*, *22*(23), 3139–3142.

Tajima, F., and S. P. Grand (1998), Variation of transition zone high-velocity anomalies and depression of 660 km discontinuity associated with subduction zones from the southern Kuriles to Izu-Bonin and Ryukyu, *J. Geophys. Res.-Sol Ea*, *103*(B7), 15015–15036.

Tosi, Y., D. A. Yuen, N. de Koler, and R. M. Wentzcovitch (2013), Mantle dynamics with pressure- and temperature-dependent thermal expansivity and conductivity, *Phys. Earth Planet. Inter.*, *217*, 48–58.

Van der Hilst, R. (1995), Complex morphology of subducted lithosphere in the mantle beneath the tonga trench, *Nature*, *374*(6518), 154–157, doi:10.1038/374154a0.

Van der Hilst, R., R. Engdahl, W. Spakman, and G. Nolet (1991), Tomographic imaging of subducted lithosphere below northwest Pacific Island arcs, *Nature*, *353*(6339), 37–43.

Wang, R. (1999), A simple orthonormalization method for stable and efficient computation of Green's functions, *Bull. seism. Soc. Am.*, *89*, 733–741.

Wang, B., and F. Niu (2010), A broad 660-km discontinuity beneath northeastern China revealed by dense regional seismic networks in China, *J. Geophys. Res., submitted*(115), XXX, doi:10.1029/2009JB006608.

Wang, K., and Z. Yao (1991), Upper mantle P velocity structure of southern China, *Chinese J. Geophys.*, *34*(3), 309 317.

Wang, Y., L. X. Wen, D. Weidner, and Y. M. He (2006), SH velocity and compositional models near the 660-km discontinuity beneath South America and northeast Asia, *J. Geophys. Res.-Sol Ea*, *111*(B7).

Wilde, S. A., X. H. Zhou, A. A. Nemchin, and M. Sun (2003), Mesozoic crust-mantle interaction beneath the North China craton: A consequence of the dispersal of Gondwanaland and accretion of Asia, *Geology*, *31*(9), 817–820.

Wu, F. Y., D. Y. Sun, H. M. Li, B. M. Jahn, and S. Wilde (2002), A-type granites in northeastern China: Age and geochemical constraints on their petrogenesis, *Chem. Geol.*, *187*, 143–173.

Xu, Y. G. (2001), Thermo-tectonic destruction of the Archaean lithospheric keel beneath the Sino-Korean Craton in China: Evidence, timing and mechanism, *Phys. Chem. Earth A*, *26*, 747–757.

Xu, Y. G. (2007), Diachronous lithospheric thinning of the North China Craton and formation of the Daxin'anling-Taihangshan gravity lineament, *Lithos*, *96*, 281–298.

Yamazaki, D., T. Inoue, M. Okamoto, and T. Irifune (2005), Grain growth kinetics of ringwoodite and its implication for rheology of the subducting slab, *Earth Planet. Sci. Lett.*, *236*(3–4), 871–881.

Ye, L., J. Li, T.-L. Tseng, and Z. Yao (2011), A stagnant slab in a water-bearing mantle transition zone beneath northeast China: Implications from regional SH waveform modelling, *Geophys. J. International*, *186*(2), 706–710, doi:10.1111/j.1365-246X.2011.05063.x.

Zhao, D. P., A. Hasegawa, and H. Kanamori (1994), Deep-structure of Japan subduction zone as derived from local, regional, and teleseismic events, *J. Geophys. Res.-Sol Ea*, *99*(B11), 22313–22329, doi:10.1029/94jb01149.

Zheng, X.-F., Z.-X. Yao, J.-H. Liang, and J. Zheng (2010), The role played and opportunities provided by IGP DMC of China National Seismic Network in Wenchuan earthquake disaster relief and researches, *Bulletin of the Seismological Society of America*, *100*(5B), 2866–2872, doi:10.1785/0120090257.

Zhu, R. X., L. Chen, F. Y. Wu, and J. L. Liu (2011), Timing, scale and mechanism of the destruction of the North China Craton *Sci. China Earth Sci.*, *54*(6), 789–797.

Zhu, R. X., Y. G. Xu, G. Zhu, H. F. Zhang, Q. K. Xia, and T. Y. Zheng (2012), Destruction of the North China Craton, *Sci. China Earth Sci.*, *55*(10), 1565–1587.

2

The Continental Collision Process Deduced from the Metamorphic Pattern in the Dabie-Hongseong and Himalayan Collision Belts

ABSTRACT

Different continental collision belts show contrasting metamorphic trend along their length, including the distribution of extreme metamorphism; i.e., ultrahigh-pressure (>100 km depth) and ultrahigh-temperature (900–1150°C) metamorphism. The different metamorphic trend can give important clues for understanding on the process of continental collision which is difficult to study because the important part of the process occurs in deep depth and most or earlier parts of the process disappear now. The present study investigates the main factors that control the metamorphic patterns along collision belts, with reference to the Dabie-Hongseong collision belt between the North and South China blocks (NCB and SCB) and the Himalayan collision belt between the Indian and Asian blocks. In the Dabie-Hongseong collision belt, collision began in the east before 245 Ma and propagated westward until ca 220 Ma. The amount of oceanic slab that subducted before collision was insufficient to pull down the continental crust to the depths of ultrahigh-pressure metamorphism in the eastern part of the belt but enough in the western part of the belt. Slab break-off also migrated from east to west, with a westward increase in the depth of break-off (from ca. 10 kbar in the east to ca. 35 kbar in the west). These lateral trends along the belt resulted in a westward change from ultrahigh-temperature (915–1160°C, 9.0–10.6 kbar) to high-pressure (835–860°C, 17.0–20.9 kbar) and finally ultrahigh-pressure metamorphism (680–880°C, 30–40 kbar). In the Himalayan collision belt, collision started from the west at ca. 55 Ma and propagated eastward. The amount of oceanic slab subducted prior to collision was sufficient to pull down the continental crust to the depths of ultrahigh-pressure metamorphism in the west. Slab break-off started in the west at ca. 46–55 Ma and propagated eastward until ca. 22–25 Ma, with an eastward decrease in the depth of slab break-off from 30–35 kbar to 17–18 kbar. Consequently, the metamorphic trend along the belt changes eastward from ultrahigh-pressure (750–770°C, 30–39 kbar) to high-pressure and finally high-pressure granulite facies metamorphism (890°C, 17–18 kbar). This study indicates that the different metamorphic pattern along the collision belt indicates different collision process and is strongly related to the amount of subducted oceanic crust between continents before collision and the depth of slab break-off. Therefore metamorphic pattern can be used to interpret both the disappeared and ongoing tectonic process during continental collision. This study also shows that the tectonic interpretation based on petrological study can contribute to better interpretation of geophysical tomography of the area where collision process is now occurring.

2.1. INTRODUCTION

Understanding of the process of continental collision is important to decipher the geologic events related to continental collision but is difficult because the important part of the process occurs in deep depth. Although there has been a significant improvement in the resolution of

Department of Earth and Environmental Sciences and Earth and Environmental Science System Research Center, Chonbuk National University, Jeonju, South Korea

Subduction Dynamics: From Mantle Flow to Mega Disasters, Geophysical Monograph 211, First Edition.
Edited by Gabriele Morra, David A. Yuen, Scott D. King, Sang-Mook Lee, and Seth Stein.
© 2016 American Geophysical Union. Published 2016 by John Wiley & Sons, Inc.

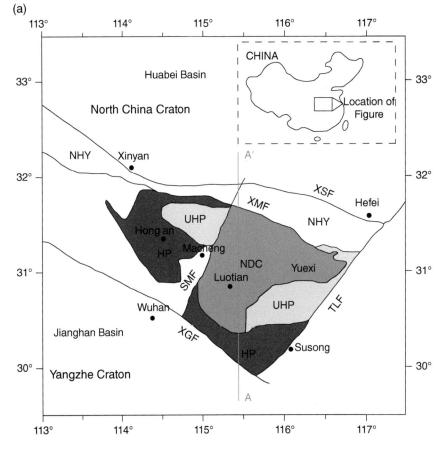

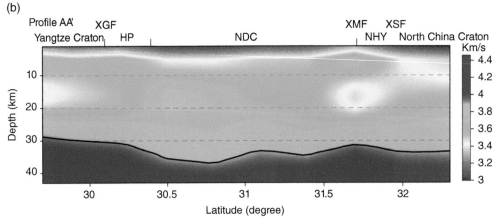

Figure 2.1 (a) Simplified geological map of the Dabie Mountains. Inset shows the location of the study area [from *Luo et al.*, 2013]. XSF = Xinyan–Shucheng Fault; XMF = Xiaotian–Mozitan Fault; SMF = Shangcheng–Macheng Fault; TLF = Tancheng–Lujiang Fault; XGF = Xiangfang–Guangji Fault; NDC = Northern Dabie complex; HP = high pressure; UHP = ultrahigh pressure. The line represents the location of the vertical cross section shown in (b). (b) Transect of shear velocities along longitude 115.4°E [from *Luo et al.*, 2013]. This transect indicates that the collision process in the Dabie area has been completed.

geophysical analysis, it is still challenging to study the geologic process that occurs in deep depth. Another problem is that geophysical observation of the collision process of most collision belts is impossible because the

collision process is already completed, as shown in Figure 2.1, with the exception of very young collision belts, such as the Himalayan. Even in the Himalayan collision belt, the youngest in the world, it is difficult to

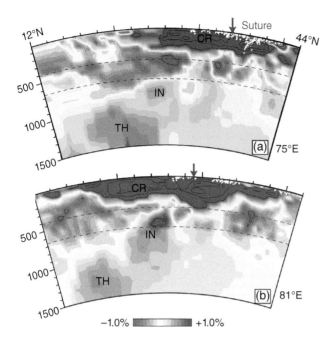

Figure 2.2 Vertical sections of P-wave tomography across the Himalayan collision belt along the longitude 75°E (a) and 81°E (b) [from *Replumaz et al.*, 2010]. This section shows that the slab break-off had already occurred along the Himalayan collision belt. The IN and TH anomalies represent the detached slabs and the CR anomaly represents the underthrusted Indian lithosphere slab.

study the whole collision process using geophysical method because the initial stage of collision process before slab break-off disappeared now as shown by the seismic tomography in Figure 2.2. Therefore, we need another research method to analyze collision processes occurring in deep depth.

One alternative is a petrological method. The collision process occurring in deep depth can be recorded in metamorphic rocks that were formed during collision and then exposed to the surface by uplifting. Therefore, the uplifted metamorphic rocks are petrological fossils that can be used to evaluate collision processes of the past.

The collision process can be recorded in metamorphic rocks in the following way. During continental collision, the pulling force exerted by the subducted oceanic slab (oceanic lithosphere) prior to collision leads to subduction of the continental slab (continental lithosphere) to a depth corresponding to high- or ultrahigh-pressure (HP or UHP, respectively) metamorphism [*Davis and Blankenburg*, 1995; *O'Brien*, 2001; *Hirajima and Nakamura*, 2003; *Massonne and O'Brien*, 2003; *Ernst*, 2006]. The oceanic slab eventually detaches from the continental slab (slab break-off) when the buoyancy force of the subducted continental slab exceeds the pulling force of the subducted oceanic slab [*Davis and Blankenburg*, 1995; *Wong et al.*, 1997; *Wrotel and Spakman*, 2000; *O'Brien*, 2001]. After slab break-off,

the continental slab moves toward the surface, and the large amount of heat supplied from the asthenosphere through the opening between the continental and detached oceanic slabs may cause ultrahigh-temperature (UHT) metamorphism [e.g., *Rajesh and Santosh*, 2004; *Oh et al.*, 2006a; *Kim et al.*, 2011].

During the uplift that follows slab break-off, HP and UHP metamorphic rocks are overprinted by epidote-amphibolite, amphibolite, or granulite facies metamorphism. As a result, metamorphism at extreme pressure-temperature conditions, including UHP (>30 kbar) and UHT (900–1150°C) metamorphism, has been reported in many continental collision belts in association with other types of metamorphism such as HP eclogite facies, granulite facies, and amphibolite facies [*Dallwitz*, 1968; *Chopin*, 1984; *Wang et al.*, 1989; *Harley*, 1988; *Ernst*, 2006; *Santosh and Sajeev*, 2006; *Rubatto et al.*, 2003]. The metamorphic ages of each step of metamorphism can be determined by age dating. Therefore, the collision process over time can be revealed by studying metamorphic evolution experienced by the uplifted HP, UHP, and UHT rocks to the surface.

However, the distribution of metamorphism, including the distribution of UHP and UHT metamorphism, differs in the various collision belts of the world. In the western part of the Himalayan collision belt, UHP eclogites occur with HP eclogites, whereas the middle/eastern part of the belt contains only HP eclogites and the easternmost part contains HP granulites [*Chopin*, 1984; *Massonne and O'Brian*, 2003; *Liu and Zhong*, 1997]. The Dabie-Hongseong collision belt, which marks the junction between the NCB and SCB, contains UHT rocks as well as HP and UHP rocks [*Wang et al.*, 1989; *Hirajima and Nakamura*, 2003; *Oh et al.*, 2005, 2006a]. In the Madurai block of southern India, part of the southern Pan-African collision belt, UHT metamorphic rocks occur with very minor HP metamorphic rocks [*Sajeev et al.*, 2004, *Sajeev et al.*, 2009]. These contrasting metamorphic patterns in the different collision belts may indicate different collision processes suggesting the possibility of using metamorphic patterns for deciphering collision processes of each collision belt.

The Permo-Triassic Dabie-Hongseong collision belt is one of the biggest, with many HP and UHP metamorphic rocks exposed on the surface with UHT rock, thus allowing relative convenience in studying past collision processes . The Himalayan collision belt is the youngest one that shows an ongoing continental collision process. Although, there are limited HP and UHP metamorphic rocks uplifted from deep depth to surface in the Himalayan belt, many geophysical studies have been done along the belt. Therefore, the Himalayan belt is the best location for comparing the ongoing collision process seen through metamorphic rocks uplifted from deep depth with the same process seen through geophysical data. In this study, the continental collision process will be analyzed by studying these two belts. In addition, 3D tectonic models of continental collision are proposed for both belts.

2.2. METAMORPHIC EVOLUTION ALONG THE DABIE-HONGSEONG COLLISION BELT IN NORTHEAST ASIA

The Triassic Dabie-Sulu collision belt between the NCB and SCB was identified by the discovery of ophiolites and UHP metamorphism (Figure 2.3; e.g., *Wang et al.*, 1989; *Zhai and Cong*, 1996; *Zhai and Liu*, 1998], and was distinguished as containing the world's largest belt of UHP metamorphic rocks. It is important to know which belts in Korea and Japan are the continuation of the Dabie-Sulu belt in order to obtain a better understanding of the tectonic processes that led to formation and exhumation of these distinctive UHP rocks. However, there are still debates concerning the tectonic relationship between Korea, China, and Japan, with a number of different tectonic models proposed to describe this relationship [Figure 2.4; *Yin and Nie*, 1993; *Ernst and Liou*, 1995; *Chang*, 1996; *Zhang*, 1997].

Yin and Nie [1993] argued that collision between the NCB and SCB began with the indentation of the northeastern south China block into the southeastern north China block in the late Permian and continued until the late Triassic. In their model, the collision belt was suggested to extend to the Imjingang belt in Korea, and the dextral Honam shear zone in Korea was regarded as a contemporaneous counterpart of the Triassic sinistral TanLu fault in China (Figure 2.4a). These faults together accommodated the northward extrusion of the block between them. However, they did not give any evidence of collision from the Imjingang belt, and the Honam shear zone can not be a counterpart of TanLu fault because it began to form from the middle Jurassic instead of Triassic [*Cho et al.*, 1999; *Lee et al.*, 2003]. *Ree et al.* [1996] reported late Permian (249 Ma) metamorphism (8.9–11.5 kbar and 640–815°C) from the Imjingang belt as evidence for HP metamorphism related to collision. However, as shown in Figure 2.5, these metamorphic pressure and temperature (P-T) conditions do not represent HP metamorphism as claimed. The high-pressure conditions reported from the belt by *Kwon et al.* [2009] and *Sajeev et al.* [2010] were wrong due to using

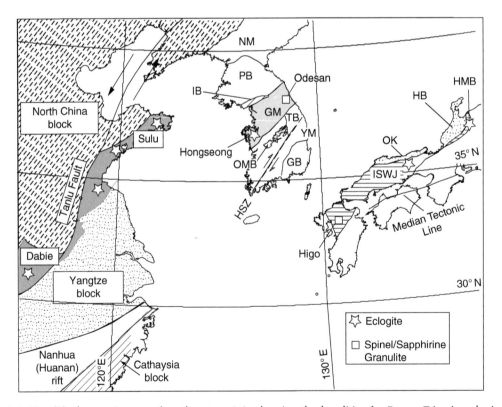

Figure 2.3 Simplified tectonic map of northeastern Asia showing the localities for Permo-Triassic eclogites and spinel granulite [from *Oh and Kusky*, 2007]. Labels: GB = Gyeongsang basin; GM = Gyeonggi Massif; HB = Hida belt; HMB = Hida marginal belt; IB = Imjingang belt; ISWJ = Inner zone of Southwest Japan; NM = Nangrim Massif; OK = Oki belt; OMB = Ogcheon metamorphic belt; PB = Pyeongnam basin; TB = Taebaeksan basin; YM = Youngnam Massif.

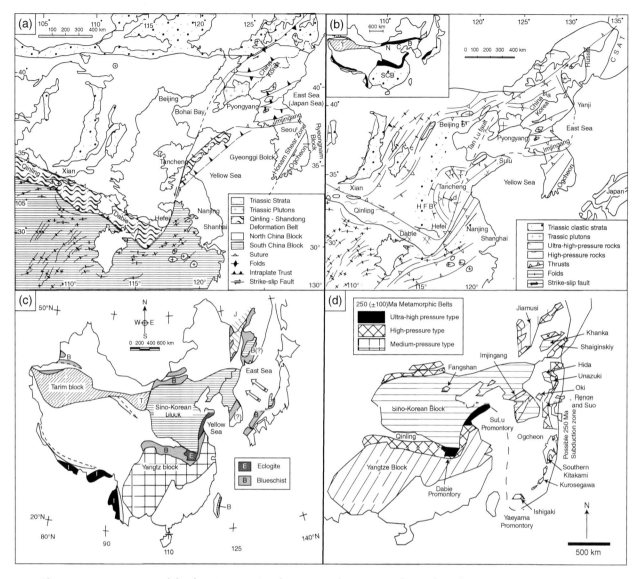

Figure 2.4 Tectonic models showing previously proposed tectonic relationships between Korea, China, and Japan. Tectonic model suggested (a) by *Yin and Nie* [1993], (b) by *Zhang* [1997], (c) by *Ernst and Liou* [1995], (d) by *Ishiwatari and Tsujimori* [2003] [from *Oh and Kusky*, 2007].

pseudosection, which did not consider plagioclase, which is essential to differentiate high-pressure metamorphism and using data that were not shown in the tables of mineral composition [Table 1 in *Kwon et al.*, 2009; Table 1 in *Sajeev et al.*, 2010]. Until now no evidence of continent collision has been found in the Imjingang belt. *Zhang* [1997] proposed a different model in which the Dabie-Sulu collision zone extends to the Yanji zone, the late Paleozoic subduction complex in North Korea, through the Imjingang belt. He further argued that the collision started from the east during the early Permian and propagated westward until the late Triassic (Figure 2.4b). These models also suggested that the Imjingang belt is the extension of the collision belt into Korea, without

giving any evidence of collision from the belt. *Ernst and Liou* [1995] regarded the Dabie-Sulu collision zone as not only crossing some parts of Korea, but also extending to the Paleozoic to Mesozoic accretionary complex in the Sikhote-Alen terrane, through the Sangun terrane in Japan (Figure 2.4c). On the other hand, *Ishiwatari and Tsujimori* [2003] suggested the collision belt extends from China to Japan directly without passing through Korea (Figure 2.4d).

The discrepancies among these different tectonic models arose mainly from a lack of core data concerning the nature of the late Paleozoic to Triassic subduction complexes and potential collision belts in Korea. Recently, Triassic eclogite facies metamorphism has been identified

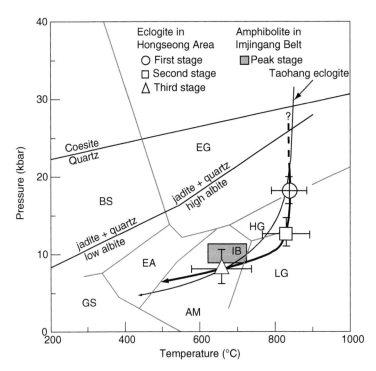

Figure 2.5 Inferred metamorphic conditions and Pressure-Temperature (P-T) paths for the Bibong eclogite in the Hongseong area [*Oh et al.*, 2005] and the Samgot amphibolite in the Imjingang belt [*Ree et al.*, 1996]. Thick and thin solid lines represent P-T paths of the Bibong eclogite and the Taohong eclogites in the Sulu belt [*Yao et al.* 2000], respectively. The petrogenetic grid used is that of *Oh and Liou* [1998]. Labels: AM = amphibolite facies; BS = blueschist facies; EA = epidote amphibolite facies; EG = eclogite facies; GS = greenschist facies; HG = high-pressure granulite facies; LG = low pressure granulite facies; IB = P-T condition for amphibolite in the Imjingang belt.

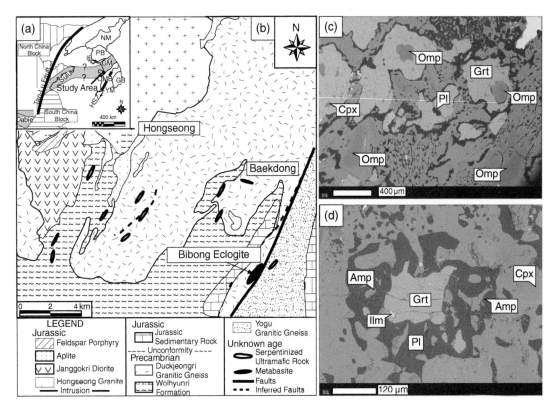

Figure 2.6 (a) Simplified tectonic map of northeast Asia and (b) geological sketch map of the Hongseong area [from *Oh et al.*, 2005]. (c) Back scattered electron image of eclogite consisting of garnet (Grt) and omphacite (Omp) with retrograde plagioclase (Pl) and clinopyroxene (Cpx). (d) Symplectite of plagioclase and amphibole (Amp) formed between garnet and clinopyroxene during the last stage of retrograde metamorphism [from *Kim et al.*, 2006]. Labels: Ilm = illmenite; GB = Gyeongsang basin; GM = Gyeonggi Massif; IB = Imjingang belt; NM = Nangrim Massif; OMB = Ogcheon metamorphic belt; PB = Pyeongnam basin; TB = Taebaeksan basin; YM = Youngnam Massif; HSZ = Honam shear zone.

(a)

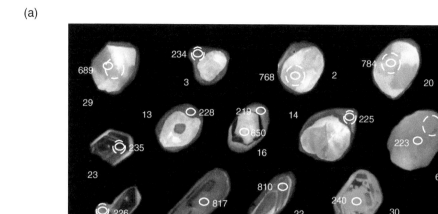

(b)

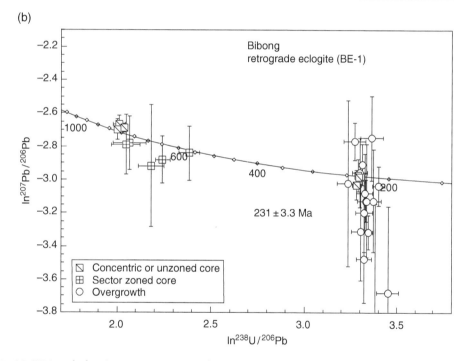

Figure 2.7 (a) SEM catholuminescence images of zircon grains from retrograded eclogite showing core with well-developed metamorphic zircon overgrowths. Numbered ellipses are locations of SHRIMP analyses with the measured age in Ma. (b) Tera-Wasserburg concordia plot of SHRIMP U-Pb isotopic analyses of zircon in the retrograded eclogite from the Bibong metabasite lenticular body [from *Kim et al.*, 2006].

from the Hongseong area in southwestern Gyeonggi Massif, Korea (Figures 2.5, 2.6, 2.7; Tables 2.1, 2.2; *Oh et al.*, 2003, 2004a, 2005], far south of the Imjingang belt. Late Paleozoic to Triassic igneous activity and UHT metamorphism related to continental collision are also reported from the Odesan area in the eastern part of Gyeonggi Massif, Korea [*Oh et al.*, 2006a]. Besides the new important findings in Korea, there has been substantial progress in elucidating an understanding of the tectonic history of east China and several new discoveries have been made from Paleozoic and Triassic HP metamorphic belts in Japan. These advances allow making a more reasonable model to describe the tectonic relationships between Korea, China, and Japan.

Table 2.1 Representative microprobe analyses of garnet and omphacite in retrograde eclogites from the Bibong area.

Rock type	BE-1				BE-2				BE-3			
Mineral	Grt (core)	Grt (rim)	Omp	Omp	Grt (core)	Grt (rim)	Omp	Omp	Grt (core)	Grt (rim)	Omp	Omp
SiO_2	38.17	37.93	53.06	53.70	37.22	37.99	52.53	53.09	37.95	37.57	52.19	52.39
TiO_2	0.12	0.05	0.21	0.07	0.16	0.04	0.23	0.18	0.11	0.13	0.20	0.20
Al_2O_3	20.60	20.73	6.56	7.01	20.86	21.09	8.76	7.22	20.87	19.74	6.94	6.94
Cr_2O_3	0.01	0.04	0.00	0.11	0.05	0.02	0.00	0.06	0.04	0.01	0.04	0.05
FeO^a	26.24	26.56	9.86	8.67	25.71	26.67	6.25	8.44	26.25	26.62	10.29	10.04
MnO	0.72	1.13	0.07	0.06	1.47	0.89	0.06	0.01	0.75	0.76	0.13	0.13
MgO	4.64	4.32	9.21	8.86	4.45	4.47	10.10	8.77	4.58	4.47	8.85	8.71
CaO	9.37	9.11	16.52	16.23	9.39	9.06	18.32	16.01	9.40	9.02	16.23	16.14
Na_2O	0.00	0.00	3.85	4.60	0.01	0.01	3.40	4.70	0.08	0.01	3.80	3.97
K_2O	0.00	0.00	0.00	0.01	0.03	0.01	0.01	0.00	0.00	0.00	0.00	0.00
Total	99.88	99.88	99.33	99.53	99.36	100.25	99.66	98.47	100.05	98.34	98.67	98.58
O	12	12	6	6	12	12	6	6	12	12	6	6
Si	3.004	2.994	1.971	1.980	2.956	2.984	1.920	1.977	2.984	3.016	1.957	1.964
Ti	0.007	0.003	0.006	0.005	0.010	0.003	0.006	0.005	0.007	0.008	0.006	0.006
Al	1.911	1.928	0.287	0.309	1.953	1.952	0.377	0.317	1.935	1.868	0.307	0.307
Cr	0.001	0.003	0.000	0.003	0.003	0.001	0.000	0.002	0.003	0.001	0.001	0.001
Fe^{3+}			0.054	0.072			0.018	0.087			0.063	0.063
Fe^{2+}	1.727	1.754	0.252	0.195	1.708	1.752	0.173	0.176	1.727	1.787	0.260	0.252
Mn	0.048	0.076	0.002	0.002	0.099	0.060	0.002	0.000	0.050	0.052	0.004	0.004
Mg	0.544	0.509	0.510	0.487	0.527	0.523	0.550	0.487	0.537	0.534	0.495	0.487
Ca	0.790	0.771	0.658	0.641	0.799	0.762	0.717	0.639	0.792	0.776	0.652	0.648
Na	0.001	0.000	0.278	0.329	0.001	0.002	0.241	0.340	0.012	0.002	0.276	0.289
K	0.000	0.000	0.000	0.001	0.003	0.001	0.000	0.000	0.000	0.000	0.000	0.000
Total	8.03	8.037	4.018	4.024	8.060	8.040	4.006	4.029	8.046	8.040	4.021	4.021
X_{alm}	0.555	0.564			0.545	0.566			0.556	0.567		
X_{prp}	0.175	0.164			0.168	0.169			0.173	0.170		
X_{sps}	0.015	0.024			0.032	0.019			0.016	0.017		
X_{grs}	0.254	0.248			0.255	0.246			0.255	0.246		
X_{jd}			0.239	0.265			0.233	0.258			0.230	0.241

Source: Kim et al. [2006].

Total iron as FeO.

$X_{alm} = Fe^{2+}/(Fe^{2+}+Mn+Mg+Ca)$.

$X_{prp} = Mg/(Fe^{2+}+Mn+Mg+Ca)$.

$X_{sps} = Mn/(Fe^{2+}+Mn+Mg+Ca)$.

$X_{grs} = Ca/(Fe^{2+}+Mn+Mg+Ca)$.

$X_{jd} = (Na^-Fe^{3+})/(Ca+Na)$.

2.3. THE DABIE-SULU COLLISION BELT BETWEEN THE NCB AND SCB IN CHINA

East China consists of two major Precambrian blocks, the NCB and SCB (Figure 2.3). The SCB is divided into the Yangtze and the Cathaysia blocks. The NCB and SCB probably converged during the early Cambrian and lay close to Australia [*Palmer*, 1974; *Burrett and Richardson*, 1980] or India [*McKenzie et al.*, 2011] by the mid and late Cambrian. *Metcalfe* [1996] proposed that the Paleo-Tethys Ocean opened in the Devonian moving the NCB and SCB northward, and the Paleo-Tethys Ocean, between two blocks, disappeared in the middle Triassic by subduction of all oceanic crust. The presence of 450–400 Ma ophiolites [*Li et al.*, 1991], arc-related 470–435 Ma metamorphism [*Kröner et al.*, 1993], and ca. 210–330 Ma eclogites [*Cong et al.*, 1992; *Zhang and You*, 1993] in the Qinling-Dabie-Sulu collision belt indicates that opening of the Paleo-Tethys had started in the mid-Ordovician instead of the Devonian. Initial stage tectonic models for the Qinling-Dabie-Sulu belt assumed that the collision zone between the NCB and SCB included only one collisional suture, but recent works have revealed a more complex situation involving at least two sutures and three tectonic plates [*Zhang et al.*, 1996; *Li et al.*, 2007]. According to these models, the Qinling-Dabie-Sulu belt includes the NCB, the Qaidam

Table 2.2 SHRIMP U-Pb zircon data from retrograde eclogite in the southwestern part of the Gyeonggi Massif, South Korea.

Grain spot	Zircon type	Pb* (ppm)	U (ppm)	Th (ppm)	Th/U	$^{206}Pb*\%$	$^{204}Pb/^{206}Pb$	±	$^{206}Pb*/^{238}U$	±	$^{207}Pb*/^{206}Pb$	±	Apparent ages (Ma) $^{206}Pb/^{238}U$	±	$^{207}Pb/^{206}Pb$	±
26.1	CZC	32.0	675	946	1.40	0.03	0.000017	0.000017	0.0368	0.0005	0.0505	0.0011	232.8	2.9	216	52
27.2	CZC	27.0	590	795	1.35	0.36	0.000197	0.000119	0.0371	0.0005	0.0507	0.0022	234.5	3.4	225	105
23.1	CZC	18.0	428	352	0.82	1.02	0.000554	0.000157	0.0372	0.0005	0.0511	0.0051	235.3	2.9	244	241
24.1	CZC	34.0	257	92	0.36	0.06	0.000035	0.000022	0.1305	0.0022	0.0678	0.0018	791	13	862	57
22.1	CZC	11.0	74	54	0.73	0.38	0.000217	0.000189	0.1339	0.0042	0.0695	0.0036	810	24	913	110
25.1	CZC	2.0	68	3	0.041	1.37	0.002313	0.000895	0.0375	0.0009	0.0483	0.0045	237.4	5.8	112	206
16.2	SZC	2.0	17	5	0.27	1.95	0.001591	0.001292	0.1062	0.0044	0.0561	0.0086	650	26	457	383
20.1	SZC	1.0	9	3	0.32	1.76	0.000020	0.000020	0.1294	0.0102	0.0615	0.0120	784	58	655	486
2.1	SZC	0.9	9	3	0.32	1.62	0.000586	0.000618	0.1266	0.0117	0.0621	0.0109	768	67	677	429
21.1	SZC	0.7	8	2	0.25	2.39	0.003064	0.002152	0.0921	0.0052	0.0583	0.0104	568	31	541	446
29.1	SZC	0.7	7	2	0.31	3.88	0.001196	0.004615	0.1127	0.0081	0.0541	0.0239	689	47	374	1519
28.1	UZC	19.0	139	45	0.33	0.29	0.000167	0.000255	0.1350	0.0022	0.0673	0.0043	817	13	847	140
27.1	UZ	2.0	76	1	0.013	1.02	0.000661	0.000524	0.0363	0.0008	0.0546	0.0071	229.9	4.8	396	323
6.1	UZ	2.0	57	0.2	0.003	2.12	0.001423	0.001016	0.0352	0.0008	0.0364	0.0039	222.7	5.1	–	–
30.1	UZ	0.9	29	1	0.035	2.04	0.001968	0.001229	0.0379	0.0024	0.0627	0.0075	240	15	697	279
1.1	UZ	0.5	17	0.1	0.007	6.67	0.005190	0.002249	0.0366	0.0019	0.0364	0.0124	232	12	–	–
12.1	UZO	2.0	62	16	0.25	1.54	0.001141	0.000828	0.0333	0.0011	0.0481	0.0060	210.8	7.0	102	272
10.1	UZO	1.0	43	1	0.029	3.57	0.000020	0.000020	0.0357	0.0010	0.0438	0.0059	225.8	6.0	–	–
11.1	UZO	0.8	27	6	0.24	8.09	0.003353	0.001557	0.0342	0.0015	0.0438	0.0158	216.8	9.2	–	–
15.1	UZO	0.7	21	0.2	0.011	4.23	0.001638	0.001673	0.0360	0.0016	0.0309	0.0093	228	10	–	–
3.1	UZO	0.6	19	0.1	0.008	4.65	0.001061	0.000884	0.0369	0.0016	0.0520	0.0121	234	10	286	461
18.1	UZO	0.5	16	1	0.055	5.50	0.003458	0.002619	0.0359	0.0016	0.0459	0.0121	227	10	–	–
14.1	UZO	0.4	14	0.1	0.008	6.51	0.004620	0.003993	0.0355	0.0019	0.0439	0.0178	225	12	–	–
13.1	UZO	0.4	14	0.2	0.013	5.67	0.000020	0.000020	0.0360	0.0017	0.0408	0.0108	228	10	–	–
17.1	UZO	0.4	13	1	0.092	7.93	0.000020	0.000020	0.0317	0.0019	0.0252	0.0173	201	12	–	–
16.1	UZO	0.4	12	0.4	0.032	5.97	0.005744	0.002512	0.0345	0.0021	0.0639	0.0185	219	13	739	739
9.1	UZO	0.3	9	0.2	0.017	4.08	0.002214	0.001853	0.0393	0.0026	0.0487	0.0315	249	16	135	3498
5.1	UZO	0.2	6	0.3	0.049	13.52	0.000020	0.000020	0.0339	0.0022	0.0136	0.0294	215	14	–	–

Source: Kim et al. (2006).
Zircon textures – UZ, unzoned; CZC, concentric zoned core; SZC, sector zoned core; UZO, unzoned overgrowth.
Pb* – radiogenic portion of Pb. Commom Pb corrected using 204Pb.

microblock, and the south China (Yangtze) block. During the early Paleozoic, the oceanic crust between the Qaidam and NCB subducted beneath the NCB, forming an active margin on the southern margin of the NCB. The NCB collided with the Qaidam microblock in the Devonian, resulting in the Shangxian-Danfeng suture, then with the SCB in the Permo-Triassic, forming the Mianxian-Lueyang suture [*Meng et al.*, 2000; *Liu et al.*, 2005; *Li et al.*, 2007 and references therein). This latter collision started in the east and propagated westward [*Zhao and Coe*, 1987; *Zhang*, 1997] until 209 Ma [*Ames et al.*, 1993]. The TanLu fault was initiated in the late Permian (258–248 Ma) as a result of the collision between the NCB and SCB and continued until the late Triassic (231–213 Ma) [*Yin and Nie*, 1993].

In the Dabie area located to the west of TanLu fault, peak metamorphic conditions of 680–880°C and 30–40 kbar were reported for the coesite-bearing Dabie eclogites, indicative of UHP metamorphism [*Hirajima and Nakamura*, 2003, and references therein). During uplift, the eclogites in the Dabie area were overprinted by amphibolite facies retrograde metamorphism (470–570°C, 6–8 kbar) and were extensively deformed [*Cong et al.*, 1995; *Carswell et al.*, 2000]. In the Sulu area located to the east of the TanLu fault, eclogites occur as lenses, blocks, and layers intercalated with the host gneiss and record UHP metamorphic conditions of 750–880°C and 26.5–41 kbar [*Hirajima and Nakamura*, 2003, and references therein). In the northern part of the Sulu area, many of the eclogites were transformed first to HP granulite (700–820°C, 10–17 kbar) and then to amphibolite (600–740°C, 8–9 kbar); however, in the southwestern part of the area, the eclogites were retrograded directly to amphibolite (510–720°C, 8–12 kbar) [*Banno et al.*, 2000; *Yao et al.*, 2000; *Zhang et al.*, 2000]. The temperatures of peak and retrograde metamorphism of the eclogites slightly increase westward from the Dabie area to the Sulu area as reported by *Wang et al.* [1992, 1995] and *Enami et al.* [1993]. The occurrence of coesite and jadeite within orthogneiss in the Dabie and Sulu areas indicates subduction of the entire unit in these areas to the depths of UHP metamorphism. UHP metamorphism in these areas has been dated to occur at ca. 210–230 Ma [*Hirajima and Nakamura*, 2003, and references therein).

2.4. THE EXTENSION OF THE DABIE-SULU BELT IN CHINA INTO THE HONGSEONG AREA IN KOREA

The southern part of the Korean peninsula consists of the Gyeonggi and Yeongnam Precambrian Massifs and two Phanerozoic belts (the Imjingang and Ogcheon belts) (Figure 2.3). The Imjingang belt consists of metasediments

of Devonian-Carboniferous age [*Ri and Ri*, 1990, 1994]. The Ogcheon belt (also called the Okcheon belt) is subdivided into the northeastern Taebaeksan basin and the southwestern Ogcheon metamorphic belt. The Ogcheon metamorphic belt consists of nonfossiliferous, low- to medium-grade metasedimentary and metavolcanic rocks from Neoproterozoic to late Paleozoic. By contrast, the Taebaeksan basin consists of fossiliferous, non- or weakly metamorphosed sedimentary rocks with Paleozoic to early Mesozoic age.

In the Hongseong area, the southwestern part of the Gyeonggi Massif of South Korea, ultramafic rocks, metabasites, and marbles are preserved as lensoid bodies within granitic gneiss [*Choi et al.*, 1998] as they are in the Sulu belt (Figure 2.6). The SHRIMP U-Pb zircon intrusion ages of the granitic gneisses have a narrow age range falling between ca 850–830 Ma and formed in the arc tectonic setting [*Kim et al.*, 2008]. The Bibong metabasite in the Hongseong also formed in the arc tectonic setting at 887 Ma and later underwent eclogite facies metamorphism (835–860°C, 17.0–20.9 kbar) during the Triassic (ca. 230 Ma) (Figures 2.5, 2.6, 2.7; Tables 2.1, 2.2; *Oh et al.*, 2009]. During uplift, the Bibong eclogite was first overprinted by granulite facies metamorphism (830–850°C, 11.5–14.6 kbar) and then by amphibolite facies metamorphism (570–740°C, 6.7–11.0 kbar) indicating isothermal decompression [Figure 2.5; *Oh et al.*, 2005; *Kim et al.*, 2006]. The isothermal decompressional P-T path represents fast uplift. A metamorphic age of ca. 230 Ma is also obtained from the rims of zircons from the Neoproterozoic granitic gneisses in the Hongseong area [*Kim et al.*, 2008]. During the decompression, symplectite formed around garnet (Figure 2.6c) and such simplectite is also observed in the other metabasite lenses in the Hongseong area. These results led *Oh et al.* [2004a, 2005] to suggest that the Hongseong area is an extension of the Dabie-Sulu collision belt in China (Figure 2.8). They also argued that collision between the NCB and SCB probably occurred earlier in Korea than in China. The finding of a gravity anomaly crossing the West Sea between Korea and China by *Choi et al.* [2006] supports the suggestion.

2.5. THE LATE-PERMIAN TO TRIASSIC COLLISION BELT IN KOREA

In the Odesan area located in the eastern part of the Gyeonggi Massif in Korea, Triassic mangerite and gabbro intruded the Paleoproterozoic migmatitic and porphyroblastic gneisses [Figures 2.9, 2.10; Table 2.3; *Oh et al.*, 2006b; *Kim et al.*, 2011]. The mangerite is porphyritic, containing alkali feldspar phenocrysts up to 3 cm long in a matrix consisting of orthopyroxene, clinopyroxene, amphibole, biotite, plagioclase, quartz,

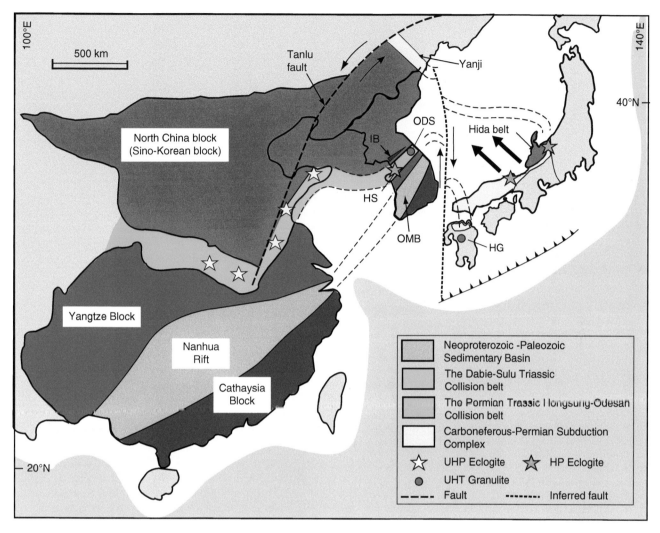

Figure 2.8 The simplified tectonic map of northeastern Asia showing the tectonic relationships between Korea, China, and Japan, and the metamorphic trend along the Dabie-Hongseong collision belt [modified after *Oh and Kusky*, 2007]. Westward along the Dabie-Hongseong collision belt, the metamorphism changes from UHT (ultrahigh-temperature metamorphism, red circle) to HP (high-pressure metamorphism, orange star) and UHP (ultrahigh-pressure metamorphism, yellow star). Labels: HP = high pressure; UHP = ultrahigh pressure; UHT = ultrahigh temperature; IB = Imjingang belt; ODS = Odesan area; HS = Hongseong; OMB = Ogcheon Metamorphic; HG = Higo.

apatite, and zircon (Figure 2.9b). *Oh et al.* [2006b] shows that the mangerites differ from mangerites formed in a typical within-plate tectonic setting in their high Mg# and Sr concentrations and negative Nb and Ta anomalies. Their LILE enrichment and negative Ti-Nb-Ta anomalies are probably inherited from a precollision subduction event, indicating a convergent margin origin for the mangerites. *Kim et al.* [2011] confirmed that the mangerite formed in the postcollisional-tectonic setting and the associated gabbro formed in the within-plate tectonic setting (Figure 2.11; Table 2.3). Both rocks show shoshonitic, calc-alkaline and high Ba-Sr geochemical characters, which are common for the post-

collision igneous rocks (Figure 2.12). Together with the geochemical data, the U-Pb zircon age of ca. 230 Ma from the mangerite and gabbro implies that the collision belt in Hongseong extends into the Odesan area [Table 2.4: *Kim et al.*, 2011]. The ca. 230 Ma postcollision igneous rocks are also found in the Yangpyeong area, which is located between the Hongseong and Odesan areas suggesting the Hongseong-Yangpyeong-Odesan collision belt within the Gyeonggi Massif [*Lee and Oh*, 2009]. The suggested Hongseong-Yangpyeong-Odesan collision belt divides the Gyeonggi Massif into the northern and southern Gyeonggi Massifs. In the northern Gyeonggi Massif, the ca. 230 Ma postcollision igneous

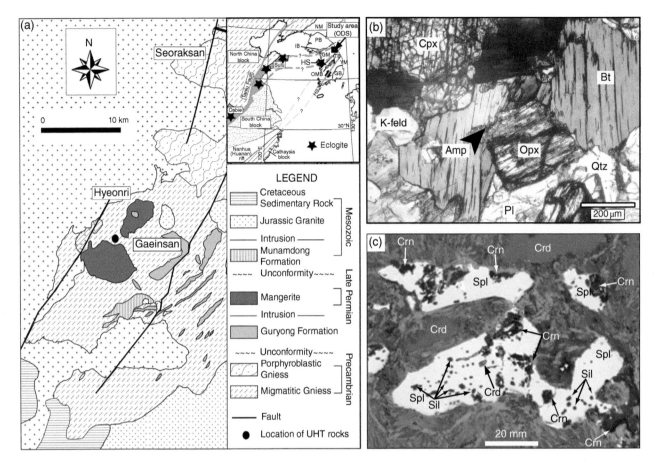

Figure 2.9 (a) Simplified tectonic map of northeast Asia and geologic map of the Odesan area [from *Oh et al.*, 2006a]. (b) Photomicrograph showing the assemblage orthopyroxene (Opx) + clinopyroxene (Cpx) + amphibole (Amp) + plagioclase (Pl) + K-feldspar (K-feld) + quartz (Qtz) within mangerite [from *Oh et al.*, 2006b]. (c) Back scattered electron image showing the assemblage spinel (Spl) + cordierite (Crd) + corundum (Crn), as well as inclusions of corundum, cordierite, and sillimanite (Sil) in spinel within spinel granulite [from *Oh et al.*, 2006a]. Labels: HS = Hongseong area; ODS = Odesan area; TB = Taebaeksan basin; GB = Gyeongsang basin; GM = Gyeonggi Massif; IB = Imjingang belt; NM = Nangrim Massif; OMB = Ogcheon metamorphic belt; PB = Pyeongnam basin; TB = Taebaeksan basin; YM = Youngnam Massif.

rocks are widely found [Figure 2.13; *Choi et al.*, 2008; *Peng et al.*, 2008; *Lee and Oh*, 2009; *Seo et al.*, 2010; *Park*, 2009; *Williams et al.*, 2009; *Kim et al.*, 2011]. During the continent-collision process, postcollision igneous rocks formed in the continent under which the other side of the continent is subducted [Figure 2.14; *Seo et al.*, 2010; *Kim et al.*, 2011]. As the SCB subducted under the NCB, the postcollision igneous rocks might have formed in the NCB. Therefore, the distribution of ca. 230 Ma postcollision igneous rocks together with eclogite indicates that the line connecting the Hongseong-Yangpyeong-Odesan areas is the collision boundary between the north and SCB in the Korean peninsula [*Oh and Kusky*, 2007; *Kim et al.*, 2011].

The Hongseong collision belt, consisting mainly of Neoproterozoic granitoids metamorphosed into gneiss at ca. 230 Ma, is terminated by NNE-striking faults and does not continue to the east side of the fault where Paleoproterozoic granitic gneiss crops out. The rocks in and around the fault seem to show sinistral strike-slip movement senses [*Fitches*, 2004] suggesting that the collision belt moved to the north in areas located to the east of the sinistral fault. The suggested Yangpyeong-Odesan belt is located to the north of the Hongseong belt, supporting the idea that the eastern part of the collision belt in Korea may be moved toward the north along the expected sinistral strike-slip fault (Figure 2.13). The fault may continue to the eastern end of the Imjingang belt where another sinistral strike-slip fault with a strong NNE-striking mylonitic foliation terminates the belt [*Jeon and Kwon*, 1999]. Although not connected, several NNE-striking faults are identified along the line connecting the two sinistral faults in eastern parts of the Hongseong and

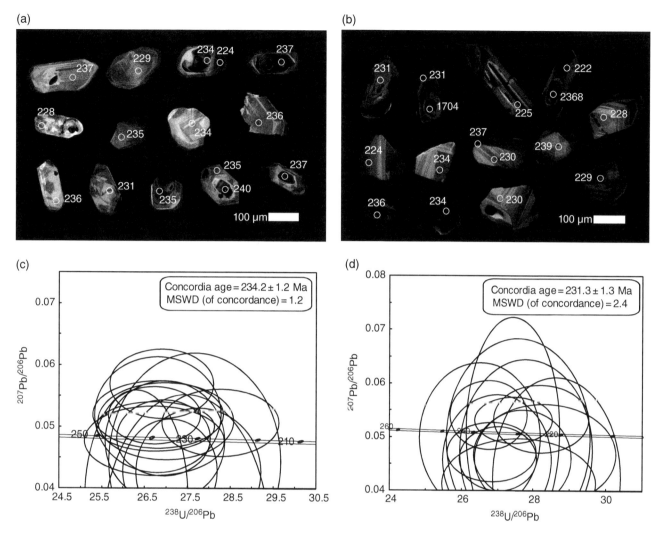

Figure 2.10 SEM cathodoluminescence (CL) images of zircon grains from (a) the mangerite and (b) gabbro. Circles indicate the locations of SHRIMP analyses (with measured ages in Ma). Concordia plots of SHRIMP U-Pb isotopic analysis of zircons from (c) mangerite and (d) gabbro in the Odesan area [*Kim et al.*, 2011].

Imjingang belts (Figure 2.13; *Korea Institute*, 2001]. The expected sinistral fault will be called "Hongseong fault" in this book.

The proposed collision belt can also be supported indirectly by the following aspects. First is the concentration of most Permo-Triassic metamorphic ages in South Korea within the Gyeonggi Massif and its neighboring two belts (the Ogcheon metamorphic belts and Imjingang belt). The 250–230 Ma metamorphic ages are obtained from the granitic gneisses and metabasites in the Gyeonggi Massif [*So et al.*, 1989; *Cho et al.*, 1996, 2001; *Cho*, 2001; *Cho and Kim*, 2003; *Sagong et al.*, 2005; *Oh et al.*, 2004a, 2005, 2006a; *Guo et al.*, 2004; *Kim*, 2011; *Lee and Oh*, 2009; *Kim*, 2011]. The 300–260 Ma metamorphic ages are reported from the Ogcheon metamorphic belt [*Kim et al.*, 2001; *Cheong*

et al., 2003; *Kim*, 2005] and 255–249 Ma metamorphic ages, from the Imjingang Belt [*Ree et al.*, 1996; *Cho et al.*, 1996, 2001]. The concentration of Permo-Triassic metamorphic ages in and around the Gyeonggi Massif might be resulted from a very strong Permo-Triassic tectonometamorphic event, such as the collision between the NCB and SCB in the Gyeonggi Massif. Especially along the Hongseong-Odesan belt, the possible boundary between the NCB and SCBs, the highest grade Permo-Triassic granulite facies metamorphism is concentrated [*Oh et al.*, 2005; *Lee and Oh.*, 2009; *Kim*, 2011].

The second line of evidence includes the intermediate-pressure (IP) metamorphism in both the Imjingang and Ogcheon metamorphic belts during Permo-Triassic time with increasing metamorphic grade toward the collision

Table 2.3 Whole rock compositions of representative mangerites and gabbros in the Odesan area, the eastern part of the Gyeonggi Massif, South Korea.

	Kim [2011]				Oh et al. [2006a]								
	Mangerite		Gabbro		Mangerite					Gabbro			
Sample	2B	2C	1C	1D	2A	2B	4A	6	7A	4B	7C	7D	8
Major elements (wt %)													
SiO_2	55.55	56.69	52.00	51.25	60.60	57.30	56.55	57.29	55.90	52.40	54.20	53.09	53.43
Al_2O_3	14.24	15.49	11.78	13.34	13.99	15.91	15.98	14.97	14.30	13.13	13.54	13.23	12.56
Fe_2O_3	0.90	0.74	1.18	1.26	N.A	N.A	N.A	N.A	N.A	N.A	N.A	N.A	N.A
FeO	6.31	5.12	8.06	7.36	5.90	6.45	6.06	6.57	6.66	9.61	8.06	8.66	9.37
MnO	0.12	0.09	0.16	0.15	0.08	0.08	0.08	0.09	0.08	0.14	0.11	0.12	0.14
MgO	5.91	4.51	8.52	8.28	3.61	4.03	4.43	5.05	4.95	8.67	6.92	7.71	8.37
CaO	5.74	5.29	8.32	8.45	3.28	3.99	4.61	5.68	4.30	7.59	5.80	7.07	7.84
Na_2O	2.24	2.42	1.97	2.21	2.80	2.98	2.78	2.70	2.51	2.19	2.43	2.17	2.15
K_2O	4.60	5.01	2.56	2.31	3.48	5.07	5.06	4.15	4.21	2.64	3.37	2.57	2.62
TiO_2	1.38	1.29	1.22	1.15	0.67	1.13	1.20	1.22	1.24	1.29	1.22	1.25	1.07
P_2O_5	0.50	0.39	0.60	0.60	0.29	0.35	0.40	0.42	0.41	0.60	0.52	0.58	0.56
LOI	1.66	1.13	2.22	1.32	4.53	2.22	1.94	0.90	4.22	1.18	3.30	3.29	1.34
Total	99.87	98.72	99.51	98.50	99.23	99.52	99.10	99.04	98.78	99.43	99.49	99.75	99.45
Trace elements (ppm)													
Sc	20	16	32	27	N.A	N.A	N.A	N.A	N.A	N.A	N.A	N.A	N.A
Be	2	2	2	2	N.A	N.A	N.A	N.A	N.A	N.A	N.A	N.A	N.A
V	146	118	208	202	79	92	105	110	114	183	150	181	176
Cr	300	240	510	480	N.A	N.A	N.A	N.A	N.A	N.A	N.A	N.A	N.A
Co	29	22	39	37	N.A	N.A	N.A	N.A	N.A	N.A	N.A	N.A	N.A
Ni	80	60	110	110	42	62	58	63	63	115	86	110	111
Cu	40	30	20	40	N.A	N.A	N.A	N.A	N.A	N.A	N.A	N.A	N.A
Zn	120	80	160	120	N.A	N.A	N.A	N.A	N.A	N.A	N.A	N.A	N.A
Ga	21	20	22	20	20	23	20	21	20	21	21	19	18
Ge	1.7	1.5	2.1	1.8	N.A	N.A	N.A	N.A	N.A	N.A	N.A	N.A	N.A
Rb	140	124	96	87	116	183	115	1142	920	121	118	104	94
Sr	1110	1184	690	929	569	709	1141	1142	920	928	997	914	774
Y	24	17	29	25	20	17	13	16	16	27	20	22	26
Zr	131	123	145	115	266	492	341	93	330	159	115	94	111
Nb	22.3	23.9	23.2	18.5	18.4	26.2	20.2	22.8	25.5	16.4	19.6	17.7	17.9

Ag	0.6	0.6	0.7	0.5	N.A	N.A	N.A	N.A	N.A	N.A	N.A	N.A	N.A
Cs	3.3	2.2	1.9	2.3	2.8	5.3	1.9	2.1	3.1	1.8	2.2	3.2	2.3
Ba	3018	3697	1710	1628	2136	2099	3672	3966	2344	2323	3885	1957	1906
La	57.7	49.1	56.6	56.6	53.3	57	40.3	44.5	48.3	57.2	48.3	53.2	53.5
Ce	112	92.7	117	114	98.4	108.3	77	85.4	93.2	117.8	98.2	104.7	108
Pr	12.1	9.8	13.3	12.6	10.7	12	8.7	9.7	10.4	13.7	11.5	12.3	12.5
Nd	47.1	36.1	51.5	45.2	38.2	42.8	32.7	37	38.9	51.5	42.8	46.4	47.7
Sm	8.5	3.09	9.52	8.52	6.4	7.1	5.6	6.5	6.8	9.5	7.7	8.3	8.5
Eu	2.62	2.8	1.84	2.18	1.6	2.3	3.2	3.1	2.5	2.4	2.5	2.4	2.1
Gd	6.51	4.54	7.26	6.53	4.9	5.4	4.5	5.1	5.1	7.8	6.1	6.7	7.2
Tb	0.91	0.62	1.03	0.94	0.8	0.8	0.7	0.87	0.7	1.2	0.9	1	1.1
Dy	4.76	3.34	5.5	5.06	4	3.8	3.1	3.6	3.7	5.8	4.4	4.8	5.3
Ho	0.87	0.62	1.01	0.94	0.8	0.7	0.6	0.6	0.7	1.1	0.8	0.9	1
Er	2.36	1.7	2.69	2.58	2.2	2	1.6	1.7	1.9	3	2.3	2.5	2.9
Tm	0.33	0.24	0.37	0.36	0.3	0.3	0.2	0.2	0.3	0.4	0.3	0.3	0.4
Yb	2.04	1.5	2.3	2.26	2.2	1.8	1.5	1.5	1.7	2.8	2	2.2	2.6
Lu	0.31	0.23	0.34	0.35	0.3	0.3	0.2	0.2	0.3	0.4	0.3	0.3	0.4
Hf	3.3	3.1	3.5	3	7.4	12.4	8	2.7	8.2	4.1	3.2	2.9	3.3
Ta	1.22	1.27	0.83	0.79	1.5	1.4	1.1	1.2	1.5	0.6	1.1	1	1
Tl	0.54	0.47	0.36	0.35	N.A	N.A	N.A	N.A	N.A	N.A	N.A	N.A	N.A
Pb	34	26	16	14	31	32	32	32	31	14	29	18	28
Th	5.31	4.77	3.41	4.93	9.4	7.5	2	2.4	6.6	1	3.9	5.5	4.8
U	1.57	1.21	0.94	1.18	2.4	2.8	0.7	1	1.6	0.5	1.3	1.6	1.3

N.A: Not analyzed.

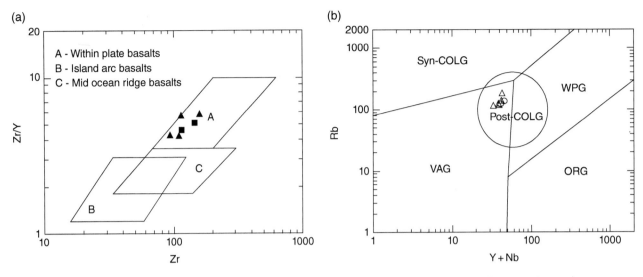

Figure 2.11 The tectonic settings for the mangerite and gabbro in the Odesan area [modified *Kim et al.*, 2011] using (a) Zr/Y vs. Zr diagram [*Pearce and Norry*, 1979] and (b) Y + Nb vs. Rb diagram [*Pearce et al.*, 1984; *Pearce*, 1996]. Symbols are the same as in Figure 2.12, open circles and filled squares represent mangerites and gabbros reported by *Kim et al.* (2011), respectively. The mangerite and gabbros reported by *Oh et al.* 2006b are shown by open and filled triangles, respectively.

boundary in both belts (northward in the Ogcheon and southward in the Imjingang). In the Ogcheon metamorphic belt, the metamorphic grade increases toward the northwest from the biotite zone to the staurolite zone [*Oh et al.*, 1995, 2004b; *Kim et al.*, 2001; *Cheong et al.*, 2003; *Kim*, 2005]. The Ogcheon metamorphic belt is located 50 km south of the Hongseong-Odesan collision belt.

These data suggest that the intermediate-P/T(IP) metamorphism in the Ogcheon metamorphic belt arose from compression caused by a distal collision event [*Oh et al.*, 2004b]. This suggestion is supported by the widespread Barrovian-type IP metamorphism formed in areas within 150 km of the collision boundary in the Himalayan collision belt [*Liou et al.*, 2004; *Patrick*, 1996]. In the Himalayan, the metamorphic grade increases toward the collision boundary during the Barrovian-type IP metamorphism. The Imjingang belt also underwent the Barrovian-type IP metamorphism during the late Permian (ca. 250–260 Ma; *Jeon and Kwon*, 1999; *Cho et al.*, 2001] and the metamorphic grade increases toward the south. As the Hongseong-Odesan collision belt is located to the south of the Imjingang belt, the Permo-Triassic metamorphism in the Imjingang belt might be also an intermediate-P/T metamorphism produced during collision in the area peripheral to the collision zone, as in the Ogcheon metamorphic belt. Therefore, the reversed direction of increasing metamorphic grade in both belts toward the Hongseong-Odesan belt during Permo-Triassic IP metamorphism supports the existence of the collision boundary within the Gyeonggi Massif located between the two belts.

Sr, Nd, and Pb isotopic compositions of the Cenozoic basalts analyzed from Baengnyongdo Island, Jeongok,

Ganseong, and Jeju Island of Korea [*Park et al.*, 2005], provide the third piece of evidence. The Cenozoic basalts in Jeju Island located south of the Hongseong-Odesan collision belt show isotopic characters similar to those in the SCB displaying Depleted Mid-ocean ridge basalt Mantle-Enriched Mantle 2(DMM-EM2) mixing, whereas the other basalts located to the north of the belt have isotopic characters similar to those in the NCB revealing mixing between DMM and EM1. *Park et al.* [2005] suggested that the subcontinental lithosphere-mantle boundary between the NCB and SCB should continue to South Korea, and the Korean continental collision zone may cross the Korean Peninsula.

2.6. THE EXTENSION OF THE DABIE-HONGSEONG COLLISION BELT INTO JAPAN AND NORTH KOREA

In previous sections, the Dabie-Sulu collision belt between the NCB and SCB is confirmed to extend into the Hongseong-Odesan collision in Korea. From here, the collision belt that extends from the Dabie in China to the Odesan area in Korea is referred as the "Dabie-Hongseong collision belt." In this section, we will check the possibility of extension of the Dabie-Hongseong collision belt into Japan.

The geology of southwestern Japan is characterized by the Hida and Oki metamorphic belts, and a series of accretionary complexes developed at a continental margin during early Paleozoic times [e.g., *Tsujimori et al.*, 2000b; Figure 2.15]. The Oki metamorphic belt forms the

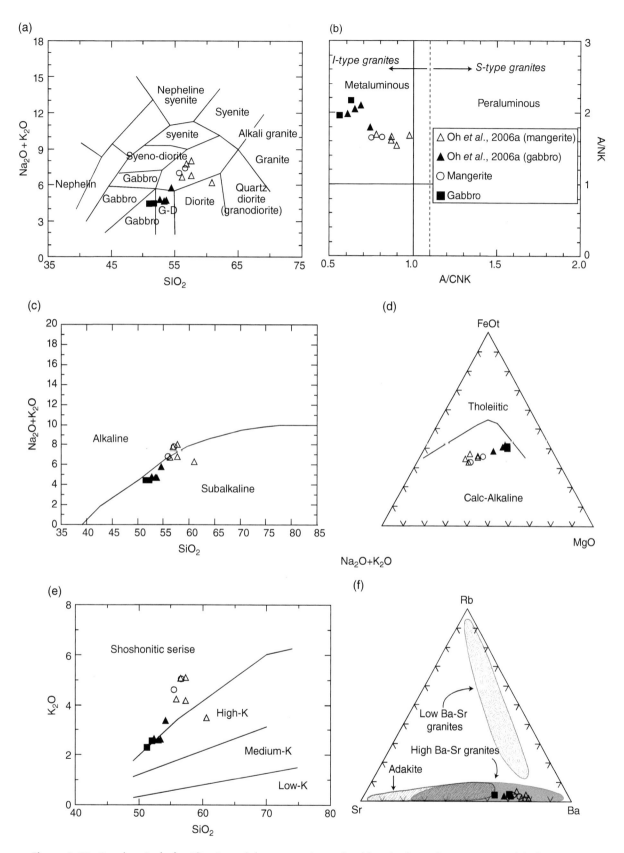

Figure 2.12 Geochemical classification of the mangerite and gabbro in the Odesan area [modified after *Kim et al.*, 2011] using (a) total alkalis vs. silica (TAS) diagram [*Cox et al.*, 1979; *Wilson*, 1989]; (b) A/NK vs. A/CNK diagram [*Maniar and Piccoli*, 1989], A = Al$_2$O$_3$, N = Na$_2$O, K = K$_2$O, and C = CaO (molar), *Chappell and White* [1974] classified I-and S-type granites; (c) total alkalis vs. silica diagram; (d) AFM diagram illustrating the dominant calc-alkaline nature of the Odesan plutonic rocks, the boundary between tholeiite and calc-alkaline series is from *Irvine and Baragar* [1971]; (e) subdivision of subalkalic rocks based on the K$_2$O vs. SiO$_2$ diagram [*Peccerillo and Taylor*, 1976]; (f) Rb-Sr-Ba ternary diagram for high and low Ba-Sr granite fields [*Tarney and Jones*, 1994].

Table 2.4 SHRIMP U-Pb zircon data from the mangerite and gabbro in the Odesan area.

Grain spot	Pb[a] (ppm)	U (ppm)	Th (ppm)	Th/U	$^{207}Pb/^{235}U$	±	$^{206}Pb/^{238}U$	±	$^{207}Pb/^{206}Pb$	±	Apparent ages (Ma) $^{206}Pb/^{238}U$	±	$^{207}Pb/^{235}U$	±
Gabbro (OD100205-1C)														
1.1	6	144	115	0.79	0.25726	0.04453	0.03645	0.00071	0.05116	0.00865	230.79	4.39	232.45	36.62
2.1	6	145	141	0.98	0.23207	0.02647	0.03704	0.00085	0.04573	0.00497	234.49	5.29	211.91	22.05
3.1	13	353	165	0.47	0.23293	0.01348	0.03700	0.00060	0.04594	0.00246	234.21	3.75	212.61	11.16
4.1	6	150	102	0.68	0.22627	0.02046	0.03533	0.00100	0.04668	0.00384	223.82	6.25	207.12	17.08
5.1	14	348	196	0.56	0.24126	0.01301	0.03725	0.00071	0.04719	0.00227	235.78	4.40	219.45	10.70
6	4	107	86	0.80	0.20379	0.01891	0.03625	0.00115	0.04123	0.00344	229.78	7.14	188.32	16.08
7	13	327	148	0.45	0.27994	0.01494	0.03737	0.00062	0.05411	0.00263	236.52	3.82	250.61	11.92
8	68	499	128	0.26	1.91303	0.04777	0.13287	0.00237	0.10442	0.00161	804.20	13.50	1085.64	16.79
9	14	371	156	0.42	0.26757	0.01033	0.03643	0.00059	0.05311	0.00176	230.69	3.65	240.75	8.31
10	7	168	144	0.86	0.26175	0.01407	0.03546	0.00068	0.05336	0.00256	224.61	4.23	236.07	11.39
11	3	86	50	0.58	0.24745	0.04002	0.03627	0.00108	0.04955	0.00768	229.68	6.69	224.50	33.11
12	4	111	84	0.75	0.23413	0.03575	0.03598	0.00097	0.04739	0.00695	227.88	6.06	213.61	29.85
13	272	704	942	1.34	6.37314	0.12250	0.30402	0.00140	0.15204	0.00481	1711.20	23.81	2028.58	17.01
14	6	168	68	0.41	0.19927	0.03772	0.03502	0.00089	0.04172	0.00768	221.92	5.54	184.51	32.45
15	5	112	88	0.79	0.25902	0.03117	0.03770	0.00098	0.04989	0.00569	238.59	6.09	233.87	25.45
16	8	197	121	0.62	0.22760	0.01926	0.03609	0.00071	0.04600	0.00367	228.58	4.44	208.22	16.05
Mangerite (OD100205-2C)														
1	6.0	141	114	0.810	0.24510	0.01395	0.03742	0.00230	0.04770	0.00093	236.82	5.81	222.59	11.44
2	8.0	219	117	0.530	0.20972	0.02191	0.03603	0.00424	0.04265	0.00077	228.15	4.81	193.32	18.56
3	12.0	287	262	0.910	0.21564	0.01811	0.03691	0.00343	0.04280	0.00061	233.66	3.80	198.27	15.24
4	15.0	381	256	0.670	0.29382	0.01361	0.03717	0.00229	0.05690	0.00068	235.27	4.25	261.56	10.73
5	5.0	130	89	0.680	0.25972	0.01509	0.03704	0.00255	0.05085	0.00089	234.49	5.54	234.44	12.23
6	22.0	563	323	0.570	0.26761	0.01124	0.03742	0.00190	0.05181	0.00061	236.81	3.81	240.78	9.05
7	19.0	443	367	0.830	0.25761	0.01055	0.03738	0.00163	0.05002	0.00079	236.29	4.88	232.67	8.55
8	18.0	448	284	0.640	0.24555	0.01404	0.03712	0.00253	0.04814	0.00061	234.95	3.82	222.95	11.51
9	8.0	193	113	0.580	0.19587	0.03818	0.03801	0.00727	0.03798	0.00071	240.49	4.44	181.63	32.95
10	9.0	200	179	0.890	0.26767	0.01370	0.03729	0.00223	0.05198	0.00087	236.04	5.40	240.83	11.04
11	5.0	132	93	0.700	0.14969	0.05008	0.03642	0.01006	0.03056	0.00107	230.60	6.65	141.63	45.22
12	11.0	294	164	0.560	0.24701	0.01248	0.03538	0.00224	0.05064	0.00070	224.11	4.33	224.15	10.21
13	5.0	131	75	0.580	0.26731	0.02139	0.03731	0.00384	0.05190	0.00084	236.14	5.23	240.54	17.28
14	6.0	149	73	0.490	0.24227	0.02168	0.03708	0.00401	0.04759	0.00078	234.69	4.88	220.28	17.88
15	5.0	135	74	0.550	0.21931	0.03693	0.03614	0.00718	0.04437	0.00107	228.83	6.66	201.33	31.23

Source: *Kim* [2011].
[a] Radiogenic Pb, corrected for common Pb using either ^{204}Pb or ^{208}Pb.

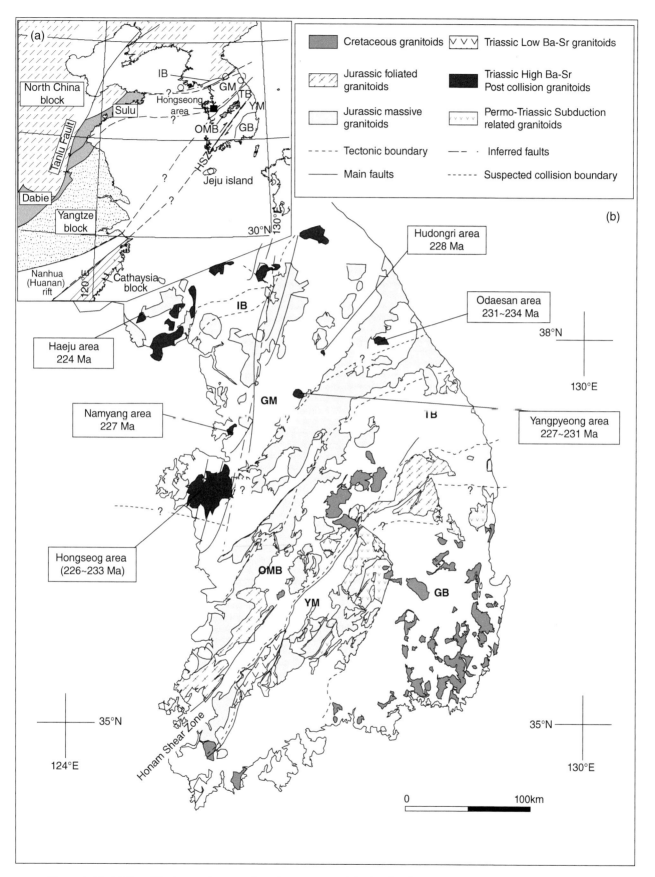

Figure 2.13 (a) Simplified tectonic map of northeast Asia and (b) the distribution of Triassic igneous rocks in South Korea [modified after *Kim et al.*, 2011]. Abbreviations are the same as those used in Figure 2.3. Ages for Triassic plutonic rocks were obtained from the following sources: *Choi et al.* [2008]; *Peng et al.* [2008]; *Williams et al.* [2009]; *Park* [2009]; *Seo et al.* [2010]; *Lee and Oh* [2009]; *Kim et al.* [2011].

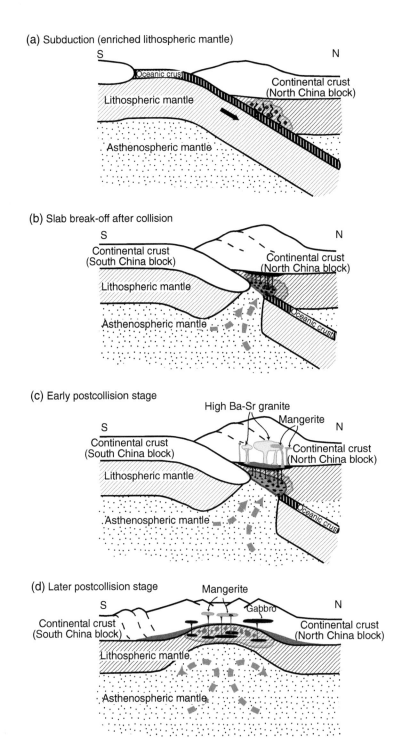

(a) Subduction (enriched lithospheric mantle)

(b) Slab break-off after collision

(c) Early postcollision stage

(d) Later postcollision stage

Figure 2.14 Tectonic model for postcollision igneous rocks [modified after *Kim et al.*, 2011]. (a) Subduction stage before continental collision. Water and crustal elements were supplied to the lithospheric mantle from the subducted oceanic crust and sediments. As a result, lithospheric mantle was enriched with water and crustal elements. (b) Slab break-off stage. During the final stages of collision, the oceanic slab broke off from the continental slab, making an opening between the continental and oceanic slabs. The heat of the asthenosphere supplied through this opening to the lithospheric mantle. Partial melting of the lithospheric mantle occurred in response to heat derived from the asthenosphere and formed gabbroic magma that underplated the lower crust. (c) Early postcollision stage. The opening became wider with more heat supply from the asthenosphere. As a result, the underplated gabbro and middle continental crust were partially melted to produce mangeritic and granitic magmas with shoshonitic character, respectively. The mangeritic magma intruded the crust together with granitic magma. In some places of the crust, these two magmas mingle together. (d) Late postcollision stage. During the late postcollision stage, the continental slab was thinning by extension and the lower continental crusts were delaminated into the mantle due to their high density obtained during thickening of crust by collision. In this stage, the mantle had uplifted to the shallow depth and underwent partial melting to produce shoshonitic gabbro and mangerite (or syenite). They intruded in the crust and mingled together in some places.

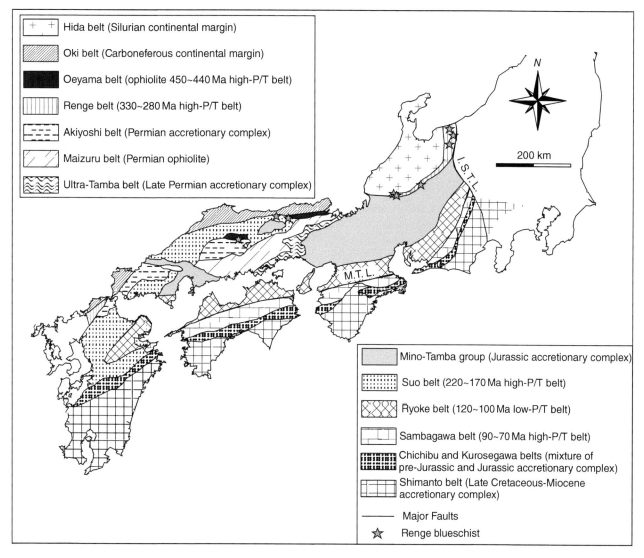

Figure 2.15 Map showing the geologic and tectonic subdivisions of southwestern Japan [modified after *Tsujimori, 2002*]. The timing of accretion generally gets younger oceanward. Labels: MTL = Median tectonic line; I.S.T.L = Itoigawa-Shizuoka tectonic time.

northernmost geotectonic unit of southwest Japan with the Hida belt [*Arakawa et al.*, 2001]. The metamorphic rocks in the belt have a Carboniferous-Permian protolith age with a ca. 250 Ma metamorphic age, and detrital zircon with Precambrian ages (up to 3000 Ma) derived from Precambrian basement [*Suzuki and Adachi*, 1994]. Amphibolites in the Oki metamorphic rocks display a chemical affinity similar to continental basalts. The Hida belt is composed of the Hida gneiss and Unazuki schist. Their protoliths were suggested to have formed in a continental shelf-platform setting [*Sohma et al.*, 1990; *Hiroi*, 1978, 1983]. Major and trace element characteristics of the igneous rocks in the Hida belt indicate their origin from a continental margin or continental arc environment [*Arakawa et al.*, 2000]. The Hida gneiss experienced

two periods of regional metamorphism: the first at ca. 350 Ma under granulite facies conditions, and the second at 240–220 Ma under amphibolite facies (IP-type metamorphism) conditions [*Arakawa et al.*, 2000]. The Unazuki schist also experienced IP metamorphism. Detrital zircons with Precambrian ages in the Hida gneisses [*Sano et al.*, 2000] point to the Hida belt being connected to Precambrian basement, while structural studies have shown that the belt and its cratonic margin were thrust onto the Permian-Jurassic accretionary complexes as a large nappe [*Komatsu and Suwa*, 1986; *Komatsu* et al., 1993]. The Hida and Oki belts in Japan were regarded as the Ordovician-Carboniferous continental margin of the NCB [*Isozaki*, 1997; *Arakawa et al.*, 2000, 2001].

In the southwestern part of Japan, four HP metamorphic belts are dispersed in subhorizontal nappes in which the older nappes typically occupy the uppermost structural positions. These belts include the ca. 450–400 Ma Oeyama belt, the ca. 330–280 Ma Renge belt, the ca. 220–170 Ma Suo belt, and the ca. 90–60 Ma Sambagawa belt [*Tsujimori et al.*, 2000b]. Ophiolitic peridotite bodies within the Oeyama belts are cut by gabbro and dolerite intrusions that display a Mid Oceanic Ridge Basalt (MORB)-like chemistry. A preliminary Sm-Nd intrusion age of ca. 560 Ma for the gabbro [*Hayasaka et al.*, 1995] indicates that the ophiolite formed in the Cambrian. The Fuko Pass metacumulate, which exhibits HP metamorphism, is exposed as a fault-bounded sheet in the Oeyama peridotite body. *Nishina et al.* [1990] obtained hornblende K-Ar ages of 426–413 Ma from the metacumulate, which implies that subduction at the paleo-Pacific margin began in the Silurian [*Tsujimori*, 1999].

The Carboniferous (330–280 Ma) regional HP Renge metamorphic belt [e.g., *Tsujimori and Itaya*, 1999] is dismembered and now distributed sporadically throughout the inner zone of southwestern Japan (Figure 2.15). The Renge belt consists of HP schists of a variety of metamorphic grades including blueschist facies metamorphic rocks and glaucophane-bearing eclogites [*Tsujimori and Itaya*, 1999; *Tsujimori et al.*, 2000a]. The Hida marginal belt is characterized by a serpentinite melange with high-P/T-type Renge blueschists and various fragments of Paleozoic accretionary complexes [*Tsujimori*, 2002]. The northern area of the Hida marginal belt is divided into eclogitic and noneclogitic units. Preliminary phengite ^{40}Ar-^{39}Ar and K-Ar ages from the eclogitic unit are 343–348 Ma [*Tsujimori et al.*, 2001] suggesting that the Hida marginal belt represents an eastern extension of the suture zone in east-central China [*Tsujimori*, 2002; *Tsujimori et al.*, 2006]. Permian accretionary complexes (the Akiyoshi, Maizuru, and Ultratemba belts) form a major component of the inner zone of southwestern Japan. These geologic settings indicate that the inner part of southwestern Japan was a continental margin where a subduction/ accretionary complex formed from the Early to Late Paleozoic. Based on these data, the inner part of southwestern Japan can be interpreted as an extension of the Dabie-Hongseong collision belt where Paleozoic subduction/accretionary complex might be formed before collision [*Oh*, 2006; *Oh and Kusky*, 2007]. *Zhang* [1997] suggested that the Yanji belt, the Carboniferous-Permian subduction complex with ophiolite, in the northeastern end of the NCB is the extension of the Dabie-Sulu belt. These data indicate that the Dabie-Hongseong belt extends into the Yanji belt through the inner part of southwestern Japan forming the Dabie-Sulu-Hongseong-Odesan-Hida-Yanji belt, which encircles the south and east borders of the NCB.

2.7. THE METAMORPHIC TREND ALONG THE DABIE-HONGSEONG COLLISION BELT

In the Dabie area, the easternmost part of the Dabie-Hongseong belt, UHP Dabie eclogite formed at 680–880°C and 30–40 kbar and were overprinted by epidote amphibolite facies retrograde metamorphism (470–570°C, 6–8 kbar) during uplift and were extensively deformed [Figure 2.16; *Cong et al.*, 1995; *Carswell et al.*, 2000; *Hirajima and Nakamura*, 2003, and references therein]. UHP eclogites in the Sulu area formed at 750–880°C and 26.5–41 kbar [*Hirajima and Nakamura*, 2003, and references therein). In the northern part of the Sulu area, many of the eclogites were transformed first to HP granulite (700–820°C, 10–17 kbar) and then to amphibolite (600–740°C, 8–9 kbar); however, in the southwestern part of the area, the eclogites were retrograded directly to amphibolite (510–720°C, 8–12 kbar) [*Banno et al.*, 2000; *Yao et al.*, 2000; *Zhang et al.*, 2000]. UHP metamorphic in these areas has been dated at ca. 210–230 Ma [*Hirajima and Nakamura*, 2003, and references therein).

HP eclogite in the Hongseong area, southwestern part of the Gyeonggi Massif, South Korea, formed at

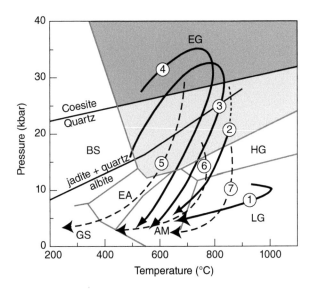

Figure 2.16 Pressure-Temperature paths along the Dabie-Hongseong (solid lines) and Himalayan collision belts (dashed lines) [from *Oh*, 2010: 1, Odesan; 2, Hongseong; 3, Sulu; 4, Dabie; 5, Kaghan; 6, Kharta; 7, Eastern Himalayan Syntaxis; *Massonne and O'Brien*, 2003; *Liu and Zhong*, 1997; *Lombardo and Rolfo*, 2000; *Liou et al.*, 1996; *Oh et al.*, 2005, 2006a]. Abbreviations in Figure 2.5 are used. Westward along the Dabie-Hongseong collision belt, the pressure and temperature conditions of peak metamorphism increases and decreases, respectively, and the grade of retrograde metamorphism decreases. The opposite trend, eastward decrease in pressure and increase in temperature, is observed along the Himalayan collision belt. The petrogenetic grid is from *Oh and Liou* [1988].

(a)

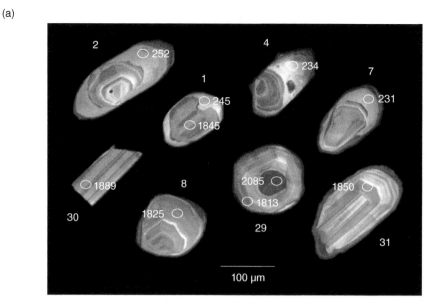

(b)

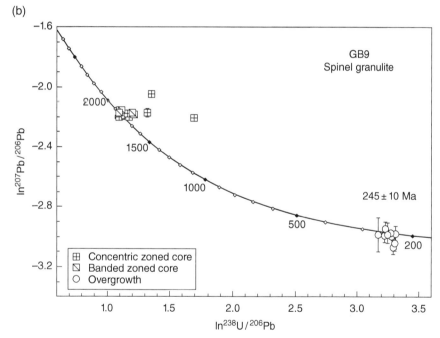

Figure 2.17 (a) SEM CL images of representative zircon grains from spinel granulite, which had undergone UHT metamorphism. Numbered ellipses mark the locations of SHRIMP analyses and the ages measured in Ma, and (b) modified Tera-Wasserburg concordia plot of SHRIMP U-Pb isotopic analyses of zircon from spinel granulite [from *Oh et al.*, 2006a].

835–860°C, 17.0–20.9 kbar during the Triassic (ca. 230 Ma). The eclogite was first overprinted by granulite facies metamorphism (830–850°C, 11.5–14.6 kbar) and then by amphibolite facies metamorphism (570–740°C, 6.7–11.0 kbar), indicating isothermal decompression [Figures 2.5, 2.16; *Oh et al.*, 2005; *Kim et al.*, 2006]. In the Odesan area, the eastern part of the Gyeonggi Massif in Korea, spinel granulites occur together with

Triassic postcollision mangerite and contain the assemblage, spinel + cordierite + corundum indicating UHT metamorphism [Figure 2.9c; *Oh et al.*, 2006a]. The P-T conditions of the UHT metamorphism are 915–1160°C and 9.0–10.6 kbar and the UHT metamorphism occurred at ca. 245 Ma (Figure 2.17; Table 2.5).

Along the Dabie-Hongseong belt, peak metamorphic conditions show a systematic change from UHT in the

Table 2.5 SHRIMP U-Pb zircon data from spinel granulite in the Odesan area, South Korea.

| Grain spot | Zircon type[a] | Pb[b] (ppm) | U (ppm) | Th (ppm) | Th/U | ^{206}Pb% | $^{204}Pb/^{206}Pb$ | ± | $^{206}Pb^{b}/^{238}U$ | ± | $^{207}Pb^{b}/^{206}Pb$ | ± | Apparent ages (Ma) | | | |
													$^{206}Pb/^{238}U$	±	$^{207}Pb/^{206}Pb$	±
GB9 – Spinel-quartz granulite																
29.1	CZC	250	967	117	0.12	0.04	0.000029	0.000008	0.2588	0.0019	0.1290	0.0014	1483.7	9.6	2085	19
10.1	CZC	64	325	112	0.35	0.31	0.000201	0.000042	0.1848	0.0012	0.1100	0.0012	1092.9	6.8	1800	20
3.1	CZC	66	198	30	0.15	0.05	0.000032	0.000025	0.3345	0.0115	0.1110	0.0024	1860	56	1815	39
20.1	CZC	60	179	40	0.23	0.03	0.000018	0.000036	0.3315	0.0054	0.1158	0.0009	1846	26	1893	14
8.1	CZC	51	155	40	0.26	0.10	0.000067	0.000026	0.3197	0.0035	0.1116	0.0011	1788	17	1825	18
29.2	CZC	51	154	67	0.44	0.03	0.000020	0.000020	0.3097	0.0053	0.1108	0.0018	1739	26	1813	29
19.1	CZC	47	149	27	0.18	0.14	0.000092	0.000027	0.3182	0.0031	0.1122	0.0012	1781	15	1835	20
22.1	CZC	38	142	22	0.16	0.07	0.000042	0.000027	0.2664	0.0064	0.1136	0.0032	1522	33	1858	52
31.1	CZC	44	135	42	0.31	0.03	0.000022	0.000022	0.3126	0.0051	0.1131	0.0018	1753	25	1850	29
1.2	BZC	83	268	101	0.38	0.05	0.000031	0.000021	0.2969	0.0047	0.1128	0.0007	1676	23	1845	11
30.1	BZC	78	228	71	0.31	0.07	0.000042	0.000021	0.3285	0.0089	0.1156	0.0018	1831	43	1889	29
21.2	BZC	28	62	99	1.61	0.12	0.000079	0.000042	0.3347	0.0061	0.1140	0.0025	1861	30	1863	41
32.1	BZC	18	54	29	0.54	0.16	0.000101	0.000084	0.3008	0.0079	0.1136	0.0021	1695	39	1858	34
5.1	DUZO	153	4328	66	0.015	0.02	0.000002	0.000002	0.0389	0.0001	0.0512	0.0003	245.8	0.7	249	13
27.1	DUZO	65	1860	28	0.015	0.03	0.000020	0.000020	0.0388	0.0002	0.0506	0.0007	245.3	1.4	222	33
26.1	DUZO	43	1270	11	0.009	0.04	0.000036	0.000031	0.0377	0.0002	0.0511	0.0010	238.8	1.2	247	46
13.1	LUZO	6	163	2	0.012	0.47	0.000831	0.000384	0.0396	0.0006	0.0527	0.0023	250.4	3.9	317	104
7.1	LUZO	5	148	0.8	0.005	0.22	0.000020	0.000020	0.0364	0.0007	0.0479	0.0027	230.7	4.2	93	126
4.1	LUZO	3	90	0.7	0.007	0.68	0.001041	0.000353	0.0370	0.0009	0.0463	0.0022	233.9	5.4	15	107
21.1	LUZO	3	65	3	0.042	2.29	0.000654	0.000345	0.0420	0.0010	0.0506	0.0060	264.9	6.1	221	253
1.1	LUZO	2	62	0.8	0.012	0.89	0.000020	0.000020	0.0388	0.0010	0.0512	0.0036	245.1	6.3	248	170
9.2	ZO	28	815	10	0.012	0.05	0.000043	0.000028	0.0376	0.0003	0.0501	0.0007	237.9	1.8	198	34
25.1	ZO	11	335	4	0.013	1.45	0.000819	0.000197	0.0364	0.0004	0.0509	0.0026	230.4	2.2	238	123
2.1	ZO	3	89	0.6	0.006	0.56	0.000534	0.000224	0.0398	0.0007	0.0502	0.0024	251.6	4.5	205	115

Source: Oh et al. [2006a].

[a] Zircon textures: C: core, O: overgrowth, CZ: concentric zoned, BZ: banded zoned, SZ: sector zoned, WZ: weakly zoned, UZ: unzoned, L: CL light, D: CL dark.

[b] Radiogenic: corrected for common Pb using either ^{204}Pb or $^{208}Pb/^{206}Pb$ and Th/U.

Table 2.6 The summary of the metamorphic evolution along the Dabie-Hongseong collision belt.

	Western area (Dabie)	Western area (Sulu)	Middle area (Hongseong)	Easternmost area (Odesan)
Peak metamorphism	UHP eclogite facies metamorphism 680–880°C, 30–40 kbar	UHP eclogite facies metamorphism 750–880°C, 27–41 kbar	Eclogite facies metamorphism 835–860°C, 17–21 kbar	Ultra-high temperature (UHT) metamorphism 915–1160°C, 9–11 kbar
Peak metamorphic age	~210–230 Ma	~210–230 Ma	~>230 Ma	~245 Ma
Retrograde metamorphism	Epidote-amphibolite facies metamorphism 470–570°C, 6–8 kbar	High-pressure granulite or amphibolite facies metamorphism 700–820°C, 10–17 kbar 510–720°C, 8–12 kbar	High-pressure granulite facies metamorphism 830–850°C, 12–15 kbar	Amphibolite facies metamorphism 571–598°C
Retrograde metamorphic age			~230 Ma	~<245 Ma
Postcollisional igneous activity		~210–230 Ma (minor igneous activity)	~230 Ma	~230 Ma
Time of collision	Before ~201–230 Ma	Before ~210–230 Ma	Before 230 Ma	Before 245 Ma

Odesan area to HP in the Hongseong area and UHP in the Dabie and Sulu areas (Figures. 2.6, 2.14, Table 2.6). The peak metamorphic temperature decreases westward from 915–1160°C to 680–880°C and the peak metamorphic pressure increases from 9.0–10.6 to 30–40 kbar, indicating a westward increase in the subducted depth of the continental slab and a westward decrease in the geothermal gradient during subduction of the continental slab. The deeper subduction in the west (i.e., the Dabie and Sulu areas) caused faster uplift, resulting in lower-grade retrograde metamorphism (epidote-amphibolite or amphibolite facies) compared with the east (the Hongseong and Odesan areas), where retrograde metamorphism was higher than the granulite facies due to the slow rate of uplift following (and as a consequence of) shallow subduction. The ages of peak metamorphism became younger toward the west, varying from ca. 245 Ma in the Odesan area to 230 Ma in the Hongseong area and ca. 210–230 Ma in the Dabie-Sulu area.

2.8. METAMORPHIC EVOLUTION ALONG THE HIMALAYAN COLLISION BELT

The Himalayan and adjacent Tibetan plateau resulted from the collision of the India plate with the Asian margin and intervening microplates and arcs. Before collision, subduction of oceanic crust produced an Andean-type continental arc (Trans-Himalayan batholith) as well as an oceanic island arc, the Kohistan-Ladakh arc, which collided with Asia at the end of the Cretaceous. During the subduction, subducted oceanic crust and sediment had undergone blueschist facies metamorphism at 80–100 Ma [*Shams*, 1972; *Honegger et al.*, 1989). India began to collide

with the Asia continent at around 50 Ma [*Searle et al.*, 1987, 1997] as documented by: (1) the end of major I-type magmatism in the Trans Himalayan batholith [*Honegger et al.*, 1982; *Petterson and Windley*, 1985]; (2) the first S-type anatectic granites and migmatites in the Lhasa block; (3) south-directed thrusting in the Asian margin and suture zone; and (4) a change of the India plate movement from ca. 15–20 cm/year down to ca. 5 cm/year at around 50 Ma [*Patriat and Achache*, 1984; *Klootwijk et al.*, 1992].

The Himalayan is subdivided into a series of narrow, parallel belts separated by major thrusts or faults [Figure 2.18; see review of *Mattauer et al.*, 1999; *Hodges*, 2000]. Ophiolites mark the sutures between Asia, India, and the Kohistan-Ladakh arc [*Gansser*, 1979]. To the south of them lie the units of the Tethyan Himalaya, weakly metamorphosed Paleozoic and Mesozoic sedimentary series of the Indian margin. The Tethyan Himalaya is separated by a normal fault, the South Tibetan Detachment (STD) from underlying high amphibolite to granulite facies gneisses belonging to the Higher Himalaya. Metamorphism in the Higher Himalaya is represented by an early Barrovian staurolite + kyanite-bearing stage (6–10 kbar), yielding ages mostly older than 30 Ma and the Barrovian metamorphism was overprinted by a younger (15–20 Ma), higher temperature but lower pressure metamorphism (4–8 kbar). It produced sillimanite-bearing assemblages, migmatites, and leucogranites [*Guillot et al.*, 1999; *Harris and Massey*, 1994; *Searle et al.*, 1997]. The Lesser Himalaya is separated from the High Himalaya by main central thrust (MCT); it experienced generally lower-grade greenschist facies metamorphism. This unit is thrust over underlying molasses sediments along the main boundary thrust (MBT), which are bounded in the south by the main frontal thrust (MFT).

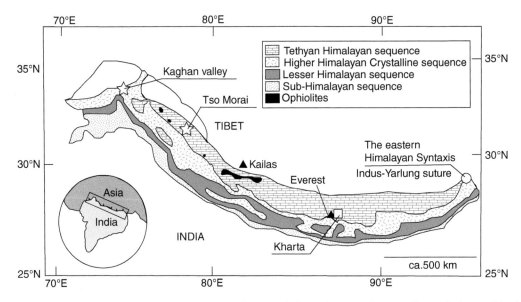

Figure 2.18 Simplified tectonic map and metamorphic trend along the Himalayan collision belt [modified after *Imayama et al.*, 2012]. Eastward along the belt, metamorphism changes from ultrahigh-pressure eclogite facies (stars) to eclogite facies (square) and high-pressure granulite facies (circle).

Evidence for high-pressure metamorphism is scarce in the Himalaya. So far, only four areas have been identified with eclogites and eclogite facies rocks [*Massonne and O'Brian*, 2003]. The areas are (1) the upper Kaghan and Neelum valleys, Pakistan [*Spencer*, 1993; *Lombardo et al.*, 2000; *Fontan et al.*, 2000; *O'Brien et al.*, 2001]; (2) the Tso Morari crystalline complex, Ladakh, India [*de Sigoyer et al.*, 1997, 2000; *Sachan et al.*, 1999, 2004]; (3) the Indus suture zone at the margin of Nanga-Parbat-Haramosh Massif [*Le Fort et al.*, 1997], and (4) the Kharta region of southern Tibet [*Lombardo et al.*, 1998; *Lombardo and Rolfo*, 2000].

2.9. THE METAMORPHISM IN THE WESTERN HIMALAYAN COLLISION BELT

The eclogite facies metamorphic rocks in the Kaghan and Neelum valleys in the Naran area, northern Pakistan [*Pognante and Spencer*, 1991; *Fontan et al.*, 2000; *O'Brien et al.*, 2001; *Treloar et al.*, 2003] lie in the most northern margin of the India plate on the immediate foot of the main mantle thrust (MMT). Coesite-bearing eclogites are common in the upper Kaghan valley [Figure 2.19; *O'Brien et al.*, 2001; *Kaneko et al.*, 2003; *Treloar et al.*, 2003]. UHP eclogites in the Kaghan record peak metamorphic conditions of 3.0 ± 0.2 GPa and $770 \pm 50°C$ [*O'Brien et al.*, 2001]. The UHP eclogites are retrograded into epidote amphibolites and/or blueschists forming magnesio-hornblende, barroisite, or glaucophane [580–610°C, 10–13 kbar; *Lombardo and Rolfo*, 2000; *O'Brien et al.*, 2001] and later overprinted locally by greenschist facies

metamorphism. The metamorphism evolution of the Kaghan eclogite is well dated. *Tonarini et al.* [1993] and *Spencer and Gebauer* [1996] first estimated the metamorphic age ranging from 50 to 40 Ma using Sm-Nd and U-Pb isotope systems. Prograde metamorphism for the quartz-bearing eclogite-facies metamorphism is dated at 50 ± 1 Ma in the upper Kaghan valley by *Kaneko et al.* [2003]. UHP metamorphism (3.0 ± 0.2 GPa) for the same rock was dated as 46.2 ± 0.7 Ma and 46.4 ± 0.1 Ma using U-Pb techniques by *Kaneko et al.* [2003] and *Parrish et al.* [2006]. Zircon and rutile from coesite-free eclogite yielded 44 Ma as the age of HP retrograde metamorphism [*Spencer and Gebauer*, 1996; *Treloar et al.*, 2003; *Parrish et al.*, 2006], implying a fast exhumation rate of ca. 3–8 cm/yr [*Guillot et al.*, 2008]. The Kaghan unit cooled below 500–400°C between 43 Ma and 40 Ma based on Rb-Sr age of phengite [*Tonarini et al.*, 1993], suggesting slowed down exhumation, ca. 0.3 cm/yr [*Guillot et al.*, 2008]. This is supported by $^{40}Ar/^{39}Ar$ ages from the surrounding rocks; 42.6 ± 1.6 Ma [*Chamberlain et al.*, 1991], 41.2 ± 2 Ma [*Smith et al.*, 1994], and 39.8 ± 1.6 Ma [*Hubbard et al.*, 1995].

The Tso Morari Massif contains the only UHP rocks in the north Himalayan Massif and outcrops south of the ITSZ in the eastern Ladakh [Figure 2.20; *Thakur*, 1983]. It has an elongated shape of 100×50 km striking northward and dipping 10° to the northwest with a thickness of less than 7 km [*de Sigoyer et al.*, 2004]. UHP parageneses occurred both in mafic and metapelitic rocks. This is further supported by the occurrence of carbonate-bearing coesite eclogite lenses in the kyanite/sillimanite-bearing rocks in the northern part of

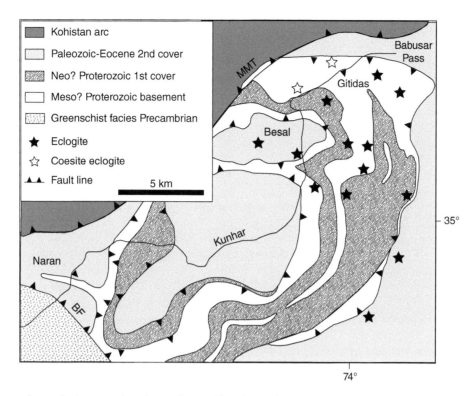

Figure 2.19 The geologic map of Kaghan valley and localities of UHP and HP eclogites [modified after *Massonne and O'Brian*, 2003; *Guillot et al.*, 2008].

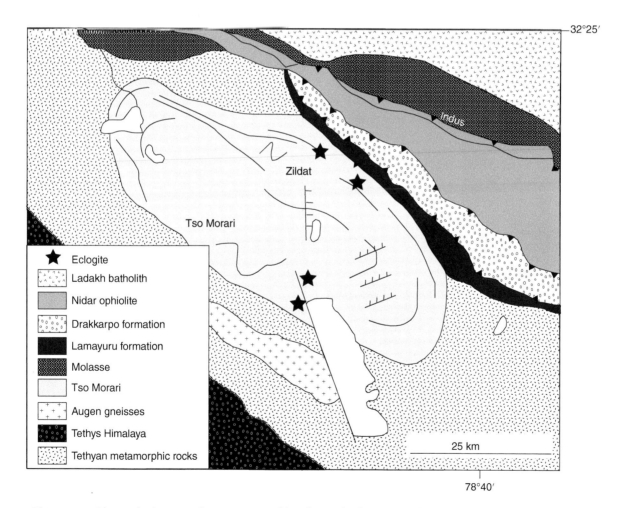

Figure 2.20 The geologic map of Tso Morari and localities of eclogites [modified after *Massonne and O'Brian*, 2003; *Guillot et al.*, 2008].

the dome suggesting pressure of >3.9 GPa and temperature of 750°C [*Mukheerjee et al.*, 2003]. *Mukheerjee et al.* [2005] reported the first occurrence of microdiamond inclusion within zircon from coesite-bearing eclogite, which suggests a pressure greater than 3.5 GPa at 750°C [*Bundy*, 1980]. The data suggest the subduction down to a depth greater than 120 km. During their exhumation up to the depth of 40–30 km, the Tso Morari rocks underwent cooling under epidote amphibolite facies condition (1.1 ± 0.3 GPa; 580 ± 50°C). In the metagraywake, staurolite, phengite, and chlorite are destabilized to form kyanite and biotite, implying heating of rocks to 630 ± 50°C at the depth of 30 km (0.9 ± 0.3 GPa) during the late exhumation [*de Sigoyer et al.*, 1997; *Guillot et al.*, 1997]. The crystallization of chlorite and white mica are common in the entire Tso Morari Massif. This marks the final deformation related to exhumation under the greenschist facies conditions.

Peak metamorphism of the Tso Morari Massif was dated as 55 ± 6 Ma using Lu-Hf, Sm-Nd, and U-Pb isotope methods [*de Sigoyer et al.*, 2000]. *Leech et al.* [2005] obtained U-Pb SHRIMP zircon ages of 53.3 ± 0.7 Ma and 50.0 ± 0.6 Ma on quartzofeldspathic gneiss from the Puga formation and interpreted the former age as the age of UHP metamorphism and the latter as the retrograde eclogite-facies metamorphism. The UHP metamorphic age is very similar to a $^{40}Ar/^{39}Ar$ phengite age of 53.8 ± 0.2 Ma obtained by *Schlup et al.* [2003]. Close ages between peak and retrograde metamorphisms suggest a very fast exhumation. Indeed, the exhumation was very fast, ca. 1.7 cm/yr, compatible with modeling of garnet diffusion profile by *O'Brien and Sachan* [2000]. The timing of amphibolite facies conditions is dated as 47 ± 0.5 Ma based on $^{40}Ar/^{30}Ar$ ages of phengite, Sm-Nd and Rb-Sr ages of amphiboles, and U-Pb zircon ages [*de Sigoyer et al.*, 2000; *Leech et al.*, 2005]. The final retrograde metamorphism under greenschist facies conditions is dated between 34 ± 2 Ma and 45 ± 2 Ma based on seven fission track analyses of zircons [*Schlup et al.*, 2003].

2.10. THE METAMORPHISM IN THE MIDEASTERN HIMALAYAN COLLISION BELT

In the mideastern Himalaya, granulitized eclogites were found in the Ama Drime Range (Kharta region) of southern Tibet and in the Arun River valley in eastern Nepal; these eclogites do not appear to have attained UHP conditions and direct correlation with eclogites in the northwest Himalaya is unclear [Figure 2.21; *Lombardo and Rolfo*, 2000; *Groppo et al.*, 2007; *Corrie et al.*, 2010]. The protolith of Ama Drime eclogites is thought to have been formed between 110 and 88 Ma based on SHRIMP zircon ages [*Rolfo et al.*, 2005]. Such

an age rules out the possibility that the age of eclogitization is Paleozoic or Precambrian and supports a Tertiary age of eclogitization. The eclogites have been strongly overprinted by subsequent metamorphism, leaving sparse evidence for eclogitization. *Groppo et al.* [2007] proposed four phases of metamorphism for the Ama Drime eclogites on the basis of microstructural observations, pseudosection analysis, and conventional thermobarometry: an initial eclogitic metamorphism (M1), high-pressure granulite-facies event (M2), and low-pressure granulite-facies event (M3), and a final cooling stage (M4).

While no direct barometric measurements can be made for eclogite-facies conditions of the granulitized eclogites, an estimate for the minimum pressures experienced by Arun eclogite can be obtained by reintegrating sodic plagioclase into pyroxene. This integration assumes: (1) at eclogite facies, all sodium was hosted in clinopyroxene; and (2) the albite and anorthite components of plagioclase texturally associated with matrix pyroxene were hosted in omphacite as the jadeite and Ca-Tschermaks components, respectively. The integration suggests jadeite content of 39.1% in omphacitic clinopyroxene, which corresponds to minimum pressure of 15 kbar at 670°C, and temperature condition is estimated via Zr-in-titanite thermometry [*Corrie et al.*, 2010]. The peak pressure condition of Ama Drime eclogites may also exceed 15 kbar and probably reach 20 kbar [*Groppo et al.*, 2007]. The P-T condition of granulite-facies metamorphism (M2), which followed the eclogite-facies metamorphism, is 750–780°C and 10–12 kbar [*Groppo et al.*, 2007; *Corrie et al.*, 2010]. The P-T condition of a later third metamorphic event (M3) represented by orthopyroxene + plagioclase overgrowth around garnet, is ca. 750°C and 4 kbar [*Groppo et al.*, 2007]. A final metamorphic event represented by the growth of amphibole in the matrix occurred at around 675°C and 6 kbar [*Corrie et al.*, 2010].

Lombardo and Rolfo [2000] considered that the eclogitic metamorphism occurred before 30 Ma based on $^{40}Ar/^{39}Ar$ ages of amphibole [*Hodges et al.*, 1994]. Attempts to date the Ama Drime eclogites have yielded U-Pb SHRIMP ages of 13–14 Ma from low-U rims of zircon, which was assumed to grow during an M3 low-P granulitic event [*Groppo et al.*, 2007]. *Cottle et al.* [2009] presented monazite and xenotime ages of 13.2 ± 1.4 Ma, which was interpreted as the timing of peak granulite metamorphism. Other published ages, including zircon Pb/U ages of 17.6 ± 0.3 Ma, were also interpreted to represent granulite metamorphism in the northern Ama Drime Range [*Li et al.*, 2003]. Lu-Hf dates from garnet separated from one retrograde eclogite from the Arun River valley indicate an age of 20.7 Ma and those from garnet amphibolite range from 13.9 to 15.1 Ma [*Corrie et al.*, 2010]. As the retrograde eclogite is now granulite rock with no

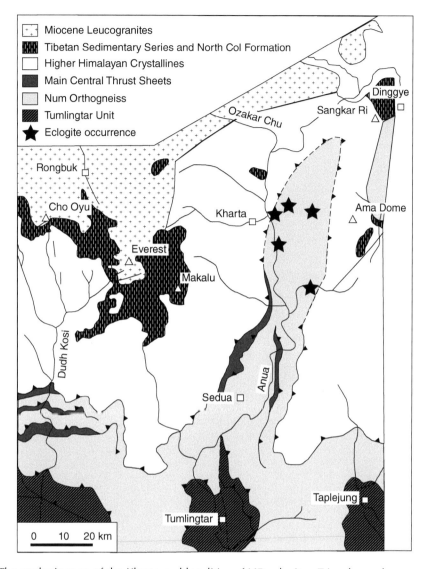

Figure 2.21 The geologic map of the Kharta and localities of HP eclogites. Triangles and squares represent the position of mountains and cities, respectively [modified after *Lombardo and Rolfo*, 2000; *Guillot et al.*, 2008; *Corrie et al.*, 2010].

omphacite, 20.7 Ma represents the time of M2 high-pressure granulite facies metamorphism indicating that eclogite formed before ca. 20 Ma. These data also indicate that 13–17 Ma ages represent the final amphibolite facies retrograde metamorphic stage. During the uplift of granulitized eclogite in the mideastern Himalaya, the pressure decrease from eclogite to high-pressure granulite facies may be about 8–10 kb equivalent to ca. 26–33 km. The uplifting rate from eclogite to amphibolite facies stages in the Kaghan and Tso Morari was 0.3 cm/yr. If we apply 0.3 cm/yr uplifting rate, the time difference between eclogite and high-P granulite facies metamorphism in the granulitized eclogite may be ca. 8.7–11 Ma. These data suggest that HP eclogite facies metamorphism occurred at ca. 30 Ma as expected by *Hodges et al.* [1994].

2.11. THE METAMORPHISM IN THE EASTERN HIMALAYAN COLLISION BELT

The Namche Barwa Syntaxis (NBS) occurs as a NE-SW striking belt with 150-km length in the eastern Himalayan collision belt (Figure 2.22). The syntaxis is characterized by a northeast-plunging antiform and is bounded by two northeast-striking strike-slip shear zones: a left-slip shear zone on the western side and a right-slip shear zone on the eastern side. These strike-slip shear zones are linked by east-west trending thrusts. The NBS is separated from the Gangdese continental margin of the Lhasa terrane by the Indus-Yarlung suture zone (IYSZ), which is marked by dismembered ophiolites and mélange. The NBS is considered an "indenter corner" of

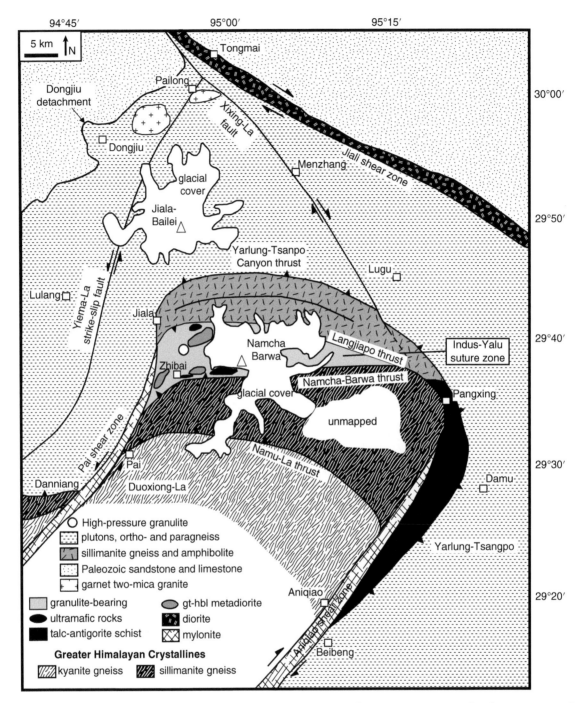

Figure 2.22 The geologic map of the Namche Barwa syntaxis based on mapping at a scale of 1:100 000 and sample location of high-pressure granulite. Triangles and squares represent the position of mountains and cities, respectively [from *Ding et al.*, 2001].

the India plate [*Koons*, 1995] and mafic boudins within the NBS, however, contain inherited Jurassic-Cretaceous zircons suggesting an affinity to the Gangdese arc [*Ding et al.*, 2001]. The granulites in the NBS record at least two episodes of granulite facies metamorphism: the first at high-pressure granulite facies metamorphism (17–18 kbar, 890°C) followed by a second low-pressure granulite facies metamorphic event at 5–10 kbar, >850–875°C [*Liu and Zhong*, 1997]. The first high-pressure granulite facies metamorphism is reported from the garnet-kyanite granulite but is not clear in mafic boudins, but low-pressure granulite facies metamorphism is clear in both rock types. The large-scale strike-slip fault in both sides of the NBS was dated as ca. 26 Ma [*Zheng and Chang*, 1979; *Zhong* and Ding, 1996], which suggests that the indentation by the collision occurred from 26 Ma. The ages of collisional granites with age between 10 and 20 Ma support the collision from ca. 26 Ma. Xu et al. [2012] carried out SHRIMP and LA-ICP-MS U-Pb dating on zircon from gneisses in the NBS and showed that granulite facies metamorphism, partial melting, and shearing could be coeval and occurred between 21.8 and 24.5 Ma. Those ages are well matched with the ages of garnet bearing two-mica leucosome [21.5–23.2 Ma; Ding et al., 2001], which might be formed by partial melting during collision. Therefore, 21.8–24.5 Ma can be regarded as the time of high-pressure granulite facies metamorphism. The timing of granulite facies metamorphism of the mafic boudin is expected to be ca. 40 Ma [Ding et al., 2001] suggesting that mafic boudin might be a part of the Gangdese arc instead of the India plate. Ca. 40 Ma may be the time the granulite facies metamorphism occurred under the Gangdese arc instead of the time of collision of India plate. The mafic boudins seem to be tectonically incorporated to NBS during collision or uplift stage. The youngest zircon age (11 Ma) and $^{40}Ar/^{39}Ar$ age (8 Ma) are interpreted to represent the timing of low-pressure granulite facies retrograde metamorphism [Ding et al., 2001]

2.12. THE METAMORPHIC PATTERN ALONG THE HIMALAYA COLLISION BELT

Both HP and UHP eclogites are reported from the Higher Himalayan crystalline nappe in the Kaghan valley and Tso Morari Massif in the western part of the collision belt [Ghazanfar and Chaudhry, 1987; Thakur, 1983]. UHP eclogites in the Kaghan record peak metamorphic conditions of 770°C and 30 kbar [O'Brien et al., 2001], which was overprinted by retrograde metamorphism of the epidote-amphibolite or blueschist facies [580–610°C, 10–13 kbar; Lombardo and Rolfo, 2000]. Sensitive high-resolution ion microprobe dating of zircon reveals that the original igneous rocks of the UHP eclogite formed at 170–253 Ma and were metamorphosed from ca. 50 Ma, thereafter being subjected to UHP metamorphism at ca. 46 Ma [Kaneko et al., 2003; Parrish et al., 2006]. The Tso Morari UHP eclogite had formed at 750°C, >39 kbar [Mukheerjee et al., 2003; Bundy, 1980] and underwent amphibolite facies retrograde metamorphism [580°C, 11 kbar) during uplift [Guillot et al., 2008]. Peak metamorphism of the Tso Morari Massif was dated at ca. 53–55 Ma [de Sigoyer et al., 2000; Leech et al., 2005] and the amphibolite retrograde metamorphism occurred at ca. 47 Ma [de Sigoyer et al., 2000; Leech et al., 2005].

Only HP eclogites have been reported from the mideastern part of the collision belt [Lombardo and Rolfo, 2000; Groppo et al., 2007; Corrie et al., 2010]. The metamorphic age of HP eclogite may be ca. 30 Ma, which is similar to the metamorphic age of the gneiss around the HP eclogite [ca. 30 Ma; Massonne and O'Brian, 2003]. The HP eclogite in the mideastern part of the Himalayan collision belt may have formed at ca. >780°C and 20 kbar and was overprinted by high-pressure granulite facies metamorphism (780–750°C, 12–10 kbar) at ca. 20 Ma [Groppo et al., 2007; Corrie et al., 2010].

HP granulite (890°C, 17–18 kbar) is reported from the NBS, at the eastern terminus of the Himalayan collision belt; the granulite was subjected to retrograde metamorphism to produce lower-pressure granulite (875–850°C, 10-5 kbar), representing near-isothermal decompression [Liu and Zhong, 1997]. The HP granulite metamorphism may have occurred at ca. 22–25 Ma.

Along the Himalayan collision belt, peak metamorphism changes eastward from UHP eclogite facies through HP eclogite facies to high-pressure granulite facies (Figure 2.16; Table 2.7) indicating a progressive eastward decrease in the depth of subduction of continental crust and an eastward increase in the geothermal gradient. Toward the east, the grade of retrograde metamorphism during uplift increases from amphibolite facies through lower granulite facies to upper granulite facies, reflecting the higher rate of uplift in the west due to deeper subduction. The peak metamorphic ages also decrease from 53–46 Ma in the west to 22–25 Ma in the east indicating propagation of collision toward the east.

2.13. DISCUSSION AND TECTONIC IMPLICATION

2.13.1. The Tectonic Evolution of the Dabie-Hongseong Collision Belt in Northeast Asia

Although paleomagnetic data indicate similar Permian paleo-latitudes for the NCB and SCB, the mean direction of paleo-north for the SCB is rotated clockwise by more than 60° with respect to that for the NCB [Zhao and Coe, 1987]. Based on these data, Zhao and Coe [1987] suggested that collision began in the easternmost part of the blocks during the late Early Permian and progressed westward as the SCB rotated clockwise relative to the NCB. Zhang [1997] supported this model based on analyses of lithofacies distribution, petrography, and intracontinental deformation in the NCB and SCB. The collision is considered to have ended in the southeast part during the Late Triassic or Early Jurassic. Zhang [1997] also suggested that the Qinling-Dabie-Tanlu-Sulu-Imjingang-Yanji belt formed by subduction and collision along the margin of the NCB. This tectonic model suggests that collision between the NCB and SCB might have started in Korea and propagated toward China, which is consistent with the finding of older collision-related metamorphic ages in the Odesan area (245 Ma) than in the Sulu and Dabie areas (210–230 Ma).

Table 2.7 The summary of the metamorphic evolution along the Himalayan collision belt.

	Western area (Kaghan, Tso Morari)	Middle eastern area (Kharta)	Easternmost area (Namcha Barwa Syntaxis)
Peak metamorphism	UHP and HP eclogite facies metamorphism 770°C, 30 kbar for Kaghan 750°C, >39 kbar for Tso Morari	Eclogite facies metamorphism > 780°C, 20 kbar	High-pressure granulite facies metamorphism 890°C, 17–18 kbar
Peak metamorphic age	46 Ma for Kaghan 53–55 Ma for Tso Morari	~30 Ma	~22–25 Ma
Retrograde metamorphism	Blueschist, epidote-amphibolite facies metamorphism 580–610°C, 10–13 kbar for Kaghan 530–630°C, 8–14 kbar for Tso Morari	High-pressure granulite facies metamorphism 780–750°C, 12–10 kbar	Low-pressure granulite facies metamorphism 875–850°C, 10–5 kbar
Retrograde metamorphic age	40–43 Ma for Kaghan 47–50 Ma for Tso Morari	~20 Ma	~8-11 Ma
Postcollisional igneous activity	~25–16 Ma (Minor igneous activity)	~20–10 Ma	~15–10 Ma
Time of collision	Before 55 Ma	Before 30 Ma	Before 25 Ma

Based on new findings, including Triassic eclogites and postcollision igneous rocks and Permian UHT metamorphism in South Korea, *Oh* [2006] and *Oh and Kusky* [2007] proposed an improved tectonic model for the development of the Qinling-Dabie-Sulu-Hongseong-Odesan-Hida-Yanji belt along the margin of the NCB (Figures 2.8, 2.23). In this model, the NCB and SCB separated from eastern Gondwana during the early Paleozoic [*Metcalfe*, 2006]. After separation, the blocks moved northward together. An active margin formed along the southern and eastern boundaries of the NCB from the Ordovician, and a passive margin developed along the northern margin of the SCB (Figure 2.23a). The Qaidam microblock existed between the NCB and SCB collided with the NCB before the main collision between the NCB and SCB (Figure 2.23a). Subduction beneath the NCB consumed the oceanic crust between the NCB an SCB and Middle to Late Paleozoic subduction complex formed in the margin of the NCB. As a result, the two blocks converged and finally collided during the Permian, on the east side of the Korean peninsula (Figure 2.23b). The collision propagated westward with rotation of the SCB (Figure 2.23c), as suggested by *Zhao and Coe* [1987]. Collision between the blocks was completed during the Early Jurassic (Figure 2.23d). Subduction and collision along the southern and eastern boundaries of the NCB resulted in the formation of the Qinling-Dabie-Sulu-Hongseong-Odesan-Hida-Yanji belt. During the collision, eastern part of the SCB indented northward forming sinistral TanLu fault. Dextral strike-slip fault, the possible counterpart of the TanLu fault, may have formed between Korea and Japan and along the fault

Japan was expected to move down southward [Figure 2.23; *Oh*, 2006; *Oh and Kusky*, 2007].

Postcollision magmatism (ca. 230 Ma) and metamorphism (245–230 Ma) occurred widely in the northern part of the Gyeonggi Massif, which locates to the north of the Hongseong-Odesan collision belt [Table 2.6; *Oh and Kusky*, 2007; *Kim et al.*, 2011]. The metamorphic grade higher than the granulite facies is concentrated along the Hongseong-Odesan collision belt. Postcollisional metamorphism and magmatism are considered to be caused by slab break-off (i.e., detachment of the oceanic slab from the continental slab) during continental collision [Figure 2.14; *Davis and Blanckenburg*, 1995].

During continental collision, the dense oceanic slab that was subducted prior to collision acted to pull down the buoyant continental slab. At depth, where the buoyancy force of the subducted continental slab exceeded the pulling force exerted by the oceanic slab, extensional deformation may have occurred at the boundary between the continental and oceanic slabs, leading to break-off of the oceanic slab along a section of certain lateral extent. Detachment of the slab would have led to an additional pulling force on the laterally neighboring, still-attached slab, leading to enhanced lateral propagation of slab break-off, thereby enlarging the gap between the continental and oceanic slabs [*Wrotel and Spakman*, 2000].

The upwelling of asthenosphere through the gap that developed at the break-off site would supply heat to the overlying metasomatized mantle lithosphere, leading to production of K-rich basaltic or andesitic magma by melting of metasomatized mantle. The basaltic magma

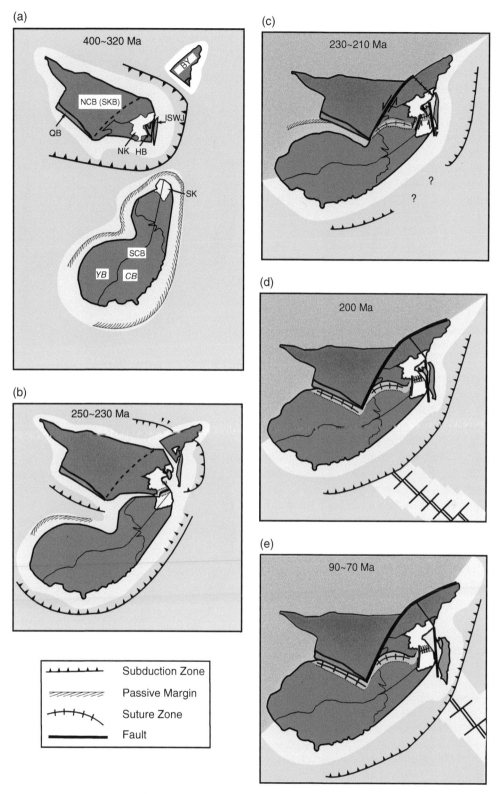

Figure 2.23 Tectonic model of the collision between the NCB and SCB [modified after *Oh and Kusky*, 2007]. (a) An active margin formed by subduction around the NCB during the mid-Paleozoic; at the same time, a passive margin developed around the northern boundary of the SCB. (b) Collision started during the Permian in the area that is now modern-day Korea. (c) Collision propagated into the area that is now the Shandong Peninsula, producing the sinistral TanLu fault and a dextral fault in the Hida belt. During collision, the Shandong and Korean peninsulas moved northward. (d) Collision ceased during the Early Jurassic. (e) Until Cretaceous, Japan moved southward. Labels: NCB = North China block; SKB = Sino-Korean block; SCB = South China block; BY = Bureya block; CB = Cathaysia block; HB = Hida belt; ISWJ = Inner zone of southwest Japan; NK = Northern Korea; QB = Qinling block; SK = Southern Korea; YB = Yangtze block.

would underplate and supply heat to the lower crust, resulting in granulite facies and UHT metamorphism and producing mangeritic or syenitic magma by partial melting of the lower crust [e.g., *Rajesh and Santosh*, 2004; *Oh et al.*, 2006b; *Kim et al.*, 2011]. However, if break-off occurs at great depth (e.g., at UHP depths), the upwelling asthenosphere may not produce a significant thermal perturbation in the distant lithosphere of the overriding plate [*Davis and Blanckenburg*, 1995]. Therefore, the postcollision magmatic activity was weak in the Dabie and Sulu belt where the break-off occurred at depth of UHP metamorphism [*Chen et al.*, 2003; *Yang et al.*, 2005].

Considering *Zhao and Coe*'s [1987] paleomagnetic data, which indicate that the angle between the long axes of the NCB and SCB was 60° before collision (with the shortest distance between the blocks in the northeastern region, as shown in Figure 2.24a), the width of ocean between the two blocks in the Dabie-Sulu area (westernmost part of the belt) and in the Hongseong area (middle part) was wider than that in the Odesan area (easternmost part) by ca. 700 and 350 km, respectively.

The trend of decreasing distance between the NCB and SCB toward the east is matched by an eastward decrease in the amount of oceanic slab subducted prior to continental collision. As a result, both the pulling force by the subducted oceanic crust and the subduction depth of the continental crust decreased toward the east causing the decrease of the depth of slab break-off toward the east. In the Odesan area, the eastern part of the Dabie-Hongseong collision belt, the depth of the slab break-off is shallower than the depth of HP eclogite facies metamorphism. In the Odesan area, due to the shallow depth of slab break-off, the heat supplied from the asthenosphere through the opening between the continental and oceanic slabs was enough to cause 245 Ma UHT metamorphism with strong Triassic postcollision magmatism. In the Hongseong area, the middle part of the Dabie-Hongseong collision belt, the HP metamorphism (older than 230 Ma) occurred showing that the slab break-off occurred at HP eclogite facies metamorphic depth. The HP eclogites were overprinted by the granulite facies metamorphism during retrograde metamorphism. In the Dabie-Sulu area, the western part of the Dabie-Hongseong collision belt, the occurrence of Triassic UHP metamorphism (average age of 220 Ma) indicates that the continental slab was subducted to the depths of UHP metamorphism (ca. 100–130 km depth) at around 220 Ma, and slab break-off occurred after UHP metamorphism. The great depth of slab break-off in the Dabie-Sulu area can be supported by minor postcollision magmatism due to weak thermal effect as explained by *Davis and Blanckenburg* [1995]. These findings indicate that the zone of oceanic slab break-off migrated westward during the period, ca. 245–220 Ma.

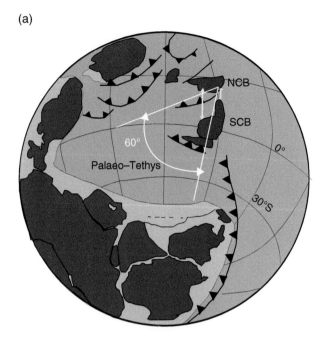

(a)

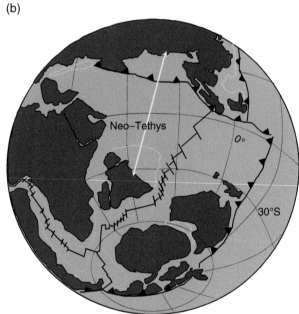

(b)

Figure 2.24 Paleogeographic map showing the relative locations of the NCB and SCB and the Asian and Indian continents [after *Li and Powell*, 2001]. (a) Before collision, the ocean between the NCB and SCB widened westward. As a result, the amount of oceanic slab subducted before continental collision was large enough to pull down the SCB to the depths of UHP metamorphism in the western part of the belt (the Dabie and Sulu areas), but not enough in the eastern parts (the Hongseong and Odesan areas). The Figure of 60° represents the angle between the long axes of the NCB and SCB at the initial stage of collision. (b) A wide ocean existing between the Asian and Indian continents caused the subduction of large amounts of oceanic crust, which acted to pull down continental crust to the depths of UHP metamorphism in the eastern Himalayan belt (where collision began).

2.13.2. The Tectonic Evolution of the Himalayan Collision Belt

The initial collision is widely thought to have started at ca. 55–52 Ma in the westernmost Himalayan belt and propagated eastward until around 41 Ma [*Rowley*, 1996; *Leech et al.*, 2005]. However, 55 Ma UHP eclogites indicate that the initial collision is older than 55 Ma. During the period of continental collision between India and Asia, India rotated counterclockwise by between 20° and 35° with progressive propagation of collision toward the east [*van der Voo et al.*, 1999].

A wide ocean existed between the Asian and Indian continents before collision along the Himalayan collision belt (Figure 2.23b); consequently, in the western part of the Himalayan collision belt where collision started, the amount of oceanic slab subducted before collision was enough to pull down the continental crust to the depths of UHP metamorphism. *Leech et al.* [2005] hypothesized that the Indian continental slab traveled rapidly along a short (and hence steep) path into the mantle. Therefore, early continental subduction, immediately following collision, occurred at a steep angle, comparable to that of modern continental subduction in the Hindu Kush, despite evidence for modern-day low-angle subduction of India beneath the Asian continent. The authors also raised the possibility that the exhumation of UHP terrains started just after oceanic slab break-off in the western Himalayan collision belt leading to the initiation of lowering subduction angle.

In the western part, slab break-off started at ca. 55–50 Ma when the continental slab reached the depth corresponding to UHP metamorphism [*de Sigoyer et al.*, 2000; *Negredo et al.*, 2007]. The peak metamorphic ages along the Himalayan collision belt suggest that slab break-off propagated to the middle eastern area of the collision belt (the Kharta area) at ca. 30 Ma and to the eastern most area (the Namche Barwa Syntaxis) at 22–25 Ma. During the eastward propagation of the slab break-off, the subduction angle decreased continuously. In the Kharta area, the middle eastern part of the belt, the reduced subduction angle after slab break-off prevented subduction of the continental crust to UHP depths and continental crust subducted only to the depth of HP eclogite facies metamorphism.

The high-pressure granulite, which indicates conditions close to UHT metamorphism, is found in the easternmost part of the belt, indicating that slab break-off in the easternmost part occurred at the shallowest depth in the Himalayan collision belt, resulting in thermal perturbations in the continental crust due to heat supplied from the asthenosphere through the gap formed by the slab break-off between the continental and oceanic slabs.

The slab break-off in the Himalayan collision belt also caused postcollision igneous activities as in the Dabie-Hongseong collision belt (Figure 2.25, Table 2.7). Intraplate postcollision magmatism in the Lhasa terrain after the Asia-India collision is represented by potassic, shoshonitic, and ultrapotassic volcanic rocks [ca. 24–8 Ma: *Chung et al.*, 2005; *Mo et al.*, 2006] and ore-bearing adakites [18–20 Ma: *Chung et al.*, 2003; *Hou et al.*, 2004; *Guo et al.*, 2007; *Gao et al.*, 2007, 2010]. The ultrapotassic postcollision rocks become younger from west to east indicating westward propagation of slab break-off, which supports the westward migration of collision between the Asia and India plate. Postcollision magmatism is almost absent in the western part of the collision belt due to the slab break-off at very deep depth. The potassic to ultra-potassic rocks are high Ba-Sr-type rocks and also have extremely high Sr and Pb and low Nd isotopic ratio. They are extremely enriched in Light Rare Earth Element (LREE) and Large Ion Lithophile Element (LILE) and depleted in Heavy Rare Earth Element (HFSE) indicating their formation by low degree partial melting of an upper mantle that was metasomatized during subduction stage before collision.

2.13.3. Tectonic Models of Continental Collision along the Dabie-Hongseong and the Himalayan Collision Belt

Although several mechanisms have been proposed to explain the uplift of subducted continental crust, including buoyancy of the crust, thrusting, and local extension [*Yang and Yu*, 2001; *Hacker et al.*, 2004; *Ernst*, 2006], buoyancy is the main factor in the uplift of subducted continental crust to midcrustal levels [*Ernst*, 2006]. On this basis, the following 3D tectonic models are proposed to explain the collision and subsequent uplift to the middle crust of the Dabie-Hongseong belt and the Himalayan collision belt (Figures 2.26, 2.27). The NCB and SCB moved northward during the Middle Paleozoic with subduction along the margin of the NCB. The two blocks first collided at their eastern margins (i.e., the Odesan area in Korea), with an angle of about 60° between the long axes of the two blocks, and the SCB began to subduct beneath the NCB due to the pulling force of the subducted oceanic slab before collision (Figure 2.26a). The collision propagated westward, accompanied by clockwise rotation of the SCB (Figure 2.26b). In the Odesan area, the amount of the subducted oceanic slab before collision was smallest along the collision belt. As a result, the SCB stopped subduction before it reached the depth of the eclogite facies, and slab break-off occurred because the buoyancy of the subducted continental slab exceeded the pulling force of the subducted oceanic slab at that depth. Consequently, UHT metamorphism and mangerite magmatism occurred during Permo-Triassic time as a result of heat supplied by the upwelling asthenosphere through the

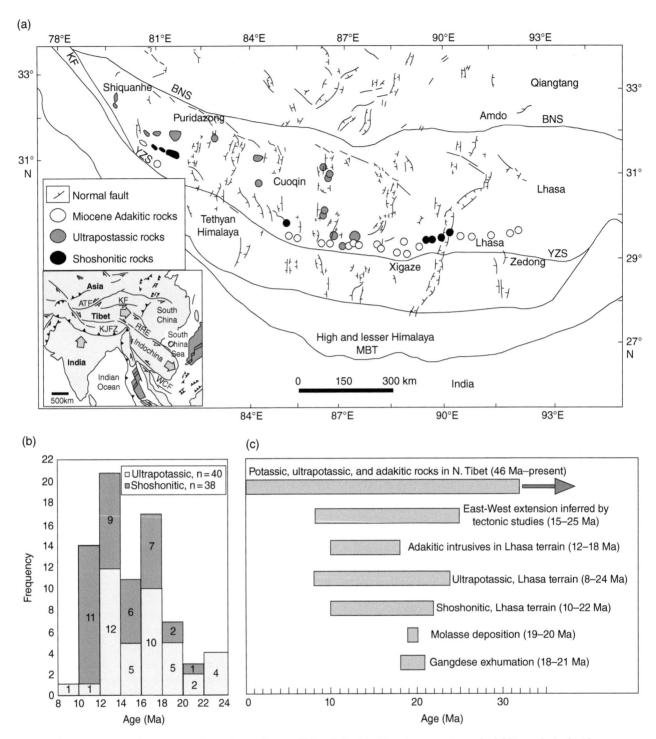

Figure 2.25 Distribution (a) and ages (b, c) of postcollisional shoshonitic, ultrapotassic, and adakitic rocks in the Lhasa terrain, southern Tibet [modified after *Zhao et al.*, 2009; *Gao et al.*, 2010]. BNS = Bangong–Nujiang Suture; KF = Karakorum fault; MBT = Main boundary thrust; YZS = Yarlung Zangbo suture.

opening between the continental slab and the detached oceanic slab. In contrast, subduction of continental crust continued in the middle and western parts of the Dabie-Hongseong belt due to relatively larger amount of oceanic crust subducted prior to collision (Figure 2.26b). The P-T conditions and metamorphic age of the Hongseong eclogites indicate that slab break-off in the Hongseong area occurred once the continental crust reached the depths of HP eclogite facies metamorphism (55–65 km) at ca. 230 Ma, indicating the westward propagation of the zone of slab break-off (Figure 2.26c). Finally, the zone of slab break-off migrated to the Dabie and Sulu areas at ca. 220 Ma,

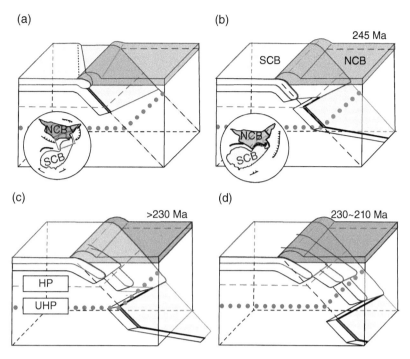

Figure 2.26 Schematic tectonic model of collision along the Dabie-Hongseong collision belt [from *Oh*, 2010]. (a) Continental collision started in the east. (b) Collision propagated westward (the Sulu and Dabie areas), accompanied by clockwise rotation of the SCB. In the east (the Odesan area), subduction of the continental block ceased due to oceanic slab break-off, which produced an opening in the subducting slab through which high heat flow occurred from the asthenosphere. (c) The zone of oceanic slab break-off propagated to the Hongseong area, terminating the subduction of the continental block after reaching the depths of HP eclogite facies metamorphism; however, subduction of the continental block continued in the Dabie and Sulu areas. (d) In the Dabie and Sulu areas, the continental block reached the depths of UHP metamorphism and then the oceanic slab began to break off. The thin blue dashed and thick red dotted lines represent the minimum depths of HP and UHP eclogite facies metamorphism, respectively. NCB = North China block; SCB = South China block.

once the continental crust had reached the depths of UHP metamorphism (100–130 km depth; Figure 2.26d).

The uplifting speed after slab break-off decreased eastward due to an eastward weakening in the buoyancy of the subducted continental crust, related to an eastward decrease in subduction depth of continental slab. This trend resulted in prolonged thermal relaxation and a higher geothermal gradient toward the east during the period of uplift following slab break-off. The higher rate of uplift in the Hongseong area compared with that in the Odesan area, meant that eclogites were overprinted by granulite facies metamorphism during uplift rather than UHT metamorphism. In the Dabie and Sulu areas, the eclogites mainly record amphibolite facies retrograde metamorphism during uplift, due to the highest rate of uplift along the Dabie-Hongseong collision belt.

The following collision model of the Himalayan collision belt is proposed based on data published in previous studies. Collision between the Indian and Asian blocks started in the west before ca. 55 Ma (Figure 2.25a). In the western part, the amount of oceanic slab subducted prior to continent collision was enough to pull the continental crust

down to the depths of UHP metamorphism, as a wide ocean existed between the Asian and Indian blocks prior to collision (Figures 2.24b, 2.27a). Following UHP metamorphism, oceanic slab break-off started at ca. 55–46 Ma in the west (Figure 2.27b) due to the very strong buoyancy of the deeply subducted continental block. In contrast, the subduction of continental crust continued at this time in the middle and eastern parts of the belt (Figure 2.27b). The zone of break-off migrated eastward, initiating a change from steep- to low-angle subduction [Figure 2.27c; *O'Brien*, 2001]. Final slab break-off may have occurred in the easternmost part of the belt at ca. 22–25 Ma (Figure 2.27d). The depth of slab break-off decreased toward east, indicating that the rate of uplift decreased eastward due to a decrease in buoyancy of the continental slab. The slower uplift resulted in a longer period of thermal relaxation and a higher geothermal gradient. In the west, the high rate of uplift resulted in the epidote amphibolite facies (580–610°C) retrograde metamorphic overprint on the UHP eclogites, whereas the relatively slow uplift in the mideastern part caused high-grade granulite (850°C) retrograde metamorphic overprint on the HP eclogites.

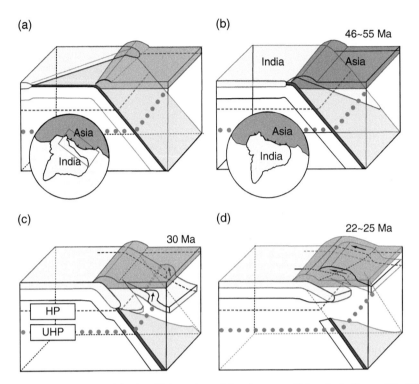

Figure 2.27 Schematic tectonic model of collision along Himalayan collision belt [from *Oh*, 2010]. (a) In the Himalayan collision belt, continental collision started from the west. (b) Collision propagated to the east (the Kharta area). In the west (the Kaghan valley), the oceanic slab began to break off due to the buoyancy of the continental crust, which reached the depths of UHP metamorphism. (c) Oceanic slab break-off propagated to the Kharta area, where the continental block had subducted to the depths of HP eclogite facies metamorphism. The continental block in the west began to be uplifted toward the surface. (d) Oceanic slab break-off also occurred in the easternmost part of the belt while UHP rocks were uplifted to shallow depths in the west. The thin blue dashed and thick red dotted lines represent the minimum depths of HP and UHP eclogite facies metamorphism, respectively.

The suggested tectonic model for the Himalayan collision belt can improve the tectonic models suggested based only in geophysical data as follows. *Replumaz et al.* [2010] reported three geophysical anomalies along the Himalayan collision belt; TH anomaly at depth deeper than 1100 Km, IN anomaly at depth between 400 and 1100 km, and CR anomaly at depth shallower than 200 km (Figure 2.28). *Negredo et al.* [2007] consider that the TH anomaly on tomographic images would correspond to the position of the subduction zone at the time of slab break-off along the Himalayan collision belt and deduced an age of slab break-off of about 44–48 Ma. *Replumaz et al.* [2010] interpreted that anomaly IN, located at depth shallower than 1100 km, represents a slab of continental material subducted after the slab break-off at 44–48 Ma. The shallower high wavespeed anomaly CR is commonly interpreted as related to the India plate [e.g., *van der Voo et al.*, 1999]. The gap between anomalies IN and CR was interpreted as an evidence of a second slab break-off event by *Replumaz et al.* [2010].

They inferred that a large portion of the northwestern margin of India began to be subducted at 35 ± 5 Ma along a 1500-km-long WNW-ESE striking zone and the subduction ended with a progressive slab break-off process. This break-off started most probably around 25 Ma at the western end of the slab and propagated eastward until complete break-off around 15 Ma. This interpretation has some problems. The petrological data explained in this study indicate that the western part of the Himalayan collision belt started to uplift from 55–45 Ma denying the initiation of subduction at 35 ± 5 Ma in the northwestern margin. Petrological data such as postcollision igneous activity from 24 Ma suggest that the break-off ended before 15 Ma.

The geophysical result can be explained better by the 3D tectonic model suggested in this study. The TH anomaly may be the detached slab fragment and represent the subduction position at 45 Ma. At this time, the slab break-off occurred in the western belt but not in other belts. Due to the movement of the Indian block toward the north, the subduction position moved to the area where IN anomaly was found. At this position, subduction occurred only at the central and eastern belts because the subduction in the western belt was finished by slab break-off at

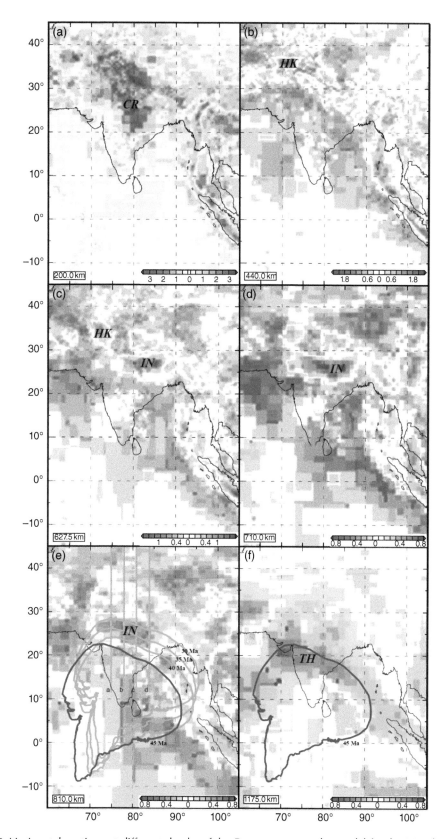

Figure 2.28 Horizontal sections at different depths of the P-wave tomography model for the Himalayan collision belt [from *Replumaz et al.*, 2010]. Anomaly HK is related to the Indian continental slab under the Hindu Kush. Anomaly CR is related to the Indian craton (section a). Sections c and d show the prominent IN anomaly, whereas section b shows that this anomaly disappears at the top of the transition zone. Blue line (in sections e and f) indicates the geometry inferred by *Negredo et al.* [2007] for continental India at about 45 Ma. This geometry of India is further rotated about poles given by *Patriat and Achache* [1984] to obtain positions at 40, 35, and 30 Ma (green lines in section e). Section f shows anomaly TH, which is interpreted as marking the location of the Tethyan oceanic subduction.

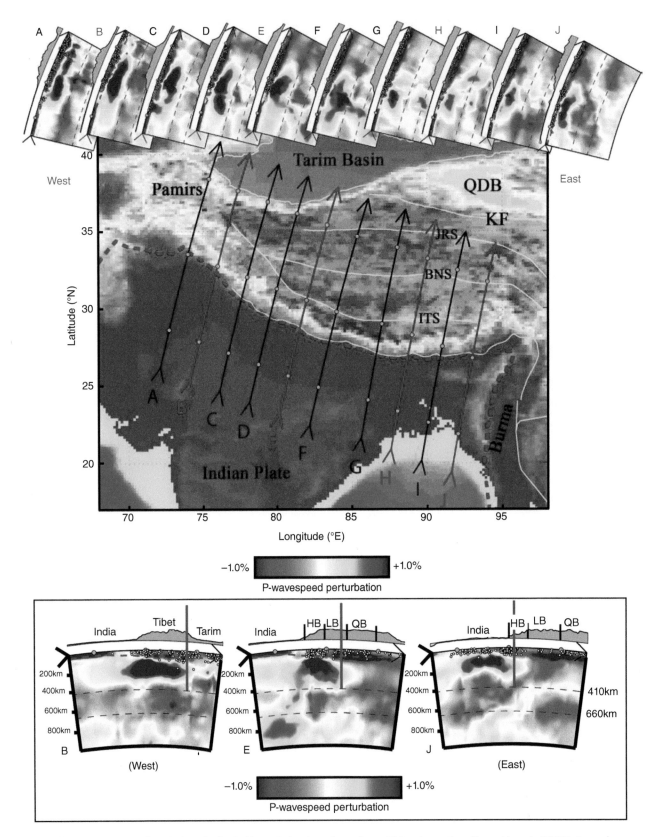

Figure 2.29 Lateral variations in the Indian subduction along the collision boundary [from *Li et al.*, 2008]. In each cross section, gray circles are earthquakes, magenta dash lines display the Moho, gray shadow on the top of each subplot is the topography, and black dash lines are the 410-km and 660-km discontinuities. The horizontal distance over which the Indian lithosphere slides northward beneath the plateau rapidly decreases from west to east. The thick blue lines in the lower Figure show the northern leading edge of the Indian subduction. Labels: HB = Himalayan block; LB = Lhasa block; QB = Qiangtang block; ITS = Indus-Tsangpo suture; BNS = Bangong-Nujiang suture; JRS = Jinsha River suture.

45 Ma. As a result, IN anomaly formed by detached slab fragment occurs only in the central and eastern belts. After oceanic slab break-off, the subducted India plate uplifted until it contact with the bottom of the lithospheric mantle. During this process, the slope of underthrusting continental lithosphere (India plate) became very gentle and some subducted parts moved up to the surface along the fault between the Asia and India plates. The uplifted India plate with gentle slope may form the CR geophysical anomaly. More detailed geophysical analysis was carried out on the CR anomaly by *Li et al.* [2008].

The new images obtained by *Li et al.* [2008] reveal that the horizontal distance over which presumed (continental) Indian lithosphere slides northward beneath the plateau decreases from west (where it underlies the Himalayas and the entire plateau) to east (where no indication is found for present-day underthrusting beyond the Himalayan block and Indus-Tsangpo suture; Figure 2.29). This is also well explained by the 3D model suggested in this paper. The subduction depth of the continental slab was longest in the west and decreased toward the east (Figure 2.29). As a result, the length of the uplifted slab decreases toward the east, which is well matched with the decrease of the horizontal distance of the Indian lithosphere toward the west.

The tectonic model for the simplified uplifting process along the Himalayan collision belt is determined by combining the result of this study and geophysical results (Figure 2.30). In the western part of the Himalayan

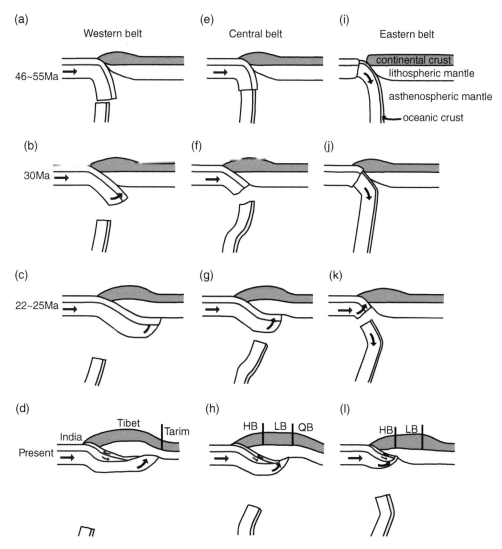

Figure 2.30 The tectonic model for uplifting of subducted continental slab after slab break-off in the western (a–d), central (e–h), and eastern Himalayan collision belts (i–l). The slab break-off first occurred in the western belt and propagated toward the east. After the slab break-off (a, f, k), the subducted continental slab uplifted until contact with the continental slab of the Asian plate (b–d, g–h, l). As the length of subducted continental slab decreased toward the east (a, f, k), the horizontal length of the uplifted continental slab also decreased toward the east (d, h, l). HB = Himalayan block; LB = Lhasa block; QB = Qiangtang block.

collision belt, the oceanic slab broke off from the continental slab at ca. 55–46 Ma (Figure 2.30a). After slab break-off, the continental slab uplifted toward the surface and the oceanic slab moved down as detached slab segments forming the TH anomaly, which now reaches to a depth deeper than 1100 km (The cross section along the line B in Figure 2.29; Figures 2.30b–d). At ca. 55–46 Ma, the continent collision started in the central belt but not in the eastern belt (Figures 2.30e, 2.30i). After 45 Ma, the slab break-off propagated toward the west and the slab break-off occurred at ca. 30 Ma in the central belt (Figure 2.30f). Due to the movement of the India plate toward Asia by forces other than the pulling force of the subducted oceanic slab, the subducted oceanic slab might be curved in the central and eastern belts before slab break-off (Figures 2.30f, 2.30j). As a result, the geophysical anomaly formed by the detached slab at the depths shallower than 1100 km (CR in Figure 2.28) locates to the north of the geophysical anomaly formed at the depths below 1100 km (TH in Figure 2.28) in the central belt. The continental slab moved up until contact with the bottom of the lithosphere of the Asian plate. However, the horizon distance of the India plate in the central belt is shorter than that in the western belt due to the shorter length of the subducted continental slab in the central belt compared to that in the western belt. The Indian lithosphere moved northward after slab break-off, which is well identified from the geophysical tomography in which the front of the Indian subhorizontal plate locates under the Lhasa block while the anomaly by the detached oceanic slab locates under the Himalayan block (The cross section along the line E in Figure 2.29; Figures 2.30g, 2.30h).

The movement of the India plate toward the north after slab break-off also can be identified in the western belt by the front of the continental slab, which is located farther north compared to the position of anomaly by the detached slab (The cross section along the line B in Figure 2.29; Figures 2.30c, 2.30d). In the eastern belt, the continental slab moved up to the surface after the slab break-off occurred at ca. 22–25 Ma. However, the length of the uplifted continental slab is shorter than those in the western and central belts, resulting in geophysical tomography that shows the shortest horizontal length of the continental slab in the eastern belt as shown in the cross section along the line J in Figure 2.29 (Figure 2.30k, l). The subducted oceanic slab might be also curved in the eastern belt before slab break-off as in the central belt. As a result, CR anomaly locates to the north of TH anomalies in the eastern belt as in the central belt. The position of geophysical anomaly by detached slab segment is similar to the front position of the underthrusted India plate suggesting that there was almost no movement of the India plate after the break-off in the eastern belt.

The study indicates that the different metamorphic patterns along the collision belt indicate different collision processes and are strongly related to the amount of subducted oceanic crust between continents before collision and the depth of slab break-off. Therefore, metamorphic patterns can be used to interpret both the past and ongoing tectonic process during continental collision. This study also shows that the tectonic interpretation based on petrological study can contribute to better understanding of geophysical tomography of the area where the collision process is now occurring.

ACKNOWLEDGMENT

I would like to express my sincere thanks to Gabriele Morra for giving me a chance to write this paper and to all who carried out coresearch on the Dabie-Hongseong collision belt. I also acknowledge students including Jiyeong Lee, Byung Choon Lee and Sang-Bong Yi in my lab for drawing Figures and making tables. This research was supported by the National Research Foundation of Korea (2010-00237347 and NRF-2014R1A2A2A 01003052).

REFERENCES

Ames, L., G. R. Tilton, and G. Zhou (1993), Timing of collision of the Sino-Korean and Yangtze cratons: U-Pb zircon dating of coesite-bearing eclogites, *Geology*, *21*, 339–342.

Arakawa, Y., T. Kouta, and Y. Kanda (2001), Geochemical characteristics of amphibolites in the Oki metamorphic rocks, Oki-Dogo Island, southwestern Japan: Mixed occurrence of amphibolites with different geochemical affinity, *J. Mineral. Petrol. Sci.*, *96*, 175–187.

Arakawa, Y., Y. Saito, and H. Amakawa (2000), Crustal development of the Hida belt, Japan: Evidence from Nd-Sr isotopic and chemical characteristics of igneous and metamorphic rocks, Tectonophysics, *328*, 183–204.

Banno, S., M. Enami, T. Hirajima, A. Ishiwatari, and Q. C. Wang (2000), Decompression P-T path of coesite eclogite to granulite from Weihai, eastern China, *Lithos*, *52*, 97–108.

Bundy, F. P. (1980), The P-T phase and reaction diagram for elemental carbon, *J. Geophys. Res.*, *85*, 6930–6936.

Burrett, C., and R. Richardson (1980), Trilobite biogeography and Cambrian tectonic models, *Tectonophysics*, *63*, 155–192.

Carswell, D. A., R. H. Wilson, and M. Zhai (2000), Metamorphic evolution, mineral chemistry and thermobarometry of schists and orthogneisses hosting ultra-high pressure eclogites in the Dabieshan of Central China, *Lithos*, *52*, 121–155.

Chamberlain, C. P., P. K. Zeitler, and E. Erickson (1991), Constraints on the tectonic evolution of the northwestern Himalaya from geochronologic and petrologic studies of Babusar Pass, *Pakistan. J. Geol.*, *99*, 829–849.

Chang, E. Z. (1996), Collisional orogene between north and south China and its eastern extension in the Korean Peninsula, *J. South. Asian Earth.*, *13*, 267–277.

Chappell, B. W., and A. J. R. White (1974), Two contrasting granite types, *Pacific Geology*, *8*, 173–174.

Chen, J. F., A. Xie, H. M. Li, X. D. Zhang, T. X. Zhou, Y. S. Park, K. S. Ahn, D. G. Chen, and X. Zhang (2003), U-Pb zircon ages for a collision-related K-rich complex at Shidao in theSulu ultrahigh pressure terrane. China, *Geochem. J.*, *37*, 35–46.

Cheong, C. S., K. Y. Cheong, H. Kim, M. S. Choi, S. Lee, and M. Cho (2003), Early Permian peak metamorphism recorded in U-Pb system of black slates from the Ogcheon metamorphic belt, South Korea, and its tectonic implication, *Chem. Geol.*, *193*, 81– 92.

Cho, D. L., and Y. Kim (2003), SHRIMP zircon age of biotite gneiss and leucogranitic gneiss in Pocheon area, Gyeonggi Massif: The age of regional metamorphic age and constraints on sedimentary age, *Abstracts volume for annual meeting of Geol. Soc. Korea*, *76* (in Korean).

Cho, D. L., K. Suzuki, M. Adachi, and U. Chwae (1996), A primary CHIME age determination of monazites from metamorphic and granitic rocks in the Gyeonggi Massif, Korea, *J. Earth Planet. Sci., Nagoya Univ.*, *43*, 49–65.

Cho, D. L., S. T. Kwon, E.Y. Jeon, and R. Armstrong (2001), SHRIMP U-Pb zircon geochronology of an amphibolite and a paragneiss from the Samgot unit, Yeoncheon complex in the Imjingang belt, Korea: Tectonic implication, *Abstracts volume for annual meeting of Geol. Soc. Korea*, *89*.

Cho, K. H., H. Takagi, and K. Suzuki (1999), CHIME monazite age of granitic rocks in the Sunchang shear zone, Korea: Timing of dextral ductile shear, *Geosci. J.*, *3*, 1–15.

Cho, M. (2001), A continuation of Chinese ultrahighpressure belt in Korea: Evidence from ion microprobe U-Pb zircon ages, *Gondwana Res.*, *4*, 708.

Choi, S., C. W. Oh, and H. Luehr (2006), Tectonic relation between northeastern China and the Korean peninsula revealed by interpretation of GRACE satellite gravity data, *Gondwana Res.*, *9*, 62–67.

Choi, S. G., S. H. Song, and S. Choi (1998), Genetic implications of the ultramafic rocks from the Hongseong area, western part of Chungnam, *Abstracts volume for annual meeting of Korean Soc. Econ. Environ. Geol.*, *46* (in Korean).

Choi, S. G., V. J. Rajesh, J. Seo, J. W. Park, C. W. Oh, S. J. Park, and S. W. Kim (2008), Petrology, geochronology and tectonic implications of Mesozoic high Ba-Sr granites in the Haemi area, Hongseong Belt, South Korea. *Island Arc*, *18*, 266–281.

Chopin, C. (1984), Coesite and pure pyrope in high grade blueschists of the Western Alps: A first recorded and some consequences, *Contrib. Mineral. Petrol.*, *86*, 107–118.

Chung, S. L., D. Liu, J. Ji, M. F. Chu, H. Y. Lee, D. J. Wen, C. H. Lo, T. Y. Lee, Q. Qian, and Q. Zhang (2003), Adakites from continental collision zones: Melting of thickened lower crust beneath southern Tibet, *Geology*, *31*, 1021–1024.

Chung, S. L., M. F. Chu, Y. Q. Zhang, Y. W. Xie, C. H. Lo, T. Y. Lee, C. Y. Lan, X. H. Li, Q. Zhang, and Y. H. Wang (2005), Tibetan tectonic evolution inferred from spatial and temporal variations in post-collisional magmatism, *Earth-Sci Rev.*, *68*, 173–196.

Cong, B., M. Zhai, D. A. Carswell, R. H. Wilson, Q. Wang, Z. Zhao, and B. F. Windley (1995), Petrogenesis of ultrahigh-pressure rocks and their country rocks in Shuanghe of Dabieshan Mountains, Central China. *Eur. J. Mineral.*, *7*, 119–138.

Cong, B., Q. Wang, R. Zhang, M. Zhai, Z. Zhao, and J. Li (1992), Discovery of the coesite-bearing granulite in the Weihai, Shandong Province, China. *Chinese Sci. Bull.*, *37*, 4–5.

Corrie, S. L., M. J. Kohn, and J. D. Vervoort (2010), Young eclogite from the Greater Himalayan Sequence, Arun Valley, eastern Nepal: P-T-t path and tectonic implications, *Earth Planet. Sci. Lett.*, *289*, 406–416.

Cottle, J. M., M. J. Jessup, D. L. Newell, M. S. A. Horstwood, S. R. Noble, R. R. Parrish, D. J. Waters, and M. P. Searle (2009). Geochronology of granulitized ecologite from the Ama Drime Massif: Implications for the tectonic evolution fo the South Tibetan Himalaya, *Tectonics*, *28*, TC1002, doi: 10.1029/2008TC002256.

Cox, K. G., J. D. Bell, and R. J. Pankhurst (1979), *The interpretation of igneous rocks*, George Allen and Unwin, London.

Dallwitz, W. B. (1968), Coexisting sapphirine and quartz in granulite from Enderby Land, Antarctica, *Nature*, *219*, 476–477.

Davis, J. H., and F. von Blankenburg (1995), Slab breakoff: A model of lithosphere detachment and its test in the magmatism and deformation of collisional orogens, *Earth Planet. Sci. Lett.*, *129*, 85–102.

De Sigoyer, J., S. Guillot, and P. Dick (2004), Exhumation processes of the high-pressure low-temperature Tso Morari dome in a convergent context (eastern-Ladakh, NW Himalaya), *Tectonics*, *23*, TC3003, doi: 10.1029/2002TC001492.

De Sigoyer, J., S. Guillot, J. M. Lardeaux, and G. Mascle (1997), Glaucophanebearing eclogites in the Tso Morari dome (eastern Ladakh, NW Himalaya), *Eur. J. Mineral.*, *9*, 1073–1083.

De Sigoyer, J., V. Chavagnac, J. Blichert-Toft, I. M. Villa, B. Luais, S. Guillot, M. Cosca, and G. Mascle (2000), Dating the Indian continental subduction and collisional thickening in the northwest Himalaya: Multichronology of the Tso Morari eclogites, *Geology*, *28*, 487–490.

Ding, L., D. L. Zhong, A. Yin, P. Kapp, and T. M. Harrisson (2001), Cenozoic structural and metamorphic evolution of the eastern Himalayan syntaxis (Namche Barwa), *Earth Planet. Sci. Lett.*, *192*, 423–438.

Enami, M., K. Suzuki, M. Zhai, and X. Zheng (1993), High-pressure eclogites in northern Jiangsu–southern Shangdong province, eastern China, *J. Metamorph. Geol.*, *11*, 589–603.

Ernst, W. G. (2006), Preservation/exhumation of ultrahigh-pressure subduction complexes. *Lithos*, *92*, 312–335.

Ernst, W. G., and J. G. Liou (1995), Contrasting plate-tectonic styles of the Qinling-Dabie-Sulu and Franciscan metamorphic belt, *Geology*, *23*, 353–356.

Fitches, B. (2004), The Yugu peridotite of South Korea: An exhumed and tectonized upper mantle fragment? In *Gondwana to Asia, International Workshop and Field Excursion, Abstr. Vol.*, *13–16*.

Fontan, D. M., C. J. Schouppe, J. Hunziker, G. Martinotti, and J. Verkaeren (2000), Metamorphic evolution and 40Ar/39Ar chronology and tectonic model for the Neelum valley, Aza Kashmir, NE Pakistan, in M. A. Khan, M. P. Searle, P. Treloar (eds.), *Tectonics of the Nanga Parbat Syntaxis and the Western Himalaya*, *170*, 431–453.

Gansser, A. (1979), Ophiolitic belts of the Himalayan and Tibetan region, in Working Group of Ophiolite Project (eds.), *International atlas of ophiolites*, Geol. Soc. Am. Map Chart Ser., MC-33, Geol. Soc. Am.,Washington, DC.

Gao, Y., Hou, Z., Kamber, B.S.,Wei, R., Meng, X., and Zhao, R. (2007), Adakite-like porphyries from the southern Tibetan continental collision zones: Evidence for slab melt metasomatism, *Contrib. Mineral. Petrol., 153*, 105–120.

Gao, Y., Z. Yang, M. Santosh, H. Zengqian, W. Ruihua, and T. Shihong (2010), Adakitic rocks from slab melt-modified sources in the continental collision zone of southern Tibet, *Lithos, 119*, 651–663.

Ghazanfar, M., and M. N. Chaudhry (1987), Geology, structure and geomorphology of Upper Kaghan Valley, northwest Himalaya, Pakistan, *Geol. Bull. Univ. Punjab, 22*, 13–56.

Groppo, C., B. Lombardo, F. Rolfo, and P. Pertusati (2007), Clockwise exhumation path of granulitized eclogites from the Ama Drime range (Eastern Himalayas), *J. Metamorph. Geol., 25*, 51–75.

Guillot, S., G. Mahéo, J. De Sigoyer, K. H. Hattori, and A. Pêcher (2008), Tethyan and Indian subduction viewed from the Himalayan high-to ultrahigh-pressure metamorphic rocks, *Tectonophysics, 451*, 225–241

Guillot, S., J. de Sigoyer, J. M. Lardeaux, and G. Mascle (1997), Eclogitic metasediments from the Tso Morari area (Ladakh, Himalaya), evidence for continental subduction during India-Asia convergence, *Contrib. Mineral. Petrol., 128*, 197–212.

Guillot, S., M. Cosca, P. Allemand, and P. Le Fort (1999), Contrasting metamorphic and geochronologic evolution along the Himalayan Belt, in A. MacFarlane, R. B. Sorkhabi, and J. Quade (eds.), *Himalaya and Tibet, Mountain Roots to Mountain Tops, Geol. Soc. Am. Spec. Pap., 328*, 117–128.

Guo, J., M. Zhai, C. W. Oh, and S. W. Kim (2004), 230 Ma Eclogite from Bibong, Hongseong area, Gyeonggi Massif, South Korea: HP metamorphism, zircon SHRIMP U-Pb ages and tectonic implication, in *Gondwana to Asia, International Workshop and Field Excursion, Abstr. Vol., 11–12*.

Guo, Z. F., M. Wilson, and J. Q. Liu (2007), Post-collisional adakites in south Tibet: Products of partial melting of sub-duction-modified lower crust, *Lithos, 96*, 205–224.

Hacker, R. B., L. Ratschbacher, and J. G. Liou (2004), Subduction, collision and exhumation in the ultrahigh-pressure Qinling-Dabie orogen, in J. Malpas, C. J. N. Fletcher, J. R. Ali, J. C. Aitchison (eds.), *Aspects of the Tectonic Evolution of China, Geol. Soc.* London, Spec. Pub., *226*, 157–175.

Harley, S. L. (1988), On the occurrence and characterization of ultrahigh-temperature (UHT) crustal metamorphism, in P. J. Treloar and P. O'Brien (eds.), *What Controls Metamorphism and Metamorphic Reactions?* Geol. Soc. London, *Spec. Pub., 138*, 75–101

Harris, N. B. W., and J. Massey (1994), Decompression and ana-texis of Himalayan metapelites, *Tectonics, 13*, 1537–1546.

Hayasaka, Y., T. Sugimoto, and T. Kano (1995), Ophiolitic complex and metamorphic rocks in the Niimi-Katsuyama area, Okayama Prefecture, in *Excursion Guidebook of 102d Annual Meeting of Geol. Soc. Japan*, 71–87.

Hirajima, T., and D. Nakamura (2003), The Dabie Shan-Sulu orogen, in D. A. Carswell and R. Compagnoni (eds.), *Ultrahigh Pressure Metamorphism*, EMU Notes in Mineralogy 5, Eur. Mineral. Union, 105–144.

Hiroi, Y. (1978), Geology of the Unazuki district in the Hida metamorphic terrain, central Japan, *J. Geol. Soc. Japan, 84*, 521–530 (in Japanese with English abstract).

Hiroi, Y. (1983), Progressive metamorphism of the Unazuki pelitic schists in the Hida terrane, central Japan, *Contrib. Mineral. Petrol., 82*, 334–350.

Hodges, K. V. (2000), Tectonics of the Himalaya and southern Tibet from two perspectives, *Geol. Soc. Am. Bull., 112*, 324–350.

Hodges, K. V., W. E. Hames, W. Olszewski, B. C. Burchfiel, L. H. Royden, and Z. Chen (1994), Thermobarometric and 40Ar/39Ar geochronologic constraints on Eohimalayan metamorphism in the Dinggyê area, southern Tibet, *Contrib. Mineral. Petrol., 117*, 151–163.

Honegger, K., P. Le Fort, G. Mascle, and J. L. Zimmerman (1989), The blueschists along the Indus Suture Zone in Ladakh, N.W. Himalaya, *J. Metamorph. Geol., 7*, 57–72.

Honegger, K., V. Dietrich, W. Frank, A. Gansser, M. Thöni, and V. Trommsdorff (1982), Magmatism and metamorphism in the Ladakh Himalayas (the Indus-Tsangpo suture zone). *Earth Planet Sci. Lett., 60*, 253–292.

Hou, Z. Q., Y. F. Gao, X. M. Qu, Z. Y. Rui, and X. X. Mo (2004), Origin of adakitic intrusives generated during mid-Miocene east-west extension in southern Tibet, *Earth Planet. Sci. Lett., 220*, 139–155.

Hubbard, M. S., D. A. Spencer, and D. P. West (1995), Tectonic exhumation of the Nanga Parbat massif, northern Pakistan, *Earth Planet. Sci. Lett., 133*, 213–225.

Imayama, T., T. Takeshita, K. Yi, D.-L. Cho, K. Kitajima, Y. Tsutsumi, M. Kayama, H. Nishido, T. Okumura, K. Yagi, T. Itaya, and Y. Sano (2012), Two-stage partial melting and contrasting cooling history within the Higher Himalayan Crystalline Sequence in the far-eastern Nepal Himalaya, *Lithos, 134–135*, 1–22.

Irvine, T. N., and W. R. A. Baragar (1971), A guide to the chemical classification of the common volcanic rocks, *Can. J. Earth Sci., 8*, 523–548.

Ishiwatari, A., and T. Tsujimori (2003), Paleozoic ophiolites and blueschists in Japan and Russian Primorye in the tectonic framework of East Asia: A synthesis, *Island Arc, 12*, 190–206.

Isozaki, Y. (1997), Contrasting two types of orogen in Permo-Triassic Japan: Accretionary versus collisional, *Island Arc, 6*, 2–24.

Jeon, E. Y., and S. T. Kwon (1999), Metamorphism of the Gonamsan area in the western Imjingang belt: Anticlockwise P-T-t path, *J. Geol. Soc. Korea, 35*, 49–72 (in Korean with English abstract).

Kaneko, Y., I. Katayama, H. Yamamoto, K. Misawa, M. Ishikawa, H. U. Rehman, A. B. Kausar, and K. Shiraishi (2003), Timing of Himalayan ultrahigh-pressure meta-morphism: Sinking rate and subduction angle of the Indian continental crust beneath Asia, *J. Metamorph. Geol , 21*, 589–599.

Kim, H., C. S. Cheong, M. Cho, G. Y. Jeong, and M. S. Choi (2001), Geochronological evidence for late Paleozoic orogeny in the Ogcheon metamorphic belt, South Korea, Abstracts with Program of Geol. Soc. Am. Annual Meeting, *33*.

Kim, S. W. (2005), Amphibole 40Ar/39Ar geochronology from the Okcheon metamorphic belt, *South Korea and its tectonic implications, Gondwana Res.*, *8*, 385–402.

Kim, S. W., C. W. Oh, I. S. Williams, D. Rubatto, I. C. Ryu, V. J. Rajesh, J. B. Kim, J. Guo, and M. Zhai (2006), Phanerozoic high-pressure eclogite and intermediate pressure granulite facies metamorphism in the Gyeonggi Massif, South Korea: Implications for the eastward extension of the Dabie-Sulu continental collision zone, *Lithos*, *92*, 357–377.

Kim, S. W., I. S. Wiliams, S. Kwon, and C. W. Oh (2008), SHRIMP zircon geochronology and geochemical characteristics of metaplutonic rocks from the south-western Gyeonggi Block, Korea: Implications for Paleoproterozoic to Mesozoic tectonic links between the Korean Peninsula and eastern China, *Precambrian Res.*, *162*, 475–497.

Kim, T. S. (2011), Igneous activity and Metamorphism ralated to the Triassic continental collision in the Odaesan area, the eastern part of the Gyeonggi Massif, South Korea. Ph.D. dissertation, Chonbuk National University, 154p.

Kim, T. S., C.W. Oh, and J. M. Kim (2011), The characteristic of mangerite and gabbro in the Odaesan area and its meaning to the Triassic tectonics of Korean peninsula, *J. Petrol. Soc. Korea*, *20*, 77–98.

Klootwijk, C. T., J. S. Gee, J. W. Peirce, G. M. Smith, and P. L. McFadden (1992), An early India-Asia contact: Paleomagnetic constraints from Ninetyeast Ridge, ODP Leg 121, *Geology*, *20*, 395–398.

Komatsu, M., and K. Suwa (1986), The structure and tectonics of pre-Jurassic serpentinite melange and overlying Hida nappe in central Japan, *Abstract volume in International Symposium on Pre-Jurassic East Asia, IGCP 224*, 97–103.

Komatsu, M., M. Nagase, K. Naito, T. Kanno, M. Ujihara, and T. Toyoshiam (1993), Structure and tectonics of the Hida massif, central Japan *Mem. Geol. Soc. Japan*, *42*, 39–62 (in Japanese with English abstract.).

Koons, P. O. (1995), Modeling the topographic evolution of collisional mountain belts, *Annual Review of Earth Planet. Sci.*, *23*, 375–408.

Korea Institute of Geoscience and Mineral Resources (KIGAM) (2001), Tectonic map of Korea (1: 1,000,000), Seoul, South Korea, Sungji Atlas Co. Ltd.

Kröner, A., G. W. Zhang, and Y. Sun (1993), Granulites in the Tongbai area, Qinling belt, China: Geochemistry, petrology, single zircon geochronology, and implications for the tectonic evolution of eastern Asia, *Tectonics*, *12*, 245–255.

Kwon, S., K. Sajeev, G. Mitra, Y. Park, S.W. Kim, and I. C. Ryu (2009), Evidence for Permo-Triassic collision in Far East Asia: The Korean collisional orogen, *Earth Planet. Sci. Lett.*, *279*, 340–349.

'Lee, S. R., B. J. Lee, D. L. Cho, W. S. Kee, H. J. Koh, B. C. Kim, K. Y. Song, J. H. Hang, and B. Y. Choi (2003), SHRIMP U-Pb zircon age from granitic rocks in Jeonju shear zone: Implications for the age of the Honam shear zone. Abstracts volume for annual joint meeting of Mineral. *Soc. Korea and Petrol. Soc. Korea*, *55* (in Korean).

Lee, S. Y., and C. W. Oh (2009), Trassic post collisional igneous activity and granulite facies metamorphic event in the Yangpyeong area, South Korea and its meaning to the tectonics of Northeast Asia, Abstracts volume for annual meeting of *Geol. Soc. Korea*, 227.

Leech, M. L., S. Singh, A. K. Jain, S. L. Klemperer, and R. M. Manickavasagam (2005), The onset of India-Asia continental collision: Early, steep subduction required by the timing of UHP metamorphism in the western Himalaya, *Earth Planet. Sci. Lett.*, *234*, 83–97.

Le Fort, P., S. Guillot, and A. Pêcher (1997), HP metamorphic belt along the Indus suture zone of NW Himalaya: new discoveries and significance, *Compte Rendus de l'Académie des Sciences, Paris*, *325*, 773–778.

Li, C., R. D. van der Hilst, A. S. Meltzer, and E. R. Engdahl (2008), Subduction of the Inidan lithosphere beneath the Tibetan Plateau and Burma, *Earth Planet. Sci. Lett.*, *274*, 157–168.

Li, D., Q. Liao, Y. Yuan, Y. Wan, D. Liu, X. Zhang, S. Yi, S. Cao, and D. Xie (2003), SHRIMP U-Pb zircon geochronology of granulites at Rimana (southern Tibet) in the central segment of the Himalayan orogen, *Chin. Sci. Bull.*, *48*, 2647–2650.

Li, S., D. Liou, Y. Chen, G. Zhang, and Z. Zhang (1991), A chronological table of the major tectonic events for Qinling-Dabie orogenic belt and its implications, in *Conference on Qinling Orogenic Belt (selected papers)*, Xi'an, China, Publishing House of Northwest University, 229–237 (in Chinese).

Li, S., T. M. Kusky, L. Wang, G. Zhang, S Lai, L. Xiaochun, D. Shuwen, and G. Zhao (2007), Collision leading to multiple-stage large-scale extrusion in the Qinling orogen: Insights from the Mianlue suture, *Gondwana Res.*, *12*, 121–143.

Li, Z. X., and C. M. A. Powell (2001), An outline of the palaeo-geographic evolution of the Australasian region since the beginning of the Neoproterozoic, *Earth Sci. Rev.*, *53*, 237–277.

Liou, J. G., R. Y. Zhang, X. Wang, E. A. Eide, W. G. Ernst, and S. Maruyama (1996), Metamorphism and tectonics of high-pressure and ultra-high-pressure belts in the Dabie-Sulu region, China, in A. Yin and M. Harrison (eds.), *The Tectonic Evolution of Asia*, Cambridge Univ. Press, New York, 300–344.

Liou, J. G., T. Tsujimori, R. Y. Zhang, I. Katayama, and S. Maruyama (2004), Global UHP metamorphism and continental subduction/collision: The Himalayan model, *Int. Geol. Rev.*, *16*, 1–27.

Liu, Y., and D. Zhong (1997), Petrology of high-pressure granulites from the eastern Himalayan syntaxis, *J. Metamorph. Geol.*, *15*, 451–466.

Liu, S., R. Steel, and G. Zhang (2005), Mesozoic sedimentary basin development and tectonic implication, northern Yangtze Block, eastern China: Record of continent-continent collision, *J. Asian Earth Sci.*, 9–27.

'Lombardo, B., and F. Rolfo (2000), Two contrasting eclogite types in the Himalayas: Implications for the Himalayan orogeny, *J. Geodyn.*, *30*, 37–60.

Lombardo, B., F. Rolfo, and R. Compagnoni (2000), Glaucophane and barroisite eclogites from the Upper Kaghan nappe: Implications for the metamorphic history of the NW Himalaya, in M. A. Khan, P. J. Treloar, M. P. Searle, and M. Q. Jan (eds.), *Tectonics of the Nanga Parbat Syntaxis and the Western Himalaya*, Geol. Soc. Spec. Pub., *170*, 411–430.

Lombardo, B., P. Pertusati, F. Rolfo, and D. Visonà (1998), First report of eclogites from the Eastern Himalaya, *Implications for the Himalayan orogeny, Memorie di Scienze Geologiche dell'Universitàdi Padova, 50*, 67–68.

Luo, Y., Y. Xu, and Y. Yang (2013), Crustal structure beneath the Dabie orogenic belt from ambient noise tomography, *Earth Planet. Sci. Lett., 313–314*, 12–22.

Massonne, H., and P. J. O'Brian (2003), The Bohemian Massif and the NW Himalaya, in D. A. Carswell and R. Compagnoni (eds.), Ultrahigh pressure metamorphism, *EMU Notes in Mineralogy, 5*, Eur. Mineral. Union, Budapest, 145–187.

Mattauer, M., P. Matte, and J. L. Olivet (1999), A 3D model of the India-Asia collision at plate scale, C. R. Sceances Acad. Sci, *Paris, Sci. Terre., 328*, 499–508.

McKenzie, N. R., N. C. Hughes, P. M. Myrow, D. Choi, and T. Y. Park (2011), Trilobites and zircons link north China with the eastern Himalaya during the Cambrian, *Geology, 39*, 591–594.

Meng, Q. R., and G. W. Zhang (2000), Geologic framework and tectonic evolution of the Qinling orogen, central China, *Tectonophysics, 323*, 183–196.

Metcalfe, I. (1996), Gondwanaland dispersion, Asian accretion and evolution of eastern Tethys, *Australian J. Earth Sci., 43*, 605–623.

Metcalfe, I. (2006), Paleozoic and Mesozoic tectonic evolution and palaeogeography of East Asian crustal fragments: The Korean Peninsula in context, *Gondwana Res., 9*, 24–46.

Mo, X., Z. Zhao, J. Deng, M. Flower, X. Yu, Z. Luo, Y. Li, S. Zhou, G. Dong, D. Zhu, and L. Wang (2006), Petrology and geochemistry of postcollisional volcanic rocks from the Tibetan plateau: Implications for lithosphere heterogeneity and collision-induced asthenospheric mantle flow, in Y. Dilek and S. Pavlides (eds.), Postcollisional tectonics and magmatism in the Mediterranean region and Asia, *Geol. Soc. Am. Spec. Paper, 409*, 507–.530, doi:10.1130/2006.2409(24).

Mukheerjee, B., H. K. Sachan, and T. Ahmad (2005), A new occurrence of microdiamond from Indus Suture zone, Himalata: a possible origin, in H. S. Memoire (ed.), Special extended Abstract Volume, Géologie Alpine, *44*, 136.

Mukheerjee, B., H. K. Sachan, Y. Ogasawaray, A. Muko, and N. Yoshioka (2003), Carbonate-bearing UHPM rocks from the Tso-Morari Region, Ladakh, India: Petrological implications, *International Geol. Rev., 45*, 49–69.

Negredo, A. M., A. Replumaz, A. Villasenor, and S. Guillot (2007), Modeling the evolution of continental subduction processes in the Pamir-Hindu Kush region, *Earth Planet. Sci. Lett., 259*, 212–225.

Nishina, K., T. Itaya, and A. Ishiwatari (1990), K-Ar ages of gabbroic rocks in the "Oeyama ophiolite," Abstracts volume for 97th annual meeting of Geol. Soc. Japan.

O'Brien, P. J. (2001), Subduction followed by Collision: Alpine and Himalayan examples, in D. C. Rubie and R. van der Hilst (eds.), Processes and consequences of deep subduction, *Phys. Earth Planet. In., 127*, 277–291.

O'Brien, P. J., and H. K. Sachan (2000), Diffusion modelling in garnet from Tso Morari eclogite and implications for exhumation models, *Earth Sci. Frontiers, 7*, 25–27.

O'Brien, P. J., N. R. Zotov, M. Law, A. Khan, and M. Q. Jan (2001), Coesite in Himalayan eclogite and implications for models of India-Asia collision, *Geology, 29*, 435–438.

Oh, C. W. (2006), A new concept on tectonic correlation between Korea, *China, and Japan: Histories from the Late Proterozoic to Cretaceous*, Gondwana Res., *9*, 47–61.

Oh, C. W. (2010), Systematic changes in metamorphic styles along the Dabie-Hongseong, and Himalayan collision belts, and their tectonic implications, *J. Asian Earth Sci., 39*, 635–644.

Oh, C. W., and J. G. Liou (1998), A petrogenetic grid for eclogite and related facies under high-pressure metamorphism, *Island Arc, 7*, 36–51.

Oh, C. W., and T. Kusky (2007), The late Permian to Triassic Hongseong-Odesan Collision Belt in South Korea, and its Tectonic Correlation with China and Japan, *Int. Geol. Rev., 49*, 636–657.

Oh, C. W., and S. T. Kim., and J. H. Lee (1995), The Metamorphic Evolution in the Southwestern Part of the Okchon Metamorphic Belt, *J. Geol. Soc. Korea, 31*, 21–31.

Oh, C. W., K. Sajeev, S. W. Kim, and Y. W. Kwon (2006a), Mangerite magmatism associated with a probable late Permian to Triassic Hongseong-Odesan collisional belt in South Korea, *Gondwana Res, 9*, 95–105.

Oh, C. W., S. G. Choi, J. Seo, V. J. Rajesh, J. H. Ree, M. Zhai, and P. Peng (2009), Neoproterozoic tectonic evolution of the Hongseong area, southwestern Gyeonggi Massif, South Korea; Implication for the tectonic evolution of Northeast Asia, *Gondwana Res., 16*, 272–284.

Oh, C. W, S. G. Choi, M. Zhai, and J. Guo (2003), The first finding of eclogite relict in the Korean Peninsula and its tectonic meaning, Abstract volume of the West Norway Eclogite Field Symposium, *107*.

Oh, C. W., S. G. Choi, S. H. Song, and S. W. Kim (2004a), Metamorphic evolution of the Baekdong metabasite in the Hongseong area, *South Korea and its relationship with the Sulu collision belt of China*, Gondwana Res., *7*, 809–816.

Oh, C. W., S. W. Kim, and I. S. Williams (2006b), Spinel granulite in Odesan area, *South Korea: tectonic implications for the collision between the North and South China blocks*, Lithos, *92*, 557–575.

Oh, C. W., S. W. Kim, I. Ryu, T. Okada, H. Hyodo, and T. Itaya (2004b), Tectono-metamorphic evolution of the Okcheon metamorphic belt, *South Korea: Tectonic implications in East Asia*, Island Arc, *13*, 387–402.

Oh, C. W., S. W. Kim, S. G. Choi, M. Zhai, J. Guo, and K. Sajeev (2005), First finding of eclogite facies metamorphic event in South Korea and its correlation with the Dabie-Sulu collision belt in China, *J. Geol., 113*, 226–232.

Palmer, A. R. (1974), Search for the Cambrian world, *Am. Sci., 62*, 216–224.

Park, K. H., J. B. Park, C. S. Cheong, and C. W. Oh (2005), Sr, Nd and Pb isotopic systematics of the Cenozoic basalts of the Korean Peninsula and their implications for the Permo-Triassic continental collision boundary, *Gondwana Res., 8*, 529–539.

Park, Y. R. (2009), Enriched geochemical and Sr-Nd isotopic characteristics of Middle Triassic plutonic rocks in Hudongri, Chuncheon: Derivation from enriched mantle, *Jour. Petrol. Soc. Korea, 18*, 255–267.

Parrish, R., S. J. Gough, M. Searle, and W. Dave (2006), Plate velocity exhumation of ultra high-pressure eclogites in the Pakistan Himalaya, *Geology, 34*, 989–992.

Patriat. P., and J. Achache (1984), India-Asia collision chronology has implications for crustal shortening and driving mechanism of plates, *Nature, 311*, 615–621.

Patrick, L. F. (1996), Evolution of the Himalaya, in A. Yin and M. Harrison (eds.), *The Tectonic Evolution of Asia*, Cambridge Univ. Press, New York, 95–109.

Pearce, J. A. (1996), Source and setting of granitic rocks, *Episodes, 19*, 120–125.

Pearce, J. A., and M. J. Norry (1979), Petrogenetic implications of Ti, Zr, Y and Nb variations in volcanic rocks, *Contrib. Mineral. Petrol., 69*, 33–47.

Pearce, J. A., N. B. Harris, and A. G. Tindle (1984), Trace element discrimination diagrams for the interpretation of granitic rocks, *J. Petrol., 25*, 957–983.

Peccerillo, A., and S. R. Taylor (1976), Geochemistry of Eocenecalc-alkaline volcanic rocks in the Kastamonu area, Northern Turkey, *Contrib. Mineral. Petrol., 58*, 63–81.

Petterson, M. G., and B. F. Windley (1985), Rb-Sr dating of the Kohistan batholith in the Trans-Himalaya of N. Pakistan and tectonic implications, *Earth Planet. Sci. Lett., 74*, 45–57.

Peng, P., M. Zhai, J. Guo, H. Zhang, and Y. Zhang (2008), Petrogenesis of Triassic post-collisional syenite plutons in the Sino-Korean craton: An example from North Korea, *Geol. Mag., 145*, 1 11.

Pognante, U., and D. A. Spencer (1991), First record of eclogites from the High Himalayan belt, Kaghan valley (northern Pakistan), *Eur. J. Mineral., 3/3*, 613–618.

Rajesh, H. M., and M. Santosh (2004), Charnockitic magmatism in southern India, *Proc. Indian, Acad. Sci (Earth Planet. Sci.), 113*, 565–585.

Ree, J. H., M. Cho, S. T. Kwon, and E. Nakamura (1996), Possible eastward extension of Chinese collision belt in South Korea: The Imjingang belt, *Geology, 24*, 1071–1074.

Replumaz, A., A. M. Negredo, A. Villasenor, and S. Guillot (2010), Indian continental subduction and slab break-off during Tertiary collision, *Terra Nova, 22*, 290–296.

Ri, J. N., and J. C. Ri (1990), *Geological Constitution of Korea 6*, Pyongyang, North Korea, Industrial Publishing House, 216 p.

Ri, J. N., and J. C. Ri (1994), Tectonic map of Korea (1:1,000,000 scale) and explanatory text: Pyongyang, North Korea, Central Geological Survey of Mineral Resources, 22 p.

Rolfo, F., W. McClelland, and B. Lombardo (2005), Geochronological constraints on the age of the eclogites-facies metamorphism in the eastern Himalaya, Memoire H.S., Special extended Abstr. Vol., *Géologie Alpine, 44*, 170.

Rowley, D. B. (1996), Age of initiation of coillision between India and Asia: A review of stratigraphic data, *Earth Planet. Sci. Lett., 145*, 1–13.

Rubatto, D., A. Liati, and D. Gebauer (2003), Dating UHP metamorphism, in D. A. Carswell and R. Compagnoni (eds.), *Ultrahigh Pressure Metamorphism*, Eur. Mineral. Union, *5*, 341–363.

Sachan, H. K., B. J. Mukherjee, R. Islam, C. S. Szabo, and F. Law (1999), Exhumation history of eclogites from the Tso Morari Crystalline Complex in eastern Ladakh: Mineralogy and fluid inclusion constraints, *J. Geol. Soc. India, 53*, 181–190.

Sachan, H. K., B. K. Mukherjee, Y. Ogasawara, S. Mayurama, H. Ishida, A. Muko, and N. Yoshioka (2004), Discovery of coesite from Indus Suture Zone (ISZ), Ladakh, India: Evidence for deep subduction, *Eur. J. Mineral., 16*, 235–240.

Sajeev, K., J. Jeong, S. Kwon, W.-S. Kee, S.W. Kim, T. Komiya, T. Itaya, H. S. Jung, and Y. Park (2010), High P-T granulite relics from the Imjingang belt, *South Korea: Tectonic significance, Gondwana Res., 17*, 75–86.

Sajeev, K., Y. Osanai, and M. Santosh (2004), Ultrahigh-temperature metamorphism followed by two-stage decompression of garnet-orthopyroxene-sillimanite granulites from Ganguvarpatti, Madurai Block, southern India, *Contrib. Mineral. Petrol., 148*, 29–46.

Sajeeva, K., B. F. Windleyb, J. A. D. Connolly, and Y. Kon (2009), Retrogressed eclogite (20 kbar, 1020°C) from the Neoproterozoic Palghat-Cauvery suture zone, southern India, *Precambrian Res., 171*, 23–36.

Sagong, H., S. T. Kwon, and J. H. Ree (2005), Mesozoic episodic magmatism in South Korea and its tectonic implication, *Tectonics, 24*, TC5002, doi:10.1029/2004TC001720.

Sano, Y., H. Hidaka, K. Terada, H. Shimizu, and M. Suzuki (2000), Ion microprobe U-Pb zircon geochronology of the Hida gneiss: Finding of the oldest minerals in Japan, *Geochem. J., 34*, 135–153.

Santosh, M., and K. Sajeev (2006), Anticlockwise evolution of ultrahigh-temperature granulites within continental collision zone in southern India, *Lithos, 92*, 447–464.

Schlup, M., A. Carter, M. Cosca, and A. Steck (2003), Exhumation history of eastern Ladakh revealed by 40Ar/39Ar and fission track ages: The Indus River-Tso Morari transect, NW Himalaya, *Journal of Geol. Soc. London, 160*, 385–399.

Searle, M. P., B. F. Windley, M. P. Coward, D. J. W. Cooper, D. Rex, L. Tingdong, X. Xuchang, V. C. Jan, V.C. Thakur, and S. Kumar (1987), The closing of Tethys and the tectonics of the Himalaya, *Geol. Soc. Am. Bull., 98*, 678–701.

Searle, M. P., R. R. Parrish, K. V. Hodges, A. Hurford, M. W. Ayres, and M. J. Whitehouse (1997), Shisha Pangma leucogranite, South Tibetan Himalaya: Field relations, geochemistry, age, origin and emplacement, *J. Geol., 105*, 295–317.

Seo, J., S. G. Choi, and C. W. Oh (2010), Petrology, geochemistry, and geochronology of the Post-collisional Triassic mangerite and syenite in the Gwangcheon area, *Hongseong Belt, SouthKorea, Gondwana Res, 18*, 479–496.

Shams, F. A. (1972), Glaucophane-bearing rocks from near Topsin, Swat, First record from Pakistan, *Pakistan J. Sci. Res., 24*, 343–345.

Smith, H. A., C. P. Chamberlain, and P. K. Zeitler (1994), Timing and duration of Himalayan metamorphism within the India plate, northwest Himalaya, Pakistan, *J. Geol., 102*, 493–508.

So, C.-S., S. H. Choi, K.Y. Lee, and K. L. Shelton (1989), Geochemical studies of hydrothermal gold deposits, *Republic of Korea: Yangpyeong-Weonju area, Journal of Korea Institute of Mining and Geology, 22*, 1–16 (in Korean with English abstract).

Sohma, T., K. Kunugiza, and M. Terabayashi (1990), Hida metamorphic belt. Excursion Guidebook in 96th Annual Meeting of the Geol. Soc. Japan, 27–53 (in Japanese with English abstract).

Spencer, D. A. (1993), Tectonics of the Higher- and Tethyan Himalaya, Upper Kaghan Valley, Northwestern Himalaya, Pakistan: Implications of an early collisional, high pressure (eclogite facies) metamorphism to the Himalayan belt, Dissertation ETH nr.10194.

Spencer, D. A., and D. Gebauer (1996), SHRIMP evidence for a Permian protolith age and a 44 Ma metamorphic age for the Himalayan eclogites (Upper Kaghan, Pakistan): Implications for the subduction of Tethys and the subdivision terminology of the NW Himalaya, 11th *Himalaya-Karakorum-Tibet Workshop, 147*.

Suzuki, K., and M. Adachi (1994), Middle Precambrian detrital monazite and zircon from the Hida gneiss on Oki-Dogo Island, *Japan: Their origin and implications for the correlation of basement gneiss of southwest Japan and Korea, Tectonophysics, 235,* 277–292.

Tarney, J., and C. E. Jones (1994), Trace element geochemistry of orogenic igneous rocks and crustal growh models, *J. Geol. Soc. London, 151,* 855– 868.

Thakur, V. C. (1983), Deformation and metamorphism of the Tso Morari crystalline complex, *Wadia Institute of Himalaya Geology, 1–8*.

Tonarini, S., I. Villa, M. Oberli, F. Meier, D. A. Spencer, U. Pognante, and J. G. Ramsay (1993), Eocene age of eclogite metamorphism in Pakistan Himalaya: Implications for India-Eurasia collision, *Terra Nova, 5,* 13–20.

Treloar, P. J., P. J. O'Brian, R. R. Parrish, and A. M. Khan (2003), Exhumation of early Tertiary, coesite-bearing eclogites from the Pakistan Himalaya, *J. Geol. Soc. London, 160,* 367–376.

Tsujimori, T. (1999), Petrogenesis of the Fuko Pass highpressure metacumulate from the Oeyama peridotite body, southwestern Japan: Evidence for early Paleozoic subduction metamorphism, *Mem. Geol. Soc. Japan, 52,* 287–302.

Tsujimori, T. (2002), Prograde and retrograde P-T paths of the late Paleozoic glaucophane-eclogite from the Renge metamorphic belt, Hida mountains, southwestern Japan, *Int. Geol. Rev., 44,* 797–818.

Tsujimori, T., A. Ishiwatari, and S. Banno (2000a), Discovery of eclogitic glaucophane schist from the Omi area, Renge metamorphic belt, the Inner zone of southwestern Japan, *J. Geol. Soc. Japan, 106,* 1–11.

Tsujimori, T., and T. Itaya (1999), Blueschist-facies metamorphism during Paleozoic orogeny in southwestern Japan: Phengite K-Ar ages of blueschist-facies tectonic blocks in a serpentinite melange beneath early Paleozoic Oeyama ophiolite, *Island Arc, 8,* 190–205.

Tsujimori, T., C. Tanaka, T. Sakurai, M. Matsumoto, Y. Miyagi, T. Mizukami, Y. Kugimiya, and M. Aoya (2000b), Illustrated introduction to eclogite in Japan, *Bulletin of Research, Institute of Natural Science, Okayama University of Science, 26,* 19–40.

Tsujimori, T., H. Hyoudo, and T. Itaya (2001), 40Ar/39Ar phengite age constrains on the exhumation of the eclogite facies rocks in the Renge metamorphic belt, SW Japan,

Abstract volume of Japan Earth Planet. Sci. Joint Meeting (CDROM).

Tsujimori, T., J. G. Liou, W. G. Ernst, and T. Itaya (2006), Triassic paragonite- and garnet-bearing epidoteamphibolite from the Hida mountains, *Japan, Gondwana Research, 9,* 167–175.

Van der Voo, R., W. Spakman, H. Bijwaard (1999), Tethyan subducted slabs under India, *Earth Planet. Sci. Lett., 171,* 7–20.

Wang, X., J. G. Liou, and H. K. Mao (1989), Coesite-bearing eclogites from the Dabie mountains, central China, *Geology, 17,* 1085–1088.

Wang, X., J. G. Liou, and S. Maruyama (1992), Coesite-bearing eclogite from the Dabie Mountains, central China: Petrogenesis, P-T paths, and implications for regional tectonics, *J. Geol., 100,* 231–250.

Wang, X., R.Y. Zhang, and J. G. Liou (1995), UHPM terrane in east central China, in R. G., Coleman and X. Wang (eds.), *Ultrahigh Pressure Metamorphism,* Cambridge Univ. Press, Cambridge, 356–390.

Williams, I. S., D. L. Cho, and S. W. Kim (2009), Geochronology, and geochemical and Nd-Sr isotopic characteristics, of Triassic plutonic rocks in the Gyeonggi Massif, South Korea: Constraints on Triassic post-collisional magmatisam, *Lithos, 107,* 239–256.

Wilson, M. (1989), *Igneous Petrogenesis,* Unwin Hyman, London, 466p

Wong, A., S. Y. M. Ton, and M. J. R. Wortel (1997), Slab detachment in continental collision zones: An analysis of controlling parameters, *Geophys. Res. Lett., 24,* 2095–2098.

Wrotel, M. J. R., and W. Spakman (2000), Subduction and slab detachment in the Mediterranean-Carpathian region, *Science, 290,* 1910–1917.

Xu, Z., S. Ji, Z. Cai, L. Zeng, , Q. Geng, and H. Cao (2012), Kinematics and dynamics of the Namche Barwa Syntaxis, eastern Himalaya: Constraints from deformation, fabrics and geochronology, *Gondwana Res., 21,* 19–36.

Yang, J. H., S. L. Chung, S. A. Wilde, F. Y. Wu, M. F. Chu, C. H. Lo, and H. R. Fan (2005), Petrogenesis of post-orogenic syenites in the Sulu Orogenic Belt, East China: Geochronological, geochemical and Nd-Sr isotopic evidence, *Chem. Geol., 214,* 99–125.

Yang, W. C., and C. Q. Yu (2001), Kinetics and dynamics of development of the Dabie-Sulu Uhpm terranes based on geophysical evidences, *Chinese J. Geophys., 44,* 340–356.

Yao, Y., K. Ye, J. Liu, B. Cong, and Q. Wang (2000), A transitional eclogite- to high pressure granulite-facies overprint on coesite-eclogite at Taohang in the Sulu ultrahigh-pressure terrane, *Eastern China, Lithos, 52*:109–120.

Yin, A., and S. Nie (1993), An indentation model for the north and south China collision and the development of the TanLu and Honam fault systems, eastern Asia, *Tectonics, 12,* 801–813.

Zhai, M., and B. Cong (1996), Major and trace element geochemistry of eclogites and related rocks, in B. Cong (ed.), *Ultrahigh-pressure metamorphic rocks in the Dabieshan-Sulu Region of China,* Science Press, Kluwer Acad. Pub., 128–160.

Zhai, M., and W. Liu (1998), The boundary between Sino-Korea Craton and Yangtze Craton and its extension to the Korean Peninsula, *J. Petrol. Soc. Korea, 7,* 15–26.

Zhang, G. W., Q. R. Meng, Z. P. Yu, Y. Sun, D. W. Zhou, and A. L. Guo (1996), Orogenesis and dynamics of the Qinling orogen, *Sci. China* (Series D), *30,* 225–234.

Zhang, K. J. (1997), North and South China collision along the eastern and southern North China margins, *Tectonophysics, 270,* 145–156.

Zhang, Z., and Z. You (1993), Development in researches on the Dabie-Shan eclogite belt, *Adv. Earth Sci., 8,* 38–43.

Zhang, Z., Z. Xu, and H. Xu (2000), Petrology of ultrahigh-pressure eclogites from Zk703 drillhole in the Donghai eastern China, *Lithos, 52,* 35–50.

Zhao, X., and R. S. Coe (1987), Paleomagnetic constraints on the collision and rotation of North and South China, *Nature, 327,* 141–144.

Zheng, X., and C. Chang (1979), A preliminary note on the tectonic features of the lower Yalu-Tsangpo river region, *Sci. Geol. Sin., 2,* 116–126.

Zhong, D., and L. Ding (1996), Discovery of high-pressure basic granulites in Namjagbarwa area, Tibet, China, *Chin. Sci. Bull., 41,* 87–88.

3

A New Tectonic Model for the Genesis of Adakitic Arc Magmatism in Cretaceous East Asia

Changyeol Lee[1] and In-Chang Ryu[2]

ABSTRACT

For decades, the migration of Cretaceous adakitic arc magmatism from South China to eastern Japan in East Asia was considered to be a result of simultaneous southwest-to-northeast migration of the subducting Izanagi-Pacific ridge, which resulted in partial melting of subducted oceanic crust. However, a recent global plate reconstruction model shows that no ridge migration occurred during the Cretaceous, which poses a problem because the very high slab surface temperatures required for the adakitic arc magmatism cannot be generated with the model. Thus, we suggest that northeast-to-southwest migration of the East Asian continental blocks during the Cretaceous resulted in apparent southwest-to-northeast migration of an intracontinental mantle plume from eastern China (Dabie-Sulu region) to far east Russia (Primorsky Krai region). The apparent migration of the mantle plume explains (1) the southwest-to-northeast migration of intracontinental adakitic rocks and A-type granitoids formed by partial melting of delaminated thickened lower mafic crust and decompression melting of the mantle plume, respectively, and (2) the simultaneous migration of adakitic arc magmatism generated by a hot lobe of the mantle plume injected into the mantle wedge, producing partial melting of subducted oceanic crust.

3.1. INTRODUCTION

Since *Defant and Drummond* [1990] first introduced the term "adakite," coined from the unusual andesite on Adak Island in the Aleutians, numerous studies about the genesis of adakite have been conducted. Initially, adakite and geochemically similar igneous rocks (adakitic rocks) were attributed to partial melting of subducted eclogitized oceanic crust [*Defant and Drummond*, 1990; *Kelemen et al.*, 2003; *Maruyama et al.*, 1997; *Peacock*, 2003]. However, a series of follow-up studies on the adakite and adakitic rocks discovered in the Andes, Philippines, China, Tibet, Japan, and Tonga showed diverse and complex geneses of the rocks:

(1) partial melting of sediments and oceanic crust relaminated from a subducted slab (cold plume) in the overlying mantle wedge [*Castro et al.*, 2013; *Gerya and Yuen*, 2003]; (2) fractional crystallization of basaltic magma under high pressure [*Castillo et al.*, 1999; *Richards and Kerrich*, 2007]; (3) partial melting of delaminated lower mafic crust in intracontinental regions [*Kay and Kay*, 1993; *Wang et al.*, 2006b; *Xu et al.*, 2002]; (4) partial melting of thickened lower mafic crust beneath overlying continents in subduction or collision zones [*Atherton and Petford*, 1993; *Chung et al.*, 2003]; and (5) direct melting of the edge of subducted oceanic crust heated by an adjacent mantle plume [*Falloon et al.*, 2008].

Through the past several decades, the adakitic arc magmatism that migrated from South China (Guangxi and Fujian) to far east Russia (Kamchatka) from the Early Cretaceous to Early Cenozoic has received considerable scientific attention [e.g., *Isozaki et al.*, 2010; *Kinoshita*, 1995, 2002; *Maruyama et al.*, 1997] (Figure 3.1a).

[1]*Faculty of Earth and Environmental Sciences, Chonnam National University, Gwangju, Republic of Korea*

[2]*Department of Geology, Kyungpook National University, Daegu, Republic of Korea*

Subduction Dynamics: From Mantle Flow to Mega Disasters, Geophysical Monograph 211, First Edition.
Edited by Gabriele Morra, David A. Yuen, Scott D. King, Sang-Mook Lee, and Seth Stein.

Southwest-to-northeast migration of the subducting Izanagi-Pacific ridge axis and the associated very high slab surface temperatures were thought to be responsible for the southwest-to-northeast migration of the adakitic arc magmatism [*Kinoshita*, 1995, 2002; *Maruyama et al.*,

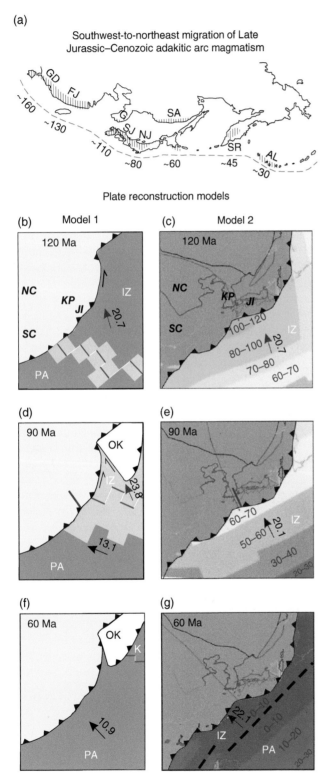

1997] (Figure 3.1b, d, and f). Thus, a plate reconstruction model including this southwest-to-northeast migration of ridge subduction (hereafter, Model 1) has been broadly accepted by a number of researchers [e.g., *Isozaki et al.*, 2010; *Kinoshita*, 1995, 2002; *Maruyama et al.*, 1997].

However, a new global plate reconstruction model with a number of improvements [*Müller et al.*, 2008; *Sdrolias and Müller*, 2006] (hereafter, Model 2) was recently introduced and successfully correlated with global sea level changes for the last 140 Myr and the tectonic evolution of Cenozoic Pacific subduction zones. A major contrast between Model 1 and 2 is that Model 2 does not include southwest-to-northeast ridge migration during the Cretaceous; subduction of the Izanagi plate was followed by Izanagi-Pacific ridge subduction at ca. 60 Ma along the trench extending from South China to Kamchatka (Figure 3.1c, e, and g). This contrast indicates that almost the entire Cretaceous subduction history is different in Model 1 and 2 and that the different motions and ages of the Izanagi and Pacific plates could result in a different tectonic history for East Asia during the Cretaceous.

Thus, a fundamental question arises naturally: which model is consistent with the southwest-to-northeast migration of the adakitic arc magmatism in East Asia during the Cretaceous? If Model 2, an improved plate reconstruction model, is correct, the rate and age of the converging oceanic plate in Model 2 should generate the very high slab surface temperatures required for adakitic arc magmatism, as did ridge subduction in Model 1. A previous study [*Lee and King*, 2010] showed that the

Figure 3.1 Schematic map of (a) the southwest-to-northeast migration of Late Jurassic–Cenozoic adakitic arc magmatism in East Asia and snapshots at 120, 90, and 60 Ma of the plate reconstruction models: Model 1 (b, d, f) and Model 2 (c, e, g). The migration map of the Late Jurassic–Cenozoic adakitic arc magmatism is modified from *Kinoshita* [1995], *Valui et al.* [2008], and *Wu et al.* [2005]. In (a), GD = Guangdong, FJ = Fujian, G = Gyeongsang, SJ = southwest Japan, NJ = northwest Japan, SA = Sikhote-Alin, SR = Sredinny, and AL = the Aleutians. The red dashed line = the Late Jurassic–Cenozoic trench; the numbers along the trench indicate representative ages of the adakitic arc magmatism. Models 1 and 2 are based on *Maruyama et al.* [1997] and *Sdrolias and Müller* [2006], respectively, using Gplate software. In (b) through (g), NC = North China, SC = South China, KP = Korean Peninsula, JI = Japanese Islands, IZ = Izanagi plates, OK = Okhotsk, and PA = Pacific plate. Units are cm/a (convergence rate) and Ma (slab age). The magenta Izanagi-Pacific ridge is fragmented by white transform faults. In Model 2, the Izanagi-Pacific ridge is identified by the black dashed line. The thin red line indicates the relative motion between the North China and Izanagi plates used to calculate the average convergence rate of the Izanagi plate during the Cretaceous.

temporal and spatial evolution of the rate and age of the converging oceanic plate were essential for the temporal and spatial evolution of adakitic arc volcanism in the Aleutians. The temporal and spatial evolution of the Cretaceous adakitic arc volcanism in East Asia also imply that consideration of the temporal and spatial evolution of the subduction parameters is essential for understanding the adakitic arc magmatism. Thus, we evaluated Model 2 by formulating two-dimensional numerical subduction models using the rates and ages of the converging oceanic plates described in Models 1 and 2. As a result, we suggest that the adakitic arc magmatism resulted from plume-slab interaction, which provided the excess heat resulting in slab melting.

3.2. NUMERICAL MODELS

As summarized above, partial melting of subducted oceanic crust is an important mechanism for generating adakitic arc magmatism. The numerical models in this study were designed to evaluate whether Model 1 and/or 2 correlates with the migration of the adakitic arc magmatism in East Asia during the Cretaceous. Because of the temporal and spatial evolution of the Izanagi and Pacific plates, time-dependent three-dimensional subduction models may ultimately be required to completely characterize the Cretaceous magmatism in East Asia. However, due to technical difficulties and computational costs, we instead used the two-dimensional numerical subduction model briefly described below. Previous studies have indicated that subduction of the southwest-to-northeast-migrating Izanagi-Pacific ridge axis generated adakitic arc magmatism in Kyushu and southern Korea from ca. 135 to 105 Ma and that the ridge axis passed beneath Kyushu at ca. 105 Ma [*Kamei*, 2004; *Kinoshita*, 2002; *Wee et al.*, 2006]. As the ages of the adakitic arc magmatism are known, we decided to model the temporal evolution of thermal structures of the subduction zone along Kyushu and southern Korea. The blue lines in Figure 3.1d and e indicate the modeling region considered in this study.

As shown in the plate reconstruction models (Figure 3.1), the convergence vectors and ages of the Izanagi and Pacific plates changed with time and are essential factors that must be included in our numerical models [e.g., *Lee and King*, 2010]. Our two-dimensional model considers only the trench-normal component of the convergence vector and excludes the trench-parallel component. Nevertheless, a previous study showed that the effects of the trench-parallel component of the convergence vector on the thermal structures of subduction zones are negligible [*Honda and Yoshida*, 2005]. Hence, our two-dimensional subduction model can approximate a three-dimensional subduction model.

In Model 1, the newly generated oceanic plate spreads obliquely from the Izanagi-Pacific ridge toward the NNW at 20.7 cm/a; thus, both the trench-normal and trench-parallel components of the Izanagi plate to the Kyushu trench were assumed to be ~13 cm/a. With the spreading oceanic plate, the Izanagi-Pacific ridge simultaneously migrates from South China to Kamchatka at ~3 cm/a (Figure 3.1b) [*Kinoshita*, 2002; *Maruyama et al.*, 1997]. After the Izanagi-Pacific ridge passes Kyushu, the Pacific plate changes its direction of motion from NNW to WNW and converges nearly perpendicular to the Kyushu trench at ~13 cm/a. Thus, the trench-normal convergence rate to the Kyushu trench can be maintained at approximately 13 cm/a during the Cretaceous. Because the spreading ridge passes along the trench from southwest to northeast, the age of the slab subducting at the Kyushu trench decreases to 105 Ma and then increases. Figure 3.2a shows the detailed evolutions of the trench-normal convergence rate and slab age at the Kyushu trench during the Cretaceous.

To represent the trench-normal convergence rate and slab age in Model 2, we used Gplates software, which was developed for quantitative and visual plate reconstruction [*Gurnis et al.*, 2012; *Williams et al.*, 2012]. Using the software, we estimated the trench-normal convergence rate by averaging the relative motion between the North China (including Korea and Kyushu) and Izanagi plates every 5 Myr. Slab age was extracted from paleomagnetic stripes every 5 Myr. The average convergence rate and slab age were approximated by using the piecewise polynomials implemented in our previous study [*Lee and King*, 2010]. Figure 3.2b shows the evolution of the trench-normal convergence rate and slab age in Model 2, which is much different from those in Model 1, described in Figure 3.2a.

The framework and rheology of our numerical models were similar to those of our previous study [*Lee and King*, 2010]. For the mantle rheology, we used a modified composite viscosity of diffusion and dislocation creep for dry olivine based on laboratory experiments on olivine rheology [*Billen and Hirth*, 2007; *Karato and Wu*, 1993]. As in previous models, wet olivine rheology was used for the corner of the mantle wedge to approximate the effect of slab dehydration on the mantle wedge [*Honda and Saito*, 2003; *Lee and King*, 2010]. Based on *Kamei* [2004], we assumed the thickness of the overriding lithosphere to be 30 km, except at the isolated corner of the mantle wedge affected by serpentinization [*Currie et al.*, 2004] (Figure 3.2c). Previous studies have indicated that slab dip varies with time due to slab buckling or rollback/advance [*Funiciello et al.*, 2003a; *Funiciello et al.*, 2003b; *Lee and King*, 2011], which poses a problem because our numerical models use a fixed slab dip. However, a slab dip of 45°, which is a "typical" slab dip of subducted slab,

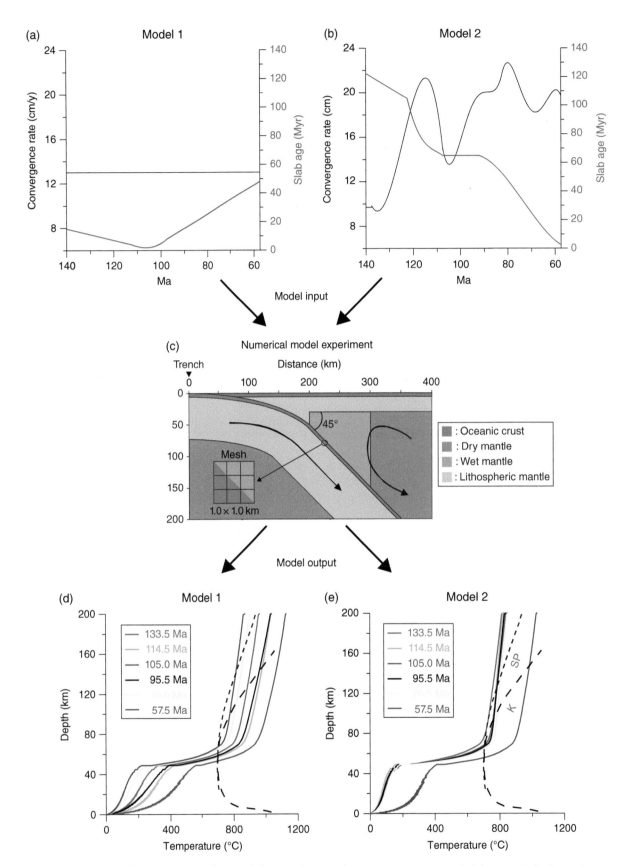

Figure 3.2 (a and b) Temporal evolution of the trench-normal convergence rates and slab ages of the incoming Izanagi plate at the trench from Models 1 and 2; (c) a schematic diagram of the numerical subduction model; and (d and e) calculated slab surface temperatures for Models 1 and 2. In (e), the dashed lines labeled SP and K correspond to the low and high end-member solidi of wet basalt from *Schmidt and Poli* [1998] and *Kessel et al.* [2005], respectively.

was assumed despite the possible temporal evolution of slab dip (Figure 3.2c). In our model, the kinematically subducted slab generates corner flow in the mantle wedge, and mantle buoyancy is neglected [e.g., *Currie et al.*, 2004; *Kneller and van Keken*, 2008; *Lee and King*, 2010].

The 200 by 400 km (1 by 2) model domain consisted of 200 by 400 four-node quadrilateral elements. Radiogenic heat productions of 3.80×10^{-11} and 7.38×10^{-12} W/kg were used for the 7-km-thick oceanic crust and elsewhere, respectively. As an initial temperature profile, a half-space cooling model for 50 Myr using a potential temperature of 1400°C was used for the whole model domain. For the mantle adiabat, a temperature gradient of 0.25°C/km was added to the model calculations a posteriori (1450°C at 200-km depth). The trench-normal convergence rate was kinematically prescribed on the subducting slabs of the Izanagi and Pacific plates by updating the convergence rate described in Figure 3.2a and b every time step. The temporal evolution of the slab age of the converging plates was prescribed as the temperature profile on the right wall by calculating the half-space cooling model every time step. The near-perpendicular ridge subduction in Model 1 could result in slab tearing or detachment and upwelling of hot asthenospheric mantle into the opened mantle wedge [e.g., *Burkett and Billen*, 2009; *Liu and Stegman*, 2012]; partial melting of subducting slab by hot asthenospheric mantle is considered to generate adakites [*Kinoshita*, 1995, 2002]. However, the age of the subducting slab close to the ridge axis is very young (<25 Ma), which results in partial melting of the oceanic crust without upwelling of hot asthenospheric mantle. Therefore, we approximated the ridge subduction by varying the slab age in the plate reconstruction model without consideration of slab tearing or detachment in our numerical models (Figure 3.2a). To solve the momentum and energy equations, we used the incompressible Boussinesq approximation implemented in ConMan, which was also used in our previous studies [*Lee and King*, 2009, 2010].

3.3. RESULTS

Figure 3.2d and e shows slab surface temperatures with depth calculated from the model experiments using the convergence rates and slab ages of Models 1 and 2, respectively. As expected, Model 1 shows that slab temperatures increase to 105 Ma due to decreasing slab age (Figure 3.2d and Model 1 in Figure 3.3). After subduction of the ridge axis at 105 Ma, the slab temperatures decrease with time due to increasing slab age. All of the slab surface temperatures are beyond the low [*Schmidt and Poli*, 1998] and high [*Kessel et al.*, 2005] end-member solidi of wet basalt, which suggests that partial melting of the subducted oceanic crust resulted in the adakitic arc magmatism in southern Korea and Kyushu. Because

dehydration reduces the water content in the subducted slab, extensive slab melting would likely occur from 135 Ma to 80 Ma, which is consistent with the ages of the adakitic arc magmatism, for example, the 114.6±9.1 Ma Jindong adakitic granites of southern Korea [*Wee et al.*, 2006], the 123, 115, and 110 Ma adakites of Kyushu [*Kinoshita*, 2002], and the 121±14 Ma sanukitic high Mg# andesites and Kunisaki diorite of Kyushu [*Kamei*, 2004]. However, except for the slab surface temperatures of 57.5 Ma, slab surface temperatures using Model 2 are only slightly higher than the solidi of wet basalt due to old and cold subducted slab (Figure 3.2e and Model 2 in Figure 3.3). Slab melting occurs at ca. 60 Ma when the Izanagi-Pacific ridge is subducted (Figure 3.3i). The experiments using Model 2 indicate that extensive adakitic arc magmatism caused by slab melting would not be expected in East Asia during the Cretaceous due to the lack of ridge subduction.

3.4. DISCUSSION

The aforementioned results clearly indicate that Model 2 does not explain the migration of adakitic arc magmatism along the ancient arc in East Asia during the Cretaceous. However, Model 2 has been improved significantly by the inclusion of (1) the back-arc basins developed in the Pacific and Indian margins [*Gaina and Müller*, 2007; *Sdrolias and Müller*, 2006], (2) the tectonic evolution of the Australian and Arctic continents [*Heine et al.*, 2004; *Sokolov et al.*, 2002], and (3) the oceanic plate from the continuously spreading Izanagi-Pacific ridge during the Cretaceous [*Müller et al.*, 2008]; the latter point contrasts with the cessation of Izanagi-Pacific ridge spreading from 110 to 80 Ma in Model 1, which is unlikely due to slab pull of the subducted slab. Therefore, Model 2 appears to be more reasonable than Model 1, even though Model 2 cannot, in and of itself, generate the partial melting of the subducted oceanic crust that explains the adakitic arc magmatism in East Asia during the Cretaceous.

As described previously, other possible mechanisms such as a cold plume, partial melting of the lower mafic crust, magma mixing, and fractionalization may result in adakitic arc magmatism [e.g., *Castillo et al.*, 1999; *Castro et al.*, 2013; *Gerya and Yuen*, 2003; *Richards and Kerrich*, 2007; *Xu et al.*, 2002]. However, none of these mechanisms can itself explain why the adakitic arc magmatism migrated along the Cretaceous arc from South China to eastern Japan. Relamination caused by a cold plume originating from the subducted slab can result in adakitic arc magmatism [*Castro et al.*, 2013; *Gerya and Yuen*, 2003]. However, no geological evidence indicating the migration of a cold plume along the Cretaceous arc has been reported. Magma mixing or fractionalization also

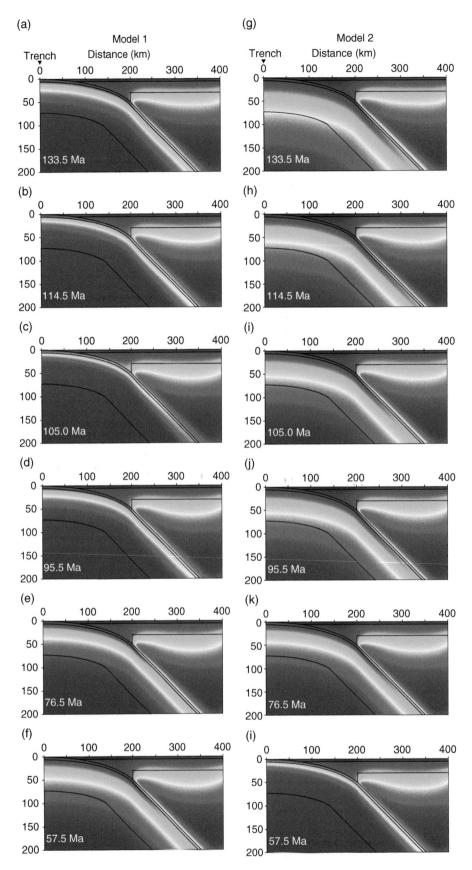

Figure 3.3 Snapshots of the calculated temperatures of the model domains corresponding to 133.5 (a, g), 114.5 (b, h), 105.0 (c, i), 95.5 (d, j), 76.5 (e, k), and 57.5 Ma (f, l) for Models 1 and 2, respectively. The black lines in the model domains indicate the lithologic boundaries described in Figure 3.2c. Increasing temperature is shown by the color changes: deep blue, blue, yellow, red, and deep red. Temporal evolutions of slab age and convergence rate significantly change the temperature profile of the subducting slab, which results in large variations in the slab surface temperature (Figure 3.2d and e). As observed in previous studies [e.g., *Currie et al.*, 2004; *Kneller and van Keken*, 2008; *Lee and King*, 2010], the overlying lithosphere gradually thickens with time.

cannot explain the migration of the adakitic arc magmatism because of a lack of geological evidence for the migration of magma mixing or fractionalization along the Cretaceous arc. Partial melting of the lower mafic crust may be possible in certain specific environments created by continental collision or underplating of subducted basaltic crust [*Atherton and Petford*, 1993; *Chung et al.*, 2003], but these are not expected to have occurred along the Cretaceous arc.

Thus, it is logical that an anomalous heat source producing partial melting of the subducted oceanic crust migrated along the Cretaceous arc from southwest to northeast. If migration of the Izanagi-Pacific ridge is excluded, the question of what could explain the anomalous heating arises. A mantle plume originating from the deeper mantle such as the core-mantle boundary or the 660-km discontinuity [*Cserepes and Yuen*, 2000; *Schubert et al.*, 2001] could be a potential candidate for the anomalous heat source because of the higher temperature (>200 K) compared to adjacent mantle [*Ribe and Christensen*, 1999]. A previous study [*Falloon et al.*, 2008] indicated that anomalous heating caused by the Samoan mantle plume resulted in partial melting of subducted oceanic crust in northern Tonga during the Quaternary; such plume-slab interaction has generated adakite in the Tonga subduction zone. *Kincaid et al.* [2013] conducted a laboratory experiment to reconcile the bifurcation of bimodal volcanism at Yellowstone (USA) and demonstrated dragging of the hot mantle plume by corner flow in the mantle wedge, indicating the importance of plume-slab interaction in arc volcanism.

Therefore, we here hypothesize that an intracontinental mantle plume that apparently migrated due to opposite-direction migration of East Asian continental blocks is responsible for the migration of adakitic arc magmatism along the Cretaceous arc. In this hypothesis, we assume that an intracontinental mantle plume existed under preexisting thickened continental crust at ca. 140 Ma. The thickened continental crust was created by curvilinear subduction and subsequent continental collision along the suture between the Sino-Korean and Yangtze cratons extending from eastern China (Dabie-Sulu region) to far east Russia (Primorsky Krai region) that occurred from the Triassic to Late Jurassic (P1 in Figure 3.4a) [*Zhang et al.*, 2009]. If the mantle plume then weakened and delaminated the thickened lower mafic crust [*Bott*, 1992; *Wilson*, 1993], it could have resulted in partial melting of the delaminated lower crust yielding adakitic rocks as well as decompression melting of the mantle plume, yielding A-type granitoids [*Betts et al.*, 2007]. This suggestion is consistent with a previous study describing the adakitic rocks and A-type granitoids (ca. 136 ± 3 Ma) observed in the Luzong, Dexing, and Yueshan areas near the Dabie-Sulu collision zone [*Wang*

(a)

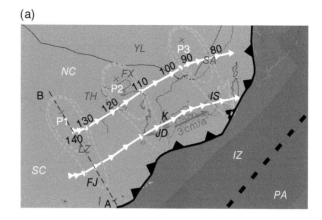

(b)

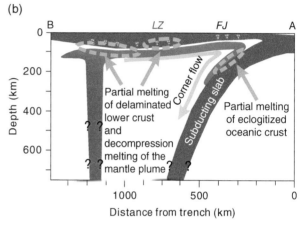

Figure 3.4 (a) Apparent southwest-to-northeast migration of the mantle plume caused by opposite migration of the China, Korea, and Japan continental blocks and (b) a schematic model showing the mantle plume and the hot lobe of the mantle plume injected into the mantle wedge by corner flow, with the resultant intracontinental adakitic rocks and A-type granitoids and adakitic arc magmatism, respectively. The white lines in (a) indicate the apparent migration of the mantle plume every 5 Myr from 140 Ma to 65 Ma; each white triangle corresponds to an apparent location of the mantle plume. The inner and outer green contours indicate the 1550°C and 1450°C isotherms, respectively, based on numerical model experiments [*Ribe and Christensen*, 1999]. Labels indicate the following: NC = North China, SC = South China, IZ = Izanagi, FJ = Fujian, and PA = Pacific plates, J = Jindong, K = Kyushu, IS = Itoigawa-Shizuoka. (b) shows a cross section crossing the intracontinent and subduction zone along the line AB. The mantle plume results in decompression melting and partial melting of the delaminated thickened lower crust created by the continental collision, yielding the A-type granitoids and adakitic rocks, respectively. The hot lobe of the mantle plume dragged into the mantle wedge results in partial melting of the subducted oceanic crust, yielding the adakitic arc magmatism.

et al., 2006a]. The relatively higher $^{87}Sr/^{86}Sr$ and lower $^{143}Nd/^{144}Nd$ in these areas [*Wang et al.*, 2006a; *Xu et al.*, 2002] than those of adakites in volcanic arcs can also be correlated with a mantle plume.

One may argue that if a mantle plume existed, typical geological expressions such as a large igneous province, radial dike swarm, and rift should exist. However, thickened continental crust due to orogeny effectively inhibits the development of large igneous provinces, radial dike swarms, and major rifts, as exemplified by the track of the A-type plutonism resulting from the mantle plume in eastern Australia [*Betts et al.*, 2007]. Additionally, a mantle plume beneath thickened lithosphere does not develop major rifting and continental splitting but rather regional rifting and regional extensional environments [*Ziegler and Cloetingh*, 2004], which is consistent with the Cretaceous extensional environments subsequent to the Dabie-Sulu collision in eastern China [*Li*, 2000; *Wang et al.*, 2006a]. A hot lobe of this mantle plume could also have been dragged by corner flow and injected into the mantle wedge. The resulting partial melting of the subducted oceanic crust explains the adakitic arc magmatism in South China (e.g., Guangxi and Central Fujian) at ca. 130 Ma [*Kinoshita*, 1995; Figure 3.4a and b].

To explain the simultaneous migration of the intracontinental adakitic rocks as well as the A-type granitoids, the mantle plume, as an anomalous heat source, would have had to migrate from southwest to northeast beneath the continental blocks. However, because mantle plumes are thought to be fixed and sustained for ~100 Myr [*Schubert et al.*, 2001; *Steinberger and O'Connell*, 1998], it is more reasonable that the East Asian continental blocks migrated with respect to the fixed mantle plume. Thus, we evaluated the migration of the East Asian continental blocks during the Cretaceous using Gplates software and found that the blocks migrated from northeast to southwest with respect to the Pacific hotspot reference frame. Given that the East Asian continental blocks migrated toward the southwest, the mantle plume under the blocks would have apparently migrated toward the northeast, passing through the Yanshan-Liaoxi and Liadong (ca. 115–105 Ma, P2 in Figure 3.4a) and then the Sikhote-Alin (far east Russia) (ca. 90–65 Ma, P3 in Figure 3.4a) regions beneath the thickened continental crust formed as a result of the continental collision. Previous petrological studies showed that the ages of the adakitic rocks and A-type granitoids in these regions are ca. 125 Ma and ca. 88–60 Ma, respectively, which is fairly consistent with the temporal migration of the mantle plume [*Valui et al.*, 2008; *Wu et al.*, 2005]. The decreasing ages of the A-type granitoids from southwest to northeast correlate with the apparent migration of the mantle plume due to the opposite migration of the East Asian continental blocks.

Along with the migration of the intracontinental adakitic rocks and A-type granitoids, the apparent migration of the mantle plume resulted in the migration of adakitic arc magmatism in southern Korea and southwest Japan. *Wee et al.* [2006] evaluated the adakitic granites (114.6 ± 9.1 Ma) in the southwestern region of the Gyeongsang basin and suggested that the origin of the adakitic granites was likely related to the interaction of slab melting and the mantle wedge, similar to the origin of the Shiraishino granodiorite (121 ± 14 Ma) [*Kamei*, 2004]. Additionally, *Kinoshita* [2002] quantitatively evaluated the adakitic arc magmatism that resulted from slab melting along southwest Japan. The evaluation showed that peak adakitic arc magmatism migrated linearly from Kyushu (ca. 105 Ma) to Itoigawa-Shizuoka (ca. 80 Ma); the migration rate of the peak adakitic magmatism was ~3 cm/a. The migration rate of the East Asian continental blocks constrained from Gplates software was also ~3 cm/a, strikingly consistent with the migration rate of peak adakitic arc magmatism (Figure 3.4a). As the Cretaceous crust in southern Korea and southwest Japan is thought to have been relatively thin (<30 km) [*Kamei*, 2004], partial melting of subducted oceanic crust, except for the lower crust, was likely responsible for the adakitic arc magmatism. Additionally, the ages of the subducted Izanagi plate in these regions were younger than those in southern China, so the subducted slab was more susceptible to partial melting. Thus, the hot lobe of the mantle plume was injected into the mantle wedge, resulting in partial melting of the subducted oceanic crust, which was responsible for the adakitic arc magmatism along southern Korea and southeast Japan.

Although the apparent migration of the Cretaceous mantle plume due to opposite migration of the East Asian continental blocks explains both the intracontinental and arc magmatism, it is worth noting some caveats of our hypothesis. First, the existence and lifespan of the mantle plume should be verified by further geological evidence including geochemical analyses. Second, the interaction between the hot mantle plume and subducted slab should be evaluated quantitatively to determine whether the dragged lobe of the mantle plume was hot enough for partial melting of the subducted oceanic crust. Third, the plate reconstruction model for Cretaceous East Asia should be improved further to address questions such as whether the collision of the Okhotomorsk block in the Izanagi plate with the Asian continental blocks from ca. 100 to 77 Ma [*Yang*, 2013] can be correlated with the concurrent migration of the adakitic arc volcanism. Therefore, further geochemical studies (e.g., evidence for the existence and lifespan of the mantle plume), numerical/laboratory experiments (e.g., thermomechanical evaluation of the plume-slab interaction, such as that by *Kincaid et al.* [2013] and *Liu and Stegman* [2012]), and plate reconstruction are required to verify our hypothesis.

3.5. CONCLUDING REMARKS

A recently introduced plate reconstruction model refutes the migration of the Izanagi-Pacific ridge formerly considered to be responsible for the adakitic arc magmatism in East Asia during the Cretaceous. However, numerical subduction models based on the plate reconstruction model cannot generate partial melting of the subducted oceanic crust. Thus, an alternative explanation is required for the southwest-to-northeast migration of the adakitic arc magmatism. Other mechanisms such as a cold plume, partial melting of delaminated lower mafic crust, magma mixing, and fractionalization also cannot explain the temporal migration of the adakitic arc magmatism. Thus, temporal migration of an anomalous heat source resulting in the migration of the partial melting of the subducted oceanic crust is required. For this migrating heat source, we hypothesize that a Cretaceous intracontinental mantle plume apparently migrated from southwest to northeast as a result of opposite migration of the East Asian continental blocks. Using Gplates software, we found that the apparent migration of the mantle plume resulting in partial melting of delaminated lower crust and decompression melting explains the trend of decreasing ages of intracontinental adakitic rocks and A-type granitoids from eastern China (e.g., Dabic-Sulu region) to far east Russia (e.g., Primorsky Krai region). A dragged hot lobe of the mantle plume injected into the mantle wedge due to corner flow resulted in partial melting of subducted oceanic crust from southwest to northeast, affecting central Fujian, southern Korea, and southeast Japan, as evidenced by the adakitic arc magmatism. The temporal evolution of the Cretaceous adakitic arc magmatism in East Asia is an excellent expression of the interaction between the mantle plume and subducted slab. Although this hypothesis is highly promising, it should be verified by further studies including geochemical, numerical/laboratory experiment, and plate reconstruction studies.

ACKNOWLEDGMENT

We thank Gabriele Morra, Taras Gerya, and an anonymous reviewer for their constructive comments, which significantly improved the manuscript. This study was supported by a National Research Foundation of Korea Grant funded by the Ministry of Education, Science, and Technology of the Government of Korea (NRF-35B-2011-1-C00043).

REFERENCES

Atherton, M. P., and N. Petford (1993), Generation of sodium-rich magmas from newly underplated basaltic crust, *Nature*, *362*, 144–146.

Betts, P. G., D. Giles, B. F. Schaefer, and G. Mark (2007), 1600–1500 Ma hotspot track in eastern Australia: Implications for Mesoproterozoic continental reconstructions, *Terra Nova*, *19*, 496–501.

Billen, M. I., and G. Hirth (2007), Rheologic controls on slab dynamics, *Geochem. Geophys. Geosys.*, *8*, Q08012.

Bott, M. H. P. (1992), Modelling the loading stresses associated with active continental rift systems, *Tectonophysics*, *215*, 99–115.

Burkett, E. R., and M. I. Billen (2009, Dynamics and implications of slab detachment due to ridge-trench collision, *J. Geophys. Res.*, *114*, B12402.

Castillo, P. R., P. E. Janney, and R. U. Solidum (1999), Petrology and geochemistry of Camiguin Island, southern Philippines: Insights to the source of adakites and other lavas in a complex arc setting, *Contrib. Mineral. Petrol.*, *134*, 33–51.

Castro, A., K. Vogt, and T. Gerya (2013), Generation of new continental crust by sublithospheric silicic-magma relamination in arcs: A test of Taylor's andesite model, *Gondwana Res.*, *23*, 1554–1566.

Chung, S.-L., D. Liu, J. Ji, M.-F. Chu, H.-Y. Lee, D.-J. Wen, C.-H. Lo, T.-Y. Lee, Q. Qian, and Q. Zhang (2003), Adakites from continental collision zones: Melting of thickened lower crust beneath southern Tibet, *Geology*, *31*, 1021–1024.

Cserepes, L., and D. A. Yuen (2000), On the possibility of a second kind of mantle plume, *Earth Planet. Sci. Lett.*, *183*, 61–71.

Currie, C. A., K. Wang, R. D., Hyndman, and J. He (2004, The thermal effects of steady-state slab-driven mantle flow above a subducting plate: the Cascadia subduction zone and back-arc, *Earth Planet. Sci. Lett.*, *223*, 35–48.

Defant, M. J., and M. S. Drummond (1990), Derivation of some modern arc magmas by melting of young subducted lithosphere, *Nature*, *347*, 662–665.

Falloon, T. J., L. V. Danyushevsky, A. J. Crawford, S. Meffre, J. D. Woodhead, and S. H. Bloomer (2008), Boninites and adakites from the northern termination of the Tonga trench: Implications for adakite petrogenesis, *J. of Petrol.*, *49*, 697–715.

Funiciello, F., C. Faccenna, D. Giardini, and K. Regenauer-Lieb (2003a), Dynamics of retreating slabs: 2. Insights from three-dimensional laboratory experiments, *J. Geophys. Res.*, *108*, 2207.

Funiciello, F., G. Morra, K. Regenauer-Lieb, and D. Giardini (2003b, Dynamics of retreating slabs: 1. Insights from two-dimensional numerical experiments, *J. Geophys. Res.*, *108*, 2206.

Gaina, C., and D. Müller (2007), Cenozoic tectonic and depth/age evolution of the Indonesian gateway and associated back-arc basins, *Earth-Sci. Rev.*, *83*, 177–203.

Gerya, T. V., and D. A. Yuen (2003), Rayleigh-Taylor instabilities from hydration and melting propel "cold plumes" at subduction zones, *Earth Planet. Sci. Lett.*, *212*, 47–62.

Gurnis, M., M. Turner, S. Zahirovic, L. DiCaprio, S. Spasojevic, R. D. Müller, J. Boyden, M. Seton, V. C. Manea, and D. J. Bower (2012), Plate tectonic reconstructions with continuously closing plates, *Computers & Geosciences*, *38*, 35–42.

Heine, C., D. Müller, and C. Gaina (2004). Reconstructing the Lost Eastern Tethys Ocean Basin: Convergence history of the

SE Asian margin and marine gateways, in P. Clift et al. (eds.) *Continent-Ocean Interactions in Southeast Asia, AGU Monograph, 149,* 37–54.

Honda, S., and M. Saito (2003), Small-scale convection under the back-arc occurring in the low viscosity wedge, *Earth Planet. Sci. Lett., 216,* 703–715.

Honda, S., and T. Yoshida (2005), Effects of oblique subduction on the 3-D pattern of small-scale convection within the mantle wedge, *Geophys. Res. Lett., 32,* L13307.

Isozaki, Y., K. Aoki, T. Nakama, and S. Yanai (2010), New insight into a subduction-related orogen: A reappraisal of the geotectonic framework and evolution of the Japanese Islands, *Gondwana Res., 18,* 82–105.

Kamei, A. (2004), An adakitic pluton on Kyushu Island, southwest Japan arc, *J. Asian Earth Sci., 24,* 43–58.

Karato, S.-I., and P. Wu (1993), Rheology of the upper mantle: A synthesis, *Science, 260,* 771–778.

Kay, R.W., and S. M. Kay (1993), Delamination and delamination magmatism, *Tectonophysics, 219,* 177–189.

Kelemen, P. B., G. M., Yogodzinski, D. W. Scholl (2003), Along-strike variation in the Aleutian island arc: Genesis of high Mg# andesite and implications for continental crust, in J. Eiler (ed.), *Inside the Subduction Factory*, Washington, DC, American Geophysical Union, 223–276.

Kessel, R., P. Ulmer, T. Pettke, M. W. Schmidt, and A. B. Thompson (2005), The water-basalt system at 4 to 6 GPa: Phase relations and second critical endpoint in a K-free eclogite at 700 to 1400°C, *Earth Planet. Sci. Lett., 237,* 873–892.

Kincaid, C., K. A. Druken, R. W. Griffiths, and D. R. Stegman (2013), Bifurcation of the Yellowstone plume driven by subduction-induced mantle flow, *Nature Geosci., 6,* 395–399.

Kinoshita, O. (1995), Migration of igneous activities related to ridge subduction in Southwest Japan and the East Asian continental margin from the Mesozoic to the Paleogene, *Tectonophysics, 245,* 25–35.

Kinoshita, O. (2002), Possible manifestations of slab window magmatisms in Cretaceous southwest Japan, *Tectonophysics, 344,* 1–13.

Kneller, E. A., and, P. E. van Keken (2008), Effect of three-dimensional slab geometry on deformation in the mantle wedge: Implications for shear wave anisotropy, *Geochem. Geophys. Geosyst., 9,* Q01003.

Lee, C., and S. D. King (2009), Effect of mantle compressibility on the thermal and flow structures of the subduction zones, *Geochem. Geophys. Geosyst., 10,* Q01006.

Lee, C., and S. D. King (2010), Why are high-Mg# andesites widespread in the western Aleutians? A numerical model approach, *Geology, 38,* 583–586.

Lee, C., and S. D. King (2011), Dynamic buckling of subducting slabs reconciles geological and geophysical observations, *Earth Planet. Sci. Lett., 312,* 360–370.

Li, X.-H. (2000), Cretaceous magmatism and lithospheric extension in Southeast China, *J. Asian Earth Sci., 18,* 293–305.

Liu, L., Earth Planet. Sci. Lett., nd D. R. Stegman (2012), Origin of Columbia River flood basalt controlled by propagating rupture of the Farallon slab, *Nature, 482,* 386–389.

Maruyama, S., Y. Isozaki, G. Kimura, and M. Terabayashi (1997), Paleogeographic maps of the Japanese Islands: Plate tectonic synthesis from 750 Ma to the present, *Island Arc, 6,* 121–142.

Müller, R. D., M. Sdrolias, C. Gaina, B., Steinberger, and C. Heine (2008), Long-term sea-level fluctuations driven by ocean basin dynamics, *Science, 319,* 1357–1362.

Peacock, S. M. (2003), Thermal structure and metamorphic evolution of subducting slab, in J. Eiler (ed.), *Inside the Subduction Factory*, Washington, DC, American Geophysical Union, 7–22.

Ribe, N. M., and U. R. Christensen (1999), The dynamical origin of Hawaiian volcanism, *Earth Planet. Sci. Lett., 171,* 517–531.

Richards, J. P., and R. Kerrich (2007), Special paper: Adakite-like rocks: Their diverse origins and questionable role in metallogenesis, *Econ. Geol., 102,* 537–576.

Schmidt, M. W., and S. Poli (1998), Experimentally based water budgets for dehydrating slabs and consequences for arc magma generation, *Earth Planet. Sci. Lett., 163,* 361–379.

Schubert, G., D. Turcotte, and P. Olson (2001), *Mantle Convection in the Earth and Planets, Cambridge Univ.* Press, Cambridge.

Sdrolias, M., and R. D. Müller (2006), Controls on back-arc basin formation, *Geochem. Geophys. Geosys., 7.*

Sokolov, S. D., G. Y. Bondarenko, O. L. Morozov, V. A. Shekhovtsov, S. P. Glotov, A. V. Ganelin, and I. R. Kravchenko-Berezhnoy (2002), South Anyui suture, northeast Arctic Russia: Facts and problems, *Geological Society of America Special Papers, 360,* 209–224.

Steinberger, B., and R. J. O'Connell (1998), Advection of plumes in mantle flow: implications for hotspot motion, mantle viscosity and plume distribution, *Geophys. J. International, 132,* 412–434.

Valui, G.A., E. Y. Moskalenko, A. A. Strizhkova, and G. R. Sayadyan (2008), Oxygen isotopes in the Cretaceous-Paleogene granites of Primorye and some problems of their genesis, *Russ. J. of Pac. Geol., 2,* 150–157.

Wang, Q., D. A. Wyman, J.-F. Xu, Z.-H. Zhao, P. Jian, X.-L. Xiong, Z.-W. Bao, C.-F. Li, and Z.-H. Bai (2006a), Petrogenesis of Cretaceous adakitic and shoshonitic igneous rocks in the Luzong area, Anhui Province (eastern China): Implications for geodynamics and Cu-Au mineralization, *Lithos, 89,* 424–446.

Wang, Q., J.-F. Xu, P. Jian, Z.-W. Bao, Z.-H. Zhao, C.-F. Li, X.-L. Xiong, and J.-L. Ma (2006b), Petrogenesis of adakitic porphyries in an extensional tectonic setting, Dexing, South China: Implications for the genesis of porphyry copper mineralization, *J. Petrol., 47,* 119–144.

Wee, S.-M., S.-G. Choi, I.-C. Ryu, and H.-J. Shin (2006), Geochemical characteristics of the Cretaceous Jindong granites in the southwestern part of the Gyeongsang Basin, Korea: Focused on adakitic signatures, *Econ. Environ. Geol., 39,* 555–566. (Korean with an English abstract)

Williams, S. E., R. D. Müller, and T. C. W. Landgrebe, and J. M. Whittaker (2012), An open-source software environment for visualizing and refining plate tectonic reconstructions using high-resolution geological and geophysical data sets, *GSA Today, 22.*

Wilson, M. (1993), Geochemical signatures of oceanic and continental basalts: A key to mantle dynamics? *J. Geol. Soc.*, *150*, 977–990.

Wu, F.-Y., J.-Q. Lin, S. A. Wilde, X. O. Zhang, and J.-H. Yang (2005), Nature and significance of the Early Cretaceous giant igneous event in eastern China, *Earth Planet. Sci. Lett.*, *233*, 103–119.

Xu, J.-F., R. Shinjo, M. J. Defant, Q. Wang, and R. P. Rapp (2002), Origin of Mesozoic adakitic intrusive rocks in the Ningzhen area of east China: Partial melting of delaminated lower continental crust? *Geology*, *30*, 1111–1114.

Yang, Y.-T. (2013), An unrecognized major collision of the Okhotomorsk Block with East Asia during the Late Cretaceous, constraints on the plate reorganization of the Northwest Pacific, *Earth-Sci. Rev.*, *126*, 96–115.

Zhang, R. Y., J. G. Liou, and W. G. Ernst (2009), The Dabie-Sulu continental collision zone: A comprehensive review, *Gondwana Res.*, *16*, 1–26.

Ziegler, P. A., and S. Cloetingh (2004), Dynamic processes controlling evolution of rifted basins, *Earth-Sci. Rev.*, *64*, 1–50.

4

Incoming Plate Variations along the Northern Manila Trench: Implications for Seafloor Morphology and Seismicity

Chuanxu Chen[1], Shiguo Wu[1], Jin Qian[2], Changlei Zhao[3], and Lingmin Cao[4]

ABSTRACT

Incoming plate variations along the whole northern Manila trench are investigated by integrating 15 reflection and 3 refraction seismic lines across the trench. Based on the along-trench variation in sediment thickness, basement topography and crust thickness, the incoming crust is divided into three segments from north to south: segment A1–the rifted basin resulting from highly thinning of the continental curst, characterized by thick sediment, ~4-km-thin crust and basement low; segment A2–high basement relief region related to moderate thinning and magmatic modification, characterized by thin sediment, ~12-km-thick crust and basement highs; segment B–normal oceanic crust, characterized by flat and moderate thick sediment and ~6-km-thick crust. The segmentation of the incoming plate correlates well with the deformation pattern of the accretionary wedges and seismically active regions, implying that the preexisting heterogeneities inherited from the rifting process of the continental margin may have an important effect on carving seafloor morphology and seismicity at the precollision zone.

4.1. INTRODUCTION

Northeastern South China Sea (SCS) is an interesting site surrounded by tectonic units from different stages of the Wilson cycle: continental breakup in the northwest; arc-continent collision in the northeast; subduction and early stage collision in the east; and continent-ocean transition in the south (Figure 4.1). It is a small but significant region that has been experienced both destruction and growth of different types of crust, making it an ideal case study to promote our understanding of (1) the evolution of arc-continent collision and the growth of continents; (2) magmatism and structural evolution of the continental margin rifting; (3) subduction and accretion of transitional crust.

The northern Manila trench is the convergent boundary between the northeastern SCS and the Philippine Sea plate (PSP). Great efforts have been made to determine the type of the northern continental margin in the SCS (magma-poor versus magma-rich) and classify the nature of the crust in the northeastern SCS along the trench (continental versus transitional versus oceanic). The northern SCS is considered to be an example of a young, "magma-moderate" rifted margin [*Lester et al.*, 2014], exhibiting characteristics of both magma-rich and magma-poor margins. This includes faulted crustal blocks stretching along the wide continental slope [*Clift et al.*, 2001; *Huang et al.*, 2005; *Zhu et al.*, 2012] and volcanic bodies intruded throughout the distal margin [*Yan et al.*, 2006; *Wang et al.*, 2006; *Zhao et al.*, 2010]. The continent-ocean boundary (COB) near the latitude 19°N (red dashed line in Figure 4.1) was suggested by

[1] *Sanya Institute of Deep-sea Science and Engineering, Chinese Academy of Sciences, Sanya, China*

[2] *Key Laboratory of Marine Geology and Environment, Institute of Oceanology, Chinese Academy of Sciences, Qingdo, China*

[3] *Research Institute of Petroleum Explorations and Development, China National Petroleum Corporation, Hangzhou, China*

[4] *South China Sea Institute of Oceanology, Chinese Academy of Sciences, Guangzhou, China*

Subduction Dynamics: From Mantle Flow to Mega Disasters, Geophysical Monograph 211, First Edition.
Edited by Gabriele Morra, David A. Yuen, Scott D. King, Sang-Mook Lee, and Seth Stein.
© 2016 American Geophysical Union. Published 2016 by John Wiley & Sons, Inc.

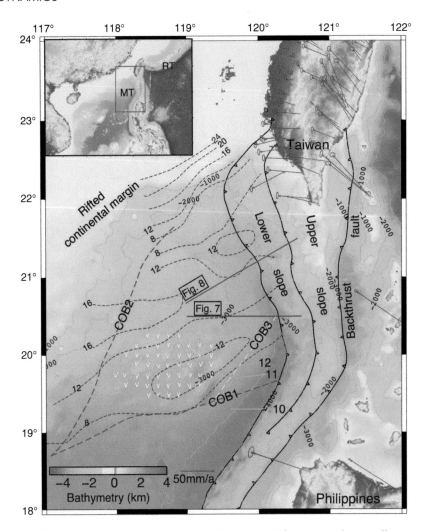

Figure 4.1 Regional bathymetric map and major tectonic features. Red arrows and error ellipses are GPS velocity field by *Sun et al.* [2011]. COB: Continent-Ocean Boundary, COB1 (red dashed line) was suggested by *Briais et al.* [1993], COB2 (blue dashed line) by *Hsu et al.* [2004] and COB3 (yellow dashed line) by *Eakin et al.* [2014]. Light grey and numbers are magnetic lineations and ishchrons by *Yeh et al.* [2010]. The yellow "v" symbols delineate the volcanic intrusion region suggested by *Hsu et al.* [2004]. The black dashed lines and numbers are approximated basement thickness based on the TAIGER data [*McIntosh et al.*, 2014]. The red bold lines denote the locations of MCS lines in Figure 4.7 and Figure 4.8. The bathymetric contour interval is 1000 meters. The inset map shows the location of the present study area (red box). RT = Ryukyu trench; MT = Manila trench.

Briais et al. [1993] based on the magnetic anomalies in the SCS, and the crust north of this boundary was previously assumed to be a magnetically quiet zone that represents a transition zone between continental and oceanic crust [*Taylor and Hayes*, 1983].

Hsu et al. [2004] placed the COB generally following the base of the continental slope (blue dashed line in Figure 4.1) based on the E-W trending magnetic polarity reversal patterns in the northernmost SCS, indicating that the nature of the crust outboard the continental shelf edge is oceanic rather than continental. However, this interpretation is controversial due to the nonuniqueness of the magnetic inversion, and these observed magnetic anomalies could also be indicative of continental crust with magmatic intrusion [*Yeh et al.*, 2012].

Recent seismic reflection images and crustal-scale velocity profiles obtained by the TAIGER (TAiwan Integrated Geodynamics Research) program revealed tilted faulting blocks in the northeastern SCS and most of the fault planes connect downward to a quasi horizontal detachment [*McIntosh et al.*, 2013; *Lester and McIntosh et al.*, 2012; *Lester et al.*, 2014; *Eakin et al.*, 2014; *Yeh et al.*, 2012], which supports the notion that it is the highly thinned continental crust in the northeast SCS rather than oceanic crust. These densely geophysical observations also revealed rough basement topography characterized by alternating basement highs and lows within the continental crust, indicating that this highly extended continental crust has experienced magmatic modification and different degrees of extension from north to south.

The seismicity offshore southwestern Taiwan is much less active, compared to eastern Taiwan, where significant collisional processes are ongoing [*Lacombe et al.*, 2001; *Chen et al.*, 2008]. There were only a few studies aimed at investigating the seismicity and seismogenic structure in the study area [*Teng et al.*, 2005; *Shyu et al.*, 2005; *Lin*, 2009]. The reason for lack of studies may include: (1) low potential for large earthquakes here when compared to eastern Taiwan; (2) no clear imaging of seismogenic megathrust, which has been widely imaged in other subduction zones; (3) incomplete records of medium to small earthquakes in offshore areas. However the occurrence of two magnitude (M_w) 7.0 offshore earthquakes (separated by approximately 8 minutes) on 26 December 2006 (known as Pingtung 2006 earthquake) alerted scientists to the possibility of devastating earthquakes and tsunamis offshore southwest Taiwan [*Wu et al.*, 2009]. This possibility was further confirmed by following historical earthquakes and tsunami investigations [*Mak and Chan*, 2007; *Lau et al.*, 2010] as well as the numerical modeling studies [*Wu and Huang*, 2009; *Liu et al.*, 2007].

These geophysical observations confirm that the nature of the incoming crust, the thickness of the sediment and basement topography are varied along-strike. In this paper, we take the whole northern Manila trench into account and integrate multi-multibeam bathymetric data, published multichannel seismic (MCS) data, wide-angle seismic data and seismicity information from earthquake catalogues to characterize the configuration of the northern Manila trench. We then discuss the morphology and seismicity variations in response to the preexisting heterogeneities within the incoming plate.

4.2. GEOLOGICAL FRAMEWORK

The Philippine Sea plate moves northwest and overrides the underlying Eurasian plate along the Manila trench. The convergence rate along the northern Manila trench ranges from ~75 mm/yr in north Philippine to ~50 mm/yr in south Taiwan, and the velocity field shows anticlockwise rotation in southern-most Taiwan [Figure 4.1, *Sun et al.*, 2011]. The segment of the northern Manila trench is the convergent area of interactions between the South China continental margin, the Taiwan-Luzon arc, and the normal SCS plate. The north-trending Luzon arc collides with the northeast-trending continental margin of the Eurasian plate at ~5–7 Ma [*Huang et al.*, 2006]. This oblique collision causes the

formation of the Taiwan island, and its southwestern propagation has been considered as a major control that results in the diachronism of the orogeny and volumetric differences of the accretionary prisms ~100 km southwest of Taiwan [*Lee et al.*, 2006; *Huang et al.*, 1997].

Sibuet et al. [2002] had previously interpreted a fossil transform fault, which was coined as LRTPB (Figure 4.2a) by *Hsu et al.* [2004]. The nature of the crust northeast of this boundary was interpreted as proto South China Sea oceanic crust or trapped Philippine Sea plate. Recently, this boundary became a questionable feature since there are no clear Moho depth differences or deformation style contrasts across this feature according to the recent geophysical observations [*Yeh et al.*, 2012; *McIntosh et al.*, 2014; *Lester et al.*, 2014; *Eakin et al.*, 2014]. The breakup of the continental margin initiated ~30 Ma in the northeastern SCS [*Lin et al.*, 2003] and generally propagated to the southwest, while the location of its southern end (i.e., the COB along the South China continental margin) is still in controversy.

There are two main debates related to the classification of the nature of the incoming plate along the northern Manila trench, which are both critical to our understanding of the tectonic responses to the subducting process. One is the aforementioned debate on the location of the COB, the other is the origin of the laterally variable basement topography within the incoming crust, starting ~200 km west of the Manila trench. *Hsu et al.* [2004] interpreted the basement highs to be volcanic intrusions and extrusions within oceanic crust, which is ascribed to postspreading volcanism or underplating. However, recent results of densely spaced seismic reflection and refraction observations are not consistent with this interpretation. Crustal-scale velocity profiles from the distal margin crust south of the shelf edge are consistent with the magmatically intruded, highly extended continental crust, rather than oceanic crust [*McIntosh et al.*, 2013; *Eakin et al.*, 2014].

The COB suggested by *Eakin et al.* [2014] is similar to *Briais et al.* [1993], except for the eastern-most segment (yellow dashed line in Figure 4.1), where distinct linear magnetic anomalies indicate that the normal ocean crust of the SCS is currently subducting. The laterally variable basement topography with several alternating highs and lows inside the continental crust south of the shelf edge are interpreted to be expressions of block faulting and different degrees of stretching with interspersed volcanic bodies. The magmatic modifications (including the extractive volcanic bodies, the sills and magmatic underplating) are ascribed to postrift magmatism that may have been induced by convective removal of continental lithosphere following breakup and the onset of seafloor spreading [*Lester et al.*, 2014].

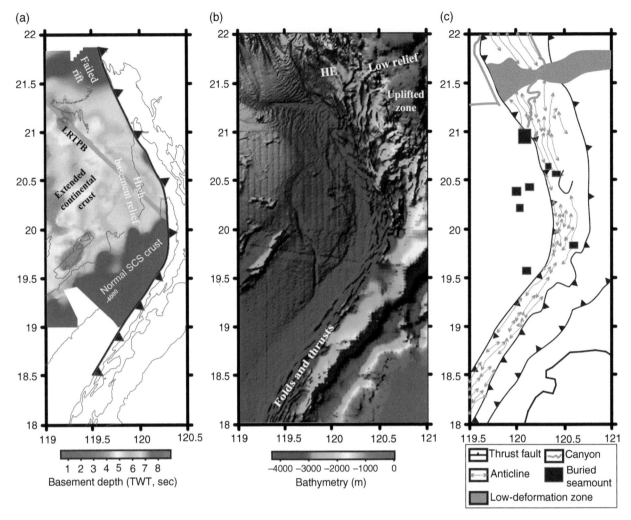

Figure 4.2 Tectonic heterogeneities of the incoming crust and its relation to morphology of the accretionary prism. (a) Basement depth of the incoming crust, modified from *Yeh et al.* [2010]. (b) Seafloor morphology revealed by high-resolution multibeam bathymetric data. (c) The tectonic interpretation of the multibeam data. The determination of the buried seamounts locations also refers to the published MCS profiles.

4.3. INCOMING PLATE VARIATION

4.3.1. Data

Information on the basement depth, sediment thickness and the crust thickness of incoming plate is obtained from two sources: (1) published seismic profiles and interpretations from the literature; and (2) previously processed seismic data released by the Academic Seismic Portal at UTIG, Marine Geoscience Data system (http://www.ig.utexas.edu/sdc). Seismic lines shown in Figure 4.3b were collected by ACT cruises, OR1-689 and OR1-693 cruises, and five seismic lines in Figure 4.3f and three wide-angle seismic lines shown in Figure 4.4a were collected by TAIGER cruises MGL0905 and MGL0908. The acquisition parameters and seismic profiles of these lines can be found in *Ku and Hsu* [2009], *Yeh and Hsu* [2004], *Yeh et al.*

[2012], and *Eakin et al.* [2014], respectively. We first digitized the two-way travel-time (TWT) of two main seismic events in the time-migrated MCS profiles, which correspond to the seafloor and the acoustic basement (Figure 4.3a), and thickness of the depth-migrated MCS profiles (Figure 4.3f). We then interpolated these lines using the Generic Mapping Tools (GMT) surface script [*Wessel and Smith*, 1998] to get a 2D image of the sediment thickness, basement topography, and crust thickness. The seafloor depth (Figure 4.3c, assuming the velocity of the seawater to be 1500 m/s) calculated by interpolating seismic lines is generally consistent with the accurate values revealed by satellite-derived bathymetric data (notice the 3000-m and 4000-m water depth contours), indicating that these 13 lines are sufficient to obtain a reliable image of the incoming plate by interpolation. However, we must note here that the main purpose for generating the 2D maps is to demonstrate lateral variations

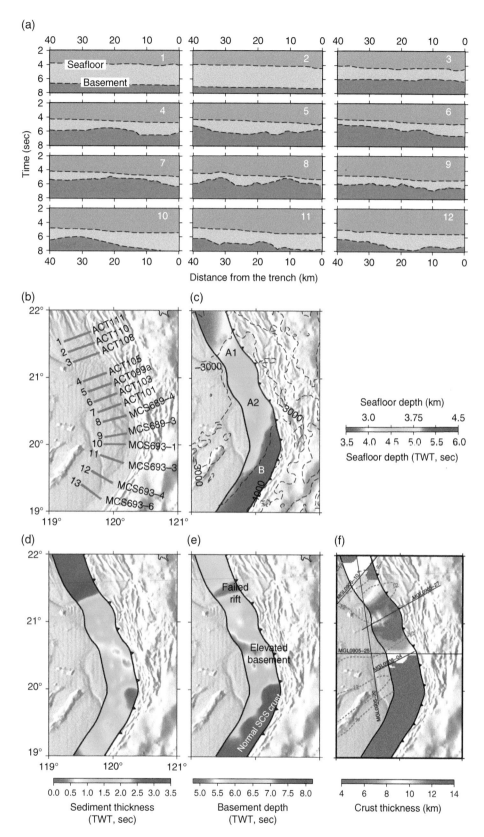

Figure 4.3 Incoming plate variation characterized by basement topography and sediment thickness. (a) Simplified multichannel seismic interpretations along seismic lines 40 km west of the trench, two dashed lines indicate the seafloor and the basement, respectively. (b) The red lines denote the seismic lines used to interpolate basement depth and the sediment thickness. The seismic profiles along these lines can be found in the papers of *Ku and Hsu* [2009]; *Yeh and Hsu* [2004]. (c)–(f) Along-strike variations of the seafloor depth, sediment thickness, basement depth, and crust thickness. HE = Hengchun embayment. A1, A2, and B denote the identifier of three segments along the trench. Black lines in (f) delineate the depth-migrated seismic lines used for interpolation. Green dashed lines and numbers delineate the approximate basement thickness as in Figure 4.1.

in parameters of the incoming plate across the study area. The actual resolution is probably not as high as the $0.1° \times 0.1°$ grid spacing, which was the parameter used in GMT surface script.

The multibeam data used to conduct seafloor morphology image were collected by Guangzhou Marine Geological Survey, Institute of Oceanology of Chinese Academy of Sciences (IOCAS) and South China Sea Institute of Oceanology (SCSIO) of Chinese Academy of Sciences during the last two decades [*Shang*, 2008]. The resolution of multibeam data is generally high in our study area, especially at the lower slope of the accretionary prisms and the South China Sea region. The resolution at the upper slope is lower due to the lack of data.

The seismicity data shown in Figure 4.4a is extracted from the International Seismological Centre (ISC) catalogue (http://www.isc.ac.uk/iscbulletin/search/catalogue/) based on the following criteria: (1) Time period: 1980 to 2012; (2) Magnitude ≥ 3.0. Hypocenter location error of the ISC catalogue is slightly larger than the EHB catalogue [*Engdahl et al.*, 1998], which has epicentral and focal depth uncertainties to roughly 15 km and 10 km, respectively [*Pesicek et al.*, 2010; *Kagan* 2003]. We think these data are sufficient to yield primary constraints on regional (tens of kilometer scale) spatiotemporal pattern of seismicity. The

focal mechanisms shown in Figure 4.4b are selected from the Global CMT catalog (http://www.globalcmt.org) [*Dziewonski et al.*, 1981; *Ekström et al.*, 2012], with recording period between January 2000 to January 2012.

Yeh and Hsu [2004] obtained similar images as Figure 4.3d and Figure 4.3e at the northernmost end of the Manila trench mainly using seismic lines of ACT cruises. Newly conducted seismic surveys during the last decade allow us to better constrain the thickness (TWT, sec) of the sediment and the depth (km) of the crust at areas south of the latitude of 20.5°N. High-resolution seafloor morphology information and additional catalogue seismicity data make it possible to investigate the possible connection between the preexisting heterogeneity within the incoming plate and accretionary wedge deformation and subduction zone seismicity.

4.3.2. Segmentation of the Incoming Plate

The geological framework of the incoming plate along the northern Manila trench can be summarized as: (1) the northern segment is continental crust that has experienced extension/hyperextension and magmatic modification; (2) the southern part is normal SCS oceanic crust. The normal SCS oceanic crust is nearly homogenous with a crustal

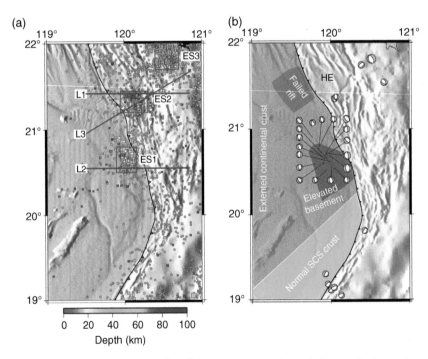

Figure 4.4 Seismicity along the northern Manila subduction zone and its relation to the incoming plate variation. (a) ISC seismicity (shown only M>=3.0) in the study area. Colored squares are epicenters of earthquakes that occurred between 2004 and 2007, while grey dots indicate all other earthquakes that occurred since 1980. Red stars indicate epicenters of 2006 Pingtung doublet earthquakes. ES1, ES2, ES3 indicate three swarms of seismicity at the outer-rise slope of the incoming plate, the lower slope, and the upper slope of the accretionary prism, respectively. (b) Focal mechanisms (blue-and-white beach balls) of the earthquakes during the period from 2000 to 2012.

thickness of ~6 km [*Eakin et al.*, 2014]. On the contrary, the northern continental segment is not homogenous due to rift-related faulting and magmatism. The heterogeneity within the incoming continental crust is revealed by the along-strike variability in basement topography, thickness of sediment and crust (Figures 4.3d, 4.3e, and 4.3f).

We divided the incoming plate along the northern Manila trench into three tectonic units from north to south based on these heterogeneities. Segment A1: NE-SW trending rifted graben (basin) characterized by basement low (~7.0 s, TWT), thick sediment (2.5–3.5 s, TWT), and ~4-km-thin continental crust. This segment is bound by deeply penetrated normal faults that connect to a low-angle detachment [*Lester et al.*, 2014]. North of this segment, ~25-km-thick continental crust stretches along the continental shelf [*McIntosh et al.*, 2014]. Segment A2: Distal continental crust is characterized by basement high (~5.0 s, TWT), thin sediment (0.5–1.5 s, TWT), and ~12–15-km-thick crust. The basement depth varies within segment A2, with the shallowest depths at latitude ~20.5°N, which is coincident with the location of sporadic volcanic bodies and buried seamounts (Figures 4.3a and 4.3c). The variable basement depth is an expression of tilted fault blocks resulting from stretching and thinning of the continental crust, also magmatic features that deform or disrupt overlying postrift strata [*Eakin et al.*, 2014; *Lester et al.*, 2014]. Segment B has normal SCS crust characterized by flat deep basement (7.5–8.0 s, TWT), moderate thick sediment (1.5–2.5 s, TWT) and typical 6-km thick ocean crust.

The boundaries at each segment are consistent with the abrupt change of the basement relief (Figures 4.2a and 4.3e). Particularly, the boundary between Area A2 and B is consistent with the existence of the COB. Because the tectonic features beneath the accretionary prism are difficult to observe seismically, especially by 2D MCS imaging. In this paper, we related accretionary prism deformation and seismic activity to the along-strike variation outboard of the prism, assuming that this variation is representative of heterogeneities of the already subducted plate beneath the prism. This assumption that the present is the key to the past in our study area is based on the following facts: (1) the breakup of the continental margin and the opening of the SCS is about 20 Ma earlier than the oblique collision between the Luzon arc and the continental margin, thus we think tectonic features of the failed rift, the evaluated basement and the normal oceanic crust have been emplaced before subduction front arrived; (2) recently seismic observations (reflection and refraction) reveal that the failed rift has been partly subducted (black dashed lines in Figure 4.1), indicating that it is reasonable to speculate the tectonic features beneath the accretionary prism based on those outboard of the trench. We think this

assumption is generally tenable, although it must still be substantiated in light of the various tectonic processes that the subducting plate may encounter during its travel at depth [*Heuret et al.*, 2012]. These processes may include that the sediment is partly underthrust deeply instead of scraped off and folded frontally into the wedge, and that the materials may be eroded from the upper plate and smooth the plate interface resulting from the tectonic erosion process.

4.4. DISCUSSION

4.4.1. Incoming Plate Variation and Its Controls on Accretionary Wedge Deformation

The properties of the incoming material (i.e., the composition, thickness, cohesion, rheology, porosity, and pore pressure) can influence the accretion process [*Davis et al.*, 1983; *Liu et al.*, 2005; *Schott and Koyi*, 2001; *Simpson*, 2010]. The existence of high bathymetric relief, for instance, seamounts, aseismic ridges, or magmatic plateaus, can make profound changes on the style of overriding plate deformation including surface uplift of the forearc region and significant deformation of the forearc basement [*Dominguez et al.*, 2000; *Rosenbaum and Mo*, 2011].

The accretionary wedge along the northern Manila trench consists of three structural belts: the lower-slope domain, the upper-slope domain, and the back-thrust domain (Figure 4.1), each exhibiting distinct character in seafloor morphology and the internal deformation [*Lin et al.*, 2009]. Besides these east to west variations, delineating different stages of the wedge growth, distinct north to south variations along the northern Manila trench are also clearly imaged by the high-resolution multibeam data (Figure 4.3b). At the southern end of the trench, the accretionary wedge is uniformly folded and thrust. The seafloor topography generally increases from west to east, which is a consequence of the typical intraoceanic subduction of the near homogeneous SCS crust.

There are three distinct bathymetric features along the northern Manila trench: the weak-deformation belt at the northern-most region, the huge uplift zone of upper slope with significant volume increase and the concave trench geometry toward the incoming plate when compared to that of the southern Manila trench (Figures 4.3b and 4.3c). The weak-deformation belt at the northernmost area of the accretionary wedge is clearly imaged by high-resolution multibeam imaging (Figure 4.3b), showing smooth seafloor tomography in the lower-slope domain and less uplift in the upper-slope domain. *Lin et al.* [2009] coined the lower slope part of this less deformation belt as Hengchun embayment (HE), which is characterized by a scallop-shaped geometry and little or no fold/thrust structures. This weakly deformed zone correlates well with the sediment sink overlain on the incoming rifted graben outboard its western edge at A1 (Figures 4.3d, 4.3e, and 4.3f).

This segment of the incoming plate is also coincident with the depositional site of a paleosubmarine fan of the Kaoping Canyon [*Chiang and Yu*, 2006], implying the sediments at A1 may contain more water. Given that the failed rift extends northeastward and is partly subducted beneath the HE [Figure 4.1, *McIntosh et al.*, 2014], we speculate that the incoming sediments that are responsible for the growth of the HE are also thick and water-rich, similar to those positioned outboard of the trench. Thus, during the growth of the HE, more water will be expelled from the fluid-rich thick sediment and will create an elevated pore pressure along the subducting interface, leading to a decrease of the shear strength and the basal friction of the decollement, ultimately, resulting in a broader protothrust zone, that is, the present HE [*Lin et al.*, 2009]. A bathymetric valley occurs inside the upper slope near latitude ~21.5°N, showing continuity to the HE weak deformation zone. There is a possibility that the low-relief belt, consisting of the HE and the valley at its east, indicates the track of the accretionary front where the westward migrating Manila trench interacts with the NE-SW emplaced failed rift.

South of the weak deformation belt, the lower slope of the accretionary wedge is strongly folded and thrust, and the upper slope is significantly uplifted with a sharp change in bathymetry and sudden volumetric increase between 21°N and 21.5°N (Figure 4.3b). Given that the dip of rift-related structure is near NW-SE and the trend of the variable basement topography is near NE-SW, we think that the subducted crust beneath this uplifted upper slope is likely the continuation of the basement high at latitude range of 20.5°N to 21°N outward of the trench. Seismic evidences from wide-angle data also confirm the existence of structural underplating beneath the upper slope (95–115 km model distance in Figure 4.6a; 125–150 km model distance in Figure 4.6c). *Lester et al.* [2013] ascribe the significant volume increase observed in the prism to structural underplating of the basement highs of subducting continental/transitional crust, which includes horst blocks of hyperextended continental crust and sporadic magmatic intrusions. The behavior and the impact of these basement highs may be analogous to the seamount subduction, but on a much larger scale. Thus, HE, with an elliptic boundary, may be considered as the "shadow zone" following the flank of a subducting basement high, which is analogous to a weak deformation zone in the wake of a subducted seamount [*Dominguez et al.*, 2000; *Rosenbaum and Mo*, 2011].

Most of trenches around the world evolve a convex arc shape toward the subducting plate, as a result of the backflow forcing during trench retreat, which can be explained by the Ping-Pong ball analogy [*Frank*, 1968]. However, concave trench curvatures are also widely observed. These exceptions are mostly consistent with the presence of high bathymetric relief, such as the seamounts, oceanic plateaus and aseismic ridges. The northern Manila Trench bends sharply at around 20°N and exhibits reentrant trench line geometry, but there is no marked high bathymetric relief adjacent to this concave curve. Numerical simulation results [*Morra et al.*, 2006; *Mahadevan et al.*, 2010] show that arc and trench curvature can be developed by along-strike variation in the rate of subduction rollback, due to internal heterogeneities within the subducted lithosphere, such as the difference in the age and buoyancy of the subducting lithosphere. The plate tectonic evolution model of SE Asia [*Hall*, 2002] showed that the Manila trench retreated westwards to the South China Sea since ~20 Ma. Compared with the southern Manila trench, most part of the incoming plate along the northern Manila trench is more buoyant continental crust rather than ocean crust. The arrival of this relatively buoyant mass at the subduction zone resists the subduction and locally reduces the rate of subduction rollback, and this in turn, generates curvatures within the subduction system and forms the concave northern Manila trench line [*Obayashi et al.*, 2009].

4.4.2. Incoming Plate Variation and Its Relation to Subduction Zone Seismicity

Figure 4.4a shows the seismicity along the northern Manila subduction zone, with the earthquake parameters extracted from the ISC catalogue. Three swarms of seismicity are clearly delineated along the northern Manila subduction zone: ES1 at the outer-rise slope, ES2 at the lower-slope region, and ES3 at the upper-slope region. The outer-rise swarm of earthquakes (ES1) is located between 20.5°N and 20.9°N, ~50 km west of the trench, which coincides with the region of high basement relief (Figures 4.3e and 4.4b). The seismogenic zone of this swarm is almost immune to large earthquakes during most time of the instrumental seismicity period since 1980, except two narrow time windows between May to June and October to November 2006 (Figures 4.5a and 4.5b). About 70 earthquakes with magnitude ≥3.0 occurred during these two seismically active periods, and the magnitude and frequency of earthquakes during the latter period are higher than that of the former one. The lower-slope swarm of earthquakes (ES2) is distributed at the south edge of the HE weak deformation zone. The timing of the vast majority of this seismicity is prior to that of the outer-rise swarm, with frequency of occurrence peaking in December 2005 (Figures 4.5a and 4.5b). Prior to the sequence of 2006 Pingtung doublet

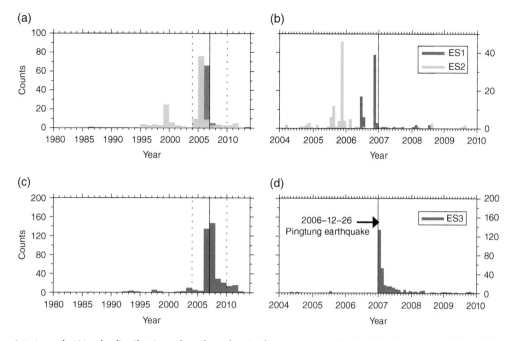

Figure 4.5 (a and c) Yearly distribution of earthquakes in three swarms of seismicity since 1980. The black dashed line delineates the time window between 2004 and 2010. (b and d) Monthly distribution of earthquakes in three swarms of seismicity between 2004 and 2010. The gray line denotes the day of the Pingtung earthquake, 26 December 2006.

earthquakes, the region of the upper-slope swarm of earthquakes (ES3) had been also seismically quiescent since 1980 (Figures 4.5c and 4.5d). Two unexpected large earthquakes (Mw = 7.0) struck southwestern Taiwan following the seismically active period of the outer-rise earthquake swarm (ES1) (note ES1 in Figure 4.5b and ES3 in Figure 4.5d). It is an interesting phenomenon that these three earthquakes swarms at the precollision zone successively become seismically active over a relatively short period (~2 years before and after 2006). It is not clear yet whether these three swarms are spatiotemporally related.

The outer-rise events occurring at shallow and intermediate depth (≤40 km) are attributed to plate bending, and the focal mechanisms of these events and their predominantly EW tension axes also suggest mechanisms of outer-rise normal faulting (Figures 4.4b and 4.6b). Viewed along a profile (L2) perpendicular to the trench axis, sparse earthquake hypocenters are concentrated in a dipping region with a width of about 20 km, which may represent a blurred image of the plate-bending-related normal fault and/or shear zone that has penetrated into the upper mantle of the incoming plate. Given the large depth uncertainties of these earthquakes, seismicity data alone are not enough to constrain the faults geometry.

Figure 4.7a shows a prestack time migrated MCS section across the ES2 and the trench, and Figure 4.7b

shows details at the outer-rise region, approximately 50 km west of the trench. From CMP 10,000 to CMP 20,000, the crust starts to bend with several bending-related normal faults and subducts eastward. The top of the basement exhibits highly variable topography with several alternating highs and lows, which may be the expression of faulted blocks of upper crust [*Eakin et al.*, 2014]. The basement highs that deform or disrupt overlying postrift strata were interpreted as volcanic zones by *Lester et al.* [2014]. These volcanic bodies are bound by the bending-related normal fault, implying that these presently deeply penetrated normal faults may have grown along the preexisting weakness, which acted as the pathway for volcanic intrusion. Figure 4.8 shows a NE-SW trench perpendicular MCS profile to the north of Figure 4.7 (see Figure 4.1 for line location). The outer-rise region of Figure 4.8 shows that the top of the basement is relatively flat and less-normal faults cut though the basement and the sedimentary layers. The seismic reflection along these two MCS lines shows chaotic and rare inner crustal reflection beneath the sediments layers. Thus, it is difficult to determine how deep these bending-related faults penetrate.

Figure 4.6b shows the velocity structure derived by *Eakin et al.* [2014] with wide-angle seismic data. Their results show that the nature of the incoming plate is thinned continental crust (~10–12 km). The seismogenic

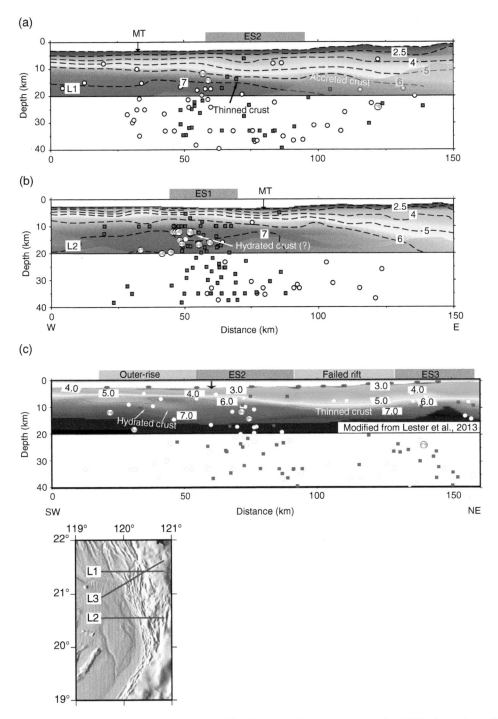

Figure 4.6 (a)–(c) Velocity structure, seismicity, and focal mechanisms (20-km swath width) along three different lines. Locations are shown in Figure 4.4a. The brown rectangle above the top of each profile denotes the projected area of each earthquake swarm. The velocity structures in (a) and (b) are obtained by interpolating velocity contour lines (black dashed lines), which are digitized from *Eakin et al.* [2014]. The transparent gray areas delineate the regions that lack ray coverage and thus have lower resolution. Red squares indicate earthquakes (shown only M>=3.0) that happened between 2004 and 2007, and white circles indicate the other events since 1980.

region of the outer-rise earthquakes is coincident with an lower crustal velocity reduction between 50 and 75 km model distance (Figure 4.6b). Additionally, this velocity reduction feature has been clearly imaged in the travel-time tomography velocity model along a ~N-S OBS (model distance between 250–300 km in Figure 4.7

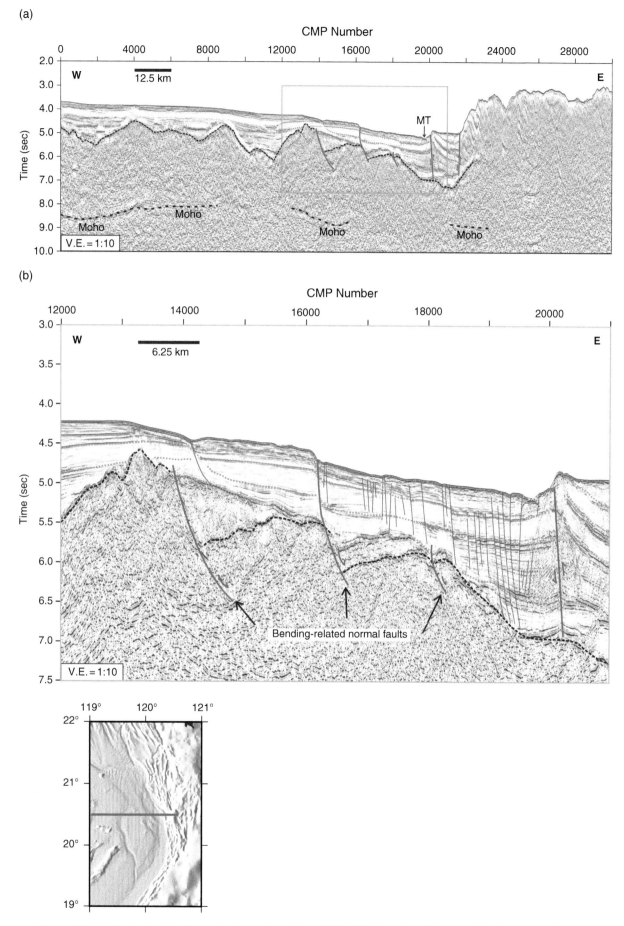

Figure 4.7 (a) Prestack migration seismic section along MGL0905_25a (see Figure 4.1 for location). Blue box is the inset shown in (b).

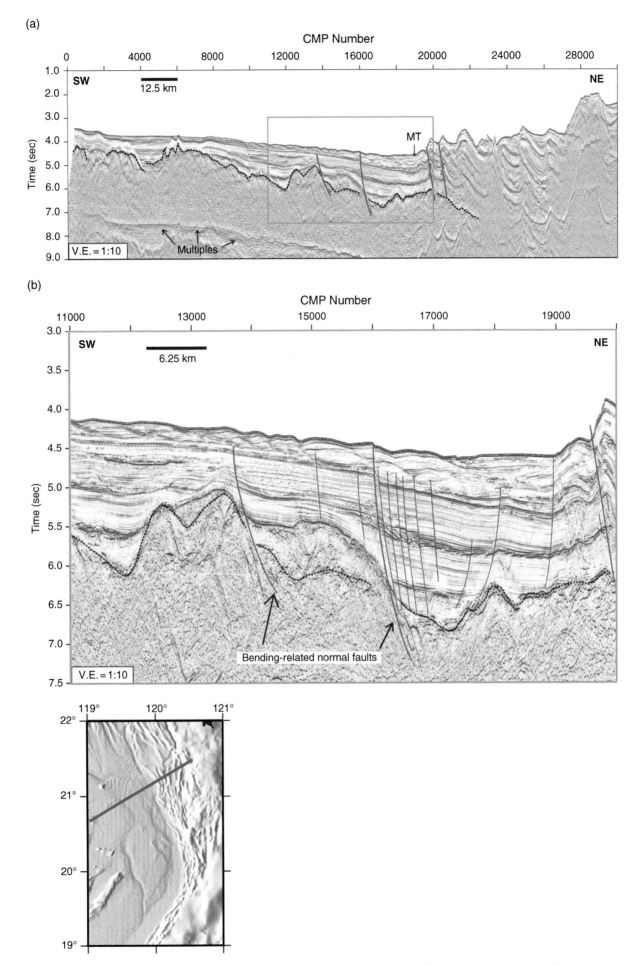

Figure 4.8 (a) Prestack migration seismic section along MGL0905_27 (see Figure 4.1 for location). Blue box is the inset shown in (b).

of *Lester et al.* [2014]), and a NE-SW OBS line to the north of transect L2 (model distance between 25 and 50 km in Figure 4.6c). Velocities are reduced by 0.3–0.5 km/s compared to the surrounding crust. Similar velocity reduction has also been observed in the trench-outer rise region offshore of Nicaragua associated with seamounts [*Ranero et al.*, 2003; *Ivandic et al.*, 2010] and of the central Chile subduction zone [*Moscoso and Contreras-Reyes*, 2012].

A plausible interpretation for the low crustal and uppermost mantle velocities reduction is that water percolation trough bending-related faults leads to mineral alteration and hydration of the oceanic crust and mantle [*Grevemeyer et al.*, 2007; *Moscoso and Contreras-Reyes*, 2012]. This crustal velocity reduction in our study area was not interpreted by previously studies, and we speculate that it may represent hydrated/fault crust resulting from infiltration of seawater through the deeply penetrated fault system. The crust/mantle hydration is generally related to normal faulting and plate bending at the outer-rise slope, and it may also be augmented by preexisting weakness inherited from the rifting phase of the Chinese margin in the case of the northern Manila subduction zone.

Another distinct spatial coincidence is between the seismogenic zone of ES1 and the uplift basement outboard the trench. The possible explanation for this coincidence is: (1) more stress accumulated at the normal faults along the outer rise due to the resistance of the basement highs; (2) The preexisting weakness, such as faulting or shear zone, which was inherited from the continental crust thinning and magmatically intrusion, has already been embedded within this segment of incoming plate prior to subduction front arrival. This preexisting weakness may have promoted the formation and growth of the bending-plate normal faults when this segment of incoming plate was positioned at the outer-rise region.

The ES2 correlates well with the southern edge of the HE weak deformation zone, and the ES3 is located offshore of southeast Taiwan. The incoming plate segment west of the HE is a local rifted basin/graben bound by deeply penetrated normal faulting related to stretching and thinning of continental crust [*Yeh et al.*, 2012; *Lester et al.*, 2014]. The crustal thickness contrast across the southern edge of this rifted basin can be as large as 8 km, and even as large as 15 km at the northern edge. Tomographic imaging derived from wide-angle seismic data shows velocity reduction within the upper crust beneath the seismogenic region of ES2 between 60 and 80 km model distance (Figure 4.6a). This velocity reduction is also imaged in the previously tomographic modeling (model distance between 90 km and 130 km in Figure 4.6c). We speculate that this low-velocity upper

crust is the seismic evidence for the existence of the rifted graben, indicating that the failed rift outboard of the prism stretched northeastward and has been subducted beneath the upper slope. Figure 4.6c shows that the ES2 and ES3 are located on two boundaries of the thinned crust, which is an indication of the NE-SW trending failed rift. Focal mechanisms of two events within ES2 and the first shock of Pingtung earthquake show normal-fault solutions and trench-normal tension axes, suggesting a plate-bending related origin.

The spatial correlation between the NE-SW-stretching failed rift and the swarms of seismicity adjacent to it might be explained by the following mechanism. The failed rift was bound by the NE-SW-trending preexisting weakness (normal faulting/ shear zone) which is inherited from continental crust thinning and rifting process. These pre-preexisting weak zones may be functioned as regions where plate-bending stress concentrate and ultimately triggered earthquakes. This mechanism also might be used to explain the possible temporal connection between these three swarms of seismicity, if these preexisting faulting/shear zones had been connected and interacted with each other.

4.5. CONCLUSION

High-resolution, multibeam bathymetric data, reflection and refraction seismic data, and seismicity information from earthquake catalogues are integrated in our work to investigate the tectonic deformation in response to incoming plate variations along the northern Manila trench. Main findings are summarized as follows:

1. Detailed seafloor morphology derived from the multibeam data shows along-strike variation of the accretionary prism deformation along the whole northern Manila trench. At the northernmost part of the trench, a weak-deformation belt stretches throughout the accretionary prism, while at the middle part, the upper slope was substantially uplifted with a significant topographic break separating the strongly folded and thrust lower slope.

2. The sediment thickness, basement topography, and crustal thickness derived from the densely distributed seismic lines delineate the incoming plate variations from north to south. These heterogeneities are spatially correlated with the deformation variations of the accretionary prisms, implying that beside the oblique convergence, the incoming plate variation also plays an important role in shaping the deformation configuration.

3. Two swarms of seismicity are spatially coincident with deeply penetrated normal-faults that are continuous with the boundaries of the rifted basin outboard the trench. A swarm of outer-rise seismicity correlates well with the elevated basement topography of the incoming crust. The preexisting heterogeneities in the incoming

plate may result in significant variations in plate bending and hydration, which can explain the observed diversity of seismicity in our study area.

ACKNOWLEDGMENT

We thank all the crew members for their decades of efforts on collecting the multibeam bathymetric and seismic data we used in presenting our work. We thank Yueming Ye and Jinling Zhang for their help on the seismic data processing. We are also grateful to Dongdong Dong and Jianke Fan for their fruitful discussions. This study was under the grant of the Knowledge Innovation Project of the Chinese Academy of Sciences (No. KZCX3-SW-229), the National Natural Science Foundation of China (41204061; 41206043). Two anonymous reviewers and our editor Gabriele Morra provided thoughtful review comments and suggestions that have improved this manuscript.

REFERENCES

Briais, A., P. Patriat, and P. Tapponnierr (1993), Updated interpretation of magnetic anomalies and seafloor spreading stages in the South China Sea: Implications for the tertiary tectonics of southeast Asia, *J. Geophys. Res.*, *98*(B4), 6299–6328.

Chen, P.-F., A. V. Newman, T. R. Wu, and C. C. Lin (2008), Earthquake probabilities and energy characteristics of seismicity offshore southeast Taiwan, *Terr. Atmos. Ocean. Sci.*, *19*, 697–703.

Chiang, C. S., and H. S. Yu (2006), Morphotectonics and incision of the Kaoping submarine canyon, SW Taiwan orogenic wedge, *Geomorphology*, *80*, 199–213.

Clift, P. D., J. Lin, and O. L. S. Party (2001), *Patterns of extension and magmatism along the continent-ocean boundary, South China margin,* in *Non-Volcanic Rifting of Continental Margins: A Comparison of Evidence from Land and Sea*, R. C. L. Wilson et al. (eds.), Geological Society, London, Special Publications, *187*, 1–8.

Davis, D., J. Suppe, and F. Dahlen (1983), Mechanics of fold-and-thrust belts and accretionary wedges, *J. Geophys. Res.*, *88*, 1153–1172.

Dominguez, S., J. Malavieille, and S. E. Lallemand (2000), Deformation of accretionary wedges in response to seamount subduction: Insights from sandbox experiments, *Tectonics*, *19*, 182–196.

Dziewonski, A. M., T.-A. Chou, and J. H. Woodhouse (1981), Determination of earthquake source parameters from waveform data for studies of global and regional seismicity, *J. Geophys. Res.*, *86*, 2825–2852.

Eakin, D. H., K. D. McIntosh, H. Van Avendonk, L. Lavier, R. Lester, C. S. Liu, and C. S. Lee (2014), Crustal-scale seismic profiles across the Manila subduction zone: The transition from intraoceanic subduction to incipient collision, *J. Geophys. Res.*, *119*, 1–17.

Ekström, G., M. Nettles, and A. M. Dziewonski (2012), The global CMT project 2004–2010: Centroid-moment tensors for 13,017 earthquakes, *Phys. Earth Planet. Inter.*, *200–201*(0), 1–9.

Engdahl, E. R., R. van der Hilst, and R. Buland (1998), Global teleseismic earthquake relocation with improved travel times and procedures for depth determination, *Bull. Seism. Soc. Am.*, *88*, 722–743.

Frank, F. (1968), Curvature of island arcs, *Nature*, *220*, 363–363.

Grevemeyer, I., C. R. Ranero, E. R. Flueh, D. Kläschen, and J. Bialas (2007), Passive and active seismological study of bending-related faulting and mantle serpentinization at the Middle America trench, *Earth Planet. Sci. Lett.*, *258*(3–4), 528–542.

Hall, R. (2002), Cenozoic geological and plate tectonic evolution of SE Asia and the SW Pacific: Computer-based reconstructions, model and animations, *J. Asian Earth Sci.*, *20*(4), 353–431.

Heuret, A., C. P. Conrad, F. Funiciello, S. Lallemand, and L. Sandri (2012), Relation between subduction megathrust earthquakes, trench sediment thickness and upper plate strain, *Geophys. Res. Lett.*, *39*, L05304, doi:10.1029/2011GL050712.

Hsu, S.-K., Y.-C. Yeh, W.-B. Doo, and C.-H. Tsai (2004), New bathymetry and magnetic lineations identifications in the northernmost South China Sea and their tectonic implications, *Mar. Geophys. Res.*, *25*, 29–44.

Huang, C., D. Zhou, Z. Sun, C. Chen, and H. Hao (2005), Deep crustal structure of Baiyun Sag, northern South China Sea revealed from deep seismic reflection profile, *Chin. Sci. Bull.*, *50*(11), 1131–1138.

Huang, C.-Y., P. B. Yuan, and S. J. Tsao (2006), Temporal and spatial records of active arc-continent collision in Taiwan: a synthesis, *Geol. Soc. Am. Bull.*, *118*, 274–288.

Huang, C. Y., W. Y. Wu, C. P. Chang, S. Tsao, P. B. Yuan, C.W. Lin and K.Y. Xia (1997), Tectonic evolution of accretionary prism in the arc-continent collision terrane of Taiwan, *Tectonophysics*, *281*, 31–51.

Ivandic, M., I. Grevemeyer, J. Bialas, and C. J. Petersen (2010), Serpentinization in the trench-outer rise region offshore of Nicaragua: constraints from seismic refraction and wide-angle data, *Geophys. J. Inter.*, *180*(3), 1253–1264.

Kagan, Y. Y. (2003), Accuracy of modern global earthquake catalogs, *Phys. Earth Planet. Inter.*, *135*(2), 173–209.

Ku, C.-Y., and S.-K. Hsu (2009), Crustal structure and deformation at the northern Manila Trench between Taiwan and Luzon islands, *Tectonophysics*, *466*, 229–240.

Lacombe, O., F. Mouthereau, J. Angelier, and B. Deffontaines (2001), Structural geodetic and seismological evidence for tectonic escape in SW Taiwan, *Tectonophysics*, *333*, 323–345.

Lau, A., A. D. Switzer, D. Dominey-Howes, J. Aitchison, and Y. Zong (2010), Written records of historical tsunamis in the northeastern South China Sea: Challenges associated with developing a new integrated database, *Nat. Hazards Earth Syst. Sci.*, *10*, 1793–1806.

Lee, Y.-H., C.-C. Chen, T.-K. Liu, H.-C. Ho, H.-Y. Lu, and W. Lo (2006), Mountain building mechanisms in the Southern Central Range of the Taiwan Orogenic Belt from accretionary wedge deformation to arc-continental collision, *Earth Planet. Sci. Lett.*, *252*, 413–422.

Lester, R., and K. McIntosh (2012), Multiple attenuation in crustal-scale imaging: examples from the TAIGER marine reflection data set, *Marine Geophys. Res.*, *33*(4), 289–305.

Lester, R., H. J. Van Avendonk, K. McIntosh, L. Lavier, C. S. Liu, T. Wang, and F. Wu (2014), Rifting and magmatism in the northeastern South China Sea from wide-angle tomography and seismic reflection imaging, *J. Geophys. Res.*, *119*, doi: 10.1002/2013JB010639.

Lester, R., K. McIntosh, H. J. A. Van Avendonk, L. Lavier, C. S. Liu, and T. K. Wang (2013), Crustal accretion in the Manila trench accretionary wedge at the transition from subduction to mountain-building in Taiwan, *Earth Planet. Sci. Lett.*, *375*, 430–440.

Lin, A. T., A. B. Watts, and S. P. Hesselbo (2003), Cenozoic stratigraphy and subsidence history of the South China Sea margin in the Taiwan region, *Basin Res.*, *15*, 453–478.

Lin, A.T., B. Yao, S.-K. Hsu, C.-S Liu, and C.-Y. Huang (2009), Tectonic features of the incipient arc-continent collision zone of Taiwan: Implications for seismicity, *Tectonophysics*, *479*, 28–42.

Liu, C.-S., B. Deffontaines, C.-Y. Lu, and S. Lallemand (2005), Deformation patterns of an accretionary wedge in the transition zone from subduction to collision offshore southwestern Taiwan, *Geophys. Res.*, *25* (Mar.), 123–137.

Liu, Y., A. Santos, S. M. Wang, Y. Shi, H. Liu, and D. A. Yuen (2007), Tsunami hazards along Chinese coast from potential earthquakes in South China Sea, *Phys. Earth Planet. Inter.*, *163*(1), 233–244.

Mahadevan, L., R. Bendick, and H. Liang (2010), Why subduction zones are curved, *Tectonics*, *29*(6), doi:10.1029/2010TC002720.

Mak, S., and L.-S. Chan (2007), Historical tsunamis in south China, *Natural Hazards*, *43*(1), 147–164.

McIntosh, K., H. Van Avendonk, L. Lavier, R. Lester, D. Eakin, F. Wu, and C.-S. Liu (2013), Inversion of a hyper-extended rifted margin in the southern central range of Taiwan, *Geology*, *41*(8), 871–874.

McIntosh, K., L. Lavier, H. van Avendonk, R. Lester, D. Eakin, and C.S. Liu (2014), Crustal structure and inferred rifting processes in the northeast South China Sea, *Marine and Petroleum Geology*, *58*, 612–626.

Morra, G., K. Regenauer-Lieb, and D. Giardini (2006), Curvature of oceanic arcs, *Geology*, *34*, 877–880.

Moscoso, E., and E. Contreras-Reyes (2012), Outer rise seismicity related to the Maule, Chile 2010 megathrust earthquake and hydration of the incoming oceanic lithosphere, *Andean Geology*, *39*(3), 564–572.

Obayashi, M., J. Yoshimitsu, and Y. Fukao (2009), Tearing of stagnant slab, *Science*, *324*, 1173–1175.

Pesicek, J. D., C. H. Thurber, H. Zhang, H. R. DeShon, E. R. Engdahl, and S. Widiyantoro (2010), Teleseismic double-difference relocation of earthquakes along the Sumatra-Andaman subduction zone using a 3-D model, *J. Geophys. Res.*, *115* (B10), doi: 10.1029/2010JB007443

Ranero, C. R., J. P. Morgan, K. McIntosh, and C. Reichert (2003), Bending, faulting, and mantle serpentinization at the Middle America trench, *Nature*, *425*, 367–373.

Rosenbaum, G., and W. Mo (2011), Tectonic and magmatic responses to the subduction of high bathymetric relief, *Gondwana Research*, *19*, 571–582.

Schott, B., and H. A. Koyi (2001), Estimating basal friction in accretionary wedges from the geometry and spacing of frontal faults, *Earth Planet. Sci. Lett.*, *194*, 221–227.

Shang, J. H., (2008), Tectonic dynamics research and subducting characteristics comparison between middle and northern part of manila subducting belt (in Chinese), Ph.D Thesis, University of Chinese Academy of Sciences, 13–17

Shyu, J. B. H., K. Sieh, Y.-G. Chen, and C.-S. Liu (2005), Neotectonic architecture of Taiwan and its implications for future large earthquakes, *J. Geophys. Res.*, *110*, B08402, doi:10.1029/2004JB003251.

Sibuet, J.-C., S.-K. Hsu, X. Le Pichon, J.-P. Le Formal, D. Reed, G. Moore, and C.-S. Liu (2002), East Asia plate tectonics since 15 Ma: constraints from the Taiwan region, *Tectonophysics*, *344*(1–2), 103–134.

Simpson, G. D. H., (2010), Formation of accretionary prisms influenced by sediment subduction and supplied by sediments from adjacent continents, *Geology*, *38*, 131–134.

Sun, J. L., H. L. Xu, and J. H. Cao (2011), Crustal movement and its dynamic mechanism of the Taiwan-Luzon convergent zone, *Chinese J. Geophys.* (in *Chinese*), *54*(12), 3016–3025.

Taylor, B., and D. Hayes (1983), Origin and history of the South China Sea basin, in D. E. Hayes (ed.), *The Tectonic and Geologic Evolution of Southeast Asian Seas and Islands*, Part 2, *Geophys. Monograph*, *23*, Washington, DC, 23–56.

Teng, L. S., R.-J. Rau, C.-T. Lee, C. S. Liu, and W.-S. Chen (2005), Faulting in southwestern Taiwan, *Western Pacific Earth Sciences (in Chinese)*, *5*, 97–128.

Wang, T. K., M.-K. Chen, C.-S. Lee, and K. Xia (2006), Seismic imaging of the transitional crust across the northeastern margin of the South China Sea, *Tectonophysics*, *412*, 237–254.

Wessel, P., and W. H. Smith (1998), New, improved version of generic mapping tools released, *Eos, Trans. Am Geophys. Un.*, *79*, 579.

Wu, T.-R., and H.-C. Huang (2009), Modeling tsunami hazards from Manila trench to Taiwan, *J. Asian Earth Sci.*, *36*(1), 21–28.

Wu, Y.-M., L. Zhao, C.-H. Chang, N.-C. Hsiao, Y.-G. Chen, and S.-K. Hsu (2009), Relocation of the 2006 Pingtung Earthquake sequence and seismotectonics in Southern Taiwan, *Tectonophysics*, *479*, 19–27.

Yan, P., H. Deng, H. Liu, Z. Zhang, and Y. Jiang (2006), The temporal and spatial distribution of volcanism in the South China Sea region, *J. Asian Earth Sci.*, *27*, 647–659.

Yeh, Y.-C., and S.-K. Hsu (2004), Crustal structures of the northernmost South China Sea: Seismic reflection and gravity modeling, *Geophys. Res.*, *25* (Mar.), 45–61.

Yeh, Y.-C., J.-C. Sibuet, S.-K. Hsu, and C.-S. Liu (2010), Tectonic evolution of the northeastern South China Sea from seismic interpretation, *J. Geophys. Res.*, *115*(B06), doi:10.1029/2009JB006354.

Yeh, Y.-C., S.-K. Hsu, W.-B. Doo, J.-C. Sibuet, C.-S. Liu, and C.-S. Lee (2012), Crustal features of the northeastern South China Sea: Insights from seismic and magnetic interpretations, *Geophys. Res.*, *33*(4, Mar.), 307–326.

Zhao, M., X. Qiu, S. Xia, H. Xu, P. Wang, T. K. Wang, C.-S. Lee, and K. Xia (2010), Seismic structure in the northeastern South China Sea: S-wave velocity and Vp/Vs ratios derived from three-component OBS data, *Tectonophysics*, *480*(1), 183–197.

Zhu, J., X. Qiu, H. Kopp, H. Xu, Z. Sun, A. Ruan, J. Sun, and X. Wei (2012), Shallow anatomy of a continental ocean transition zone in the northern South China Sea from multichannel seismic data, *Tectonophysics*, *554–557*(0), 18–29.

5

Source of the Cenozoic Volcanism in Central Asia

Gabriele Morra[1], David A. Yuen[2], Sang-Mook Lee[3], and Siqi Zhang[4]

ABSTRACT

Cenozoic intraplate volcanism in central Asia is characterized by individual small volcanic provinces spread over three broad regions, thousand of kilometers long and hundred of kilometers wide, whose origin remains controversial. Seismological observation of high-velocity anomalies at the base of the upper mantle, where the subducted Pacific plate has been sliding for tens of millions of years, prompted the "Big Mantle Wedge" hypothesis that some of the surface volcanism emerges from the rise of partially molten mantle rocks originating above the slab. Constraints from mineral physics, petrology, and geophysics have suggested that the wadsleyite layer at the top of the transition zone can store large amounts of volatiles. We show that while a 1-km single diapir would rise too slowly and diffuse to the surrounding mantle, a sufficiently large group of diapirs would reach the surface because its rising speed increases with the square of the cluster size. We relate the regular distribution of the surface volcanism, characterized by cones that are about 200-km distant, to the thickness of the volatile-rich layer from which diapirs rise. Doing so we find that a 20-km thick layer of diapirs on top of the transition zone would explain most of the observed volcanism. Our models suggest that decarbonation-dehydration melting in the transition zone is the main cause of the upwellings and of the surface volcanism in central Asia.

5.1. INTRODUCTION

Regularly spaced volcanism has been observed in continental rifting [Bonatti, 1985; Corti et al., 2003], island arcs [Marsh, 1979; Zhu et al., 2011], and along midocean ridges [Whitehead et al., 1984]. These regularities have been frequently explained by buoyant instabilities of partially molten rocks that rise from the shallow or deep

mantle to the Earth's surface, cause partial melts, and originate surface volcanism.

Cenozoic intraplate volcanism in central Asia is characterized by small alkaline [Peng et al., 1986] individual volcanic provinces, typically below $30 \, km^3$ [Zhou and Armstrong, 1982; Miyashiro, 1986; Barry et al., 2007], spread over several regions with distances of around 1000 km long with a breadth of hundred of kilometers. Many hypotheses have been put forward to explain its origin, ranging from mantle plumes [Windley and Allen, 1993], crustal melts [Yarmolyuk, 1991] lithosphere extension or delamination caused by the India-Asia collision, to thermal blanketing [Petit et al., 2002]. An overview of the location and morphology of the main volcanic regions are shown in Figure 5.1, modified after [Yin, 2010], where the central and eastern volcanism is grouped in three elongated regions: A, B, and C. Of the three, the most enigmatic is region A because it is the most westward, while the regions B and C are close to the Pacific and

[1]Department of Physics and School of Geosciences University of Louisiana at Lafayette, Lafayette, Louisiana, USA

[2]School of Environment Studies, China University of Geosciences, Wuhan, China; Minnesota Supercomputing Institute Department of Earth Sciences, University of Minnesota, Minneapolis, Minnesota, USA

[3]School of Earth and Environmental Sciences Seoul National University, Gwanak-Gu, South Korea

[4]Geophysics and Geodynamics Group Macquarie University, Sydney, Australia

Subduction Dynamics: From Mantle Flow to Mega Disasters, Geophysical Monograph 211, First Edition.
Edited by Gabriele Morra, David A. Yuen, Scott D. King, Sang-Mook Lee, and Seth Stein.
© 2016 American Geophysical Union. Published 2016 by John Wiley & Sons, Inc.

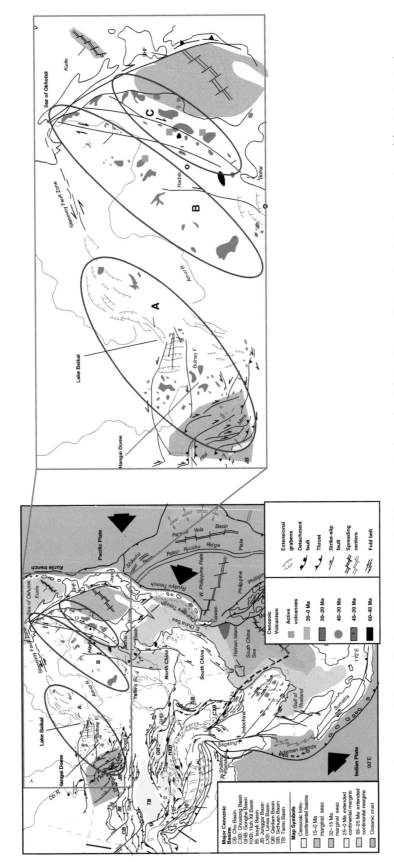

Figure 5.1 Summary of the distribution of the region with high magmatic throughput during Cenozoic in eastern Asia. An overview of the Asian tectonics is shown in the large map, modified after *Yin* [2010], while the lateral box shows a zoom into the volcanically active regions. Black arrows represent the motion of oceanic plates. Cenozoic volcanism can be at a first order divided in three broad elongated regions. The Hangai dome region is indicated in the zoomed version. More detailed information is shown in Figure 5.3.

Philippine subduction regions, suggesting a subduction-related origin [*Zhao et al.*, 2011].

Among the magmatism of the region A, the most studied areas are around the Baikal lake [*Lebedev et al.*, 2006] and Mongolia [*Barry et al.*, 2007], in particular the Hangai dome [*Hunt et al.*, 2012]. Here detailed petrological studies have shown no age progression, small volumes, lack of mantle xenoliths above 1100°C [*Ionov*, 2002], and lack of high heat flow, estimated around 50–60 mW/m² [*Windley and Allen*, 1993]. These observations, combined with the absence of a major basalt province indicate the absence of a plume rooted in the lower mantle.

Two main hypotheses remain open, one invoking continental rifting or lithosphere delamination, due to the stresses induced by the indentation of the Indian plate and formation of the Himalaya [*Liu et al.*, 2001; *Ren et al.*, 2007]. This hypothesis has the advantage of explaining the SW-NE anisotropic shape of the volcanically active regions and the northeast-trending normal faults and northwest-trending thrusts [*Barry et al.*, 2007; *Hunt et al.*, 2012], but it does not explain the igneous gap in central Asia for example, Junggar basin, Tian Shan, northern Altai, Tarim basin, Qilian Shan–Nan Shan region, Qaidam basin, and the western part of South China [*Yin*, 2010]. How the far field regional stress would cause delamination is also unclear. Some recent research has proposed that a reactivation of suture zones might explain some of the intracontinental volcanism [*Gorczyk and Vogt*, 2013; *Gorczyk et al.*, 2013]. Indeed reconstruction of the late Mesozoic volcanism in the Great Xing'an range suggests the presence of a suture zone called Mongol-Okhotsk, which has been related to Mesozoic (160–120 Ma) volcanism [*Wang et al.*, 2006], but it is unclear which trigger would have reactivated this region after 80–100 Myrs from the delamination. Furthermore, intraplate volcanism cannot fully explain the three stripes A, B, and C, nor the compositional similarity of the basalts [*Barry et al.*, 2007].

The second hypothesis links the evolution of the surface volcanism to numerous upwellings from the stagnant subducted Pacific slab [e.g., *Lei and Zhao*, 2005; *Shen et al.*, 2008]. This so-called "Big Mantle Wedge" hypothesis arises from the observation that the Pacific plate has been continuously subducting during the past 60 Myrs, and possibly sliding for tens of million of years through the transition zone below a stable Asian plate. This hypothesis is supported by a number of seismological [*Zhao et al.*, 2009; *Tang et al.*, 2014], geochemical [*Kuritani et al.*, 2013] and geodynamical studies [*Richard and Iwamori*, 2010], however it leaves open how hydrous and carbon-rich phases would remain stable at transition zone mantle conditions and be released after 10–20 Myrs from their subduction.

Figure 5.2 shows that the regions displaying volcanism only partially overlap with the stagnant slab as imaged by mantle tomography since only regions B and C are above stagnant slab, while region A is beyond the edge of the imaged slab. The presence of a volcanism near stagnant slab edges has been already observed for the Columbia basalt, in the Central-Western Mediterranean region and the proto-Philippine plate [*Faccenna et al.*, 2010]. A check-board test by *Li and van der Hilst* [2010] has shown that tomographic inversions of the transition zone above latitude 45°N is unreliable using P-wave travel time, which adds uncertainty in the identification of the exact slab edge. This implies that we can potentially only detect the geometry of the stagnant slab at present time, while during the Cenozoic it might have extended farther than today.

In this work, we focus on whether the Big Mantle Wedge hypothesis is a viable candidate for explaining volcanism in regions B, C, and in particular A. Using numerical models, we investigate which mechanism would generate the geomorphological features of the surface volcanism in the Cenozoic. Geological field data [*Barry et al.*, 2007; *Hunt et al.*, 2012] suggest that the detailed distribution of the volcanic cones at the surface is compatible with a regular spacing of about 200 km (Figure 5.3). We show that such a pattern spontaneously emerges from the instability of a ~20-km thick layer of randomly distributed partially molten diapirs, and how only such a cluster would be able to rise to the surface and win equilibration. In the last sections we discuss the conditions and implications of the second hypothesis for the circulation of carbonates and hydrous rocks in the transition zone.

5.2. PETROLOGICAL SETTING

A radical new paradigm has been introduced by *Bercovici and Karato* [2003], who have proposed that the ability of wadsleyite to store much more water than the other mantle minerals might imply the existence of a ~10-km-thin water-filter layer of molten silicate just above the 410-km phase transition. This layer is predicted to be stable since melt phases are denser than solid ones at very high pressures. While recent measurements of electric conductivity in the upper mantle support the existence of the hypothesized higher water content [e.g., *Karato*, 2006; *Kelbert et al.*, 2009], a different way to test this hypothesis is to numerically model the small features of the plumes rising from the lying slabs and from the hydrated mantle in the transition zone, and predict observable geological features. These "cold" plumes, in fact, are different from the ones arising from the deep mantle, and their structures have to reflect the size and morphology of the original thin crust above the subducted lithosphere [*Faccenna et al.*, 2010].

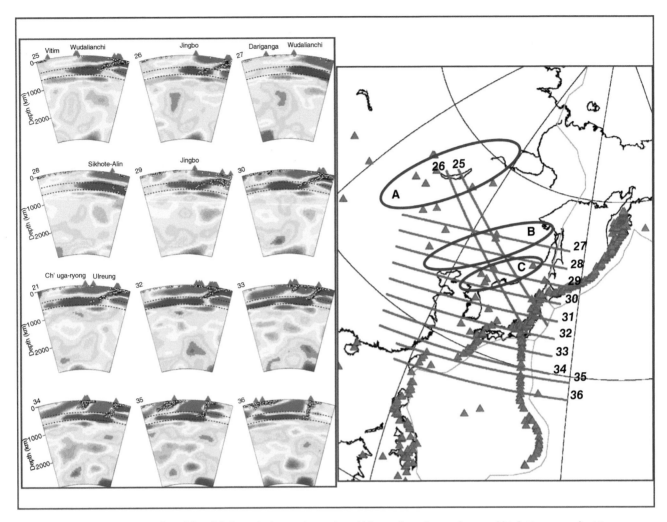

Figure 5.2 Tomography of the slab lying below Asia. Red and blue colors denote low and high P-wave velocities, respectively. Red triangles show active and Cenozoic volcanoes. The two dashed lines denote the 410 km and 660 km discontinuities. White dots are earthquakes that occurred within a 100 km width of each profile. Image modified after *Zhao et al.* [2011].

Ivanov and Litasov [2013] showed that if the carbonic sediments lying above the slab can reach the transition zone, they will form metastable phases that, upon reaching melting condition, can ascend either through the silicate matrix or the mobilized mantle rock. From the residual carbonatic layer above the slab in the transition zone, small volumes of partially molten rock would form diapirs at kilometer scale [*Litasov et al.*, 2013], they would mobilize the surrounding mantle and allow a sufficiently fast upwelling in order to overcome equilibration. We show in the Section 5.6 that our numerical models are compatible with this hypothesis when compared with the observed Cenozoic volcanism in Asia. This suggests that not only hydrated rocks could be responsible for the rock mobility in the upper mantle, but that a combination of carbonates and hydrous minerals might explain the observed volcanism.

5.3. BUOYANCY DRIVEN VISCOUS INSTABILITIES

There has been a trove of theoretical knowledge garnered on the development of secondary upwellings from an (inviscid or viscous) unstable buoyant fluid reservoir. We review here quickly a sampling of results related to size and morphology of buoyancy driven instability of cylinders and planes, and their relationship with the structure and the effective viscosity of the original source.

Biot [1961] first showed that a layer of buoyant material sandwiched between two heavier layers develops an instability that depends only on the thickness of the weaker layer and the viscosity ratio between the two materials. *Lister* [1989] showed that the withdrawal of a buoyant fluid from a sink in a heavier fluid is controlled by the surface tension between the two fluids. If the surface tension is null (or if the size of the unstable layers is

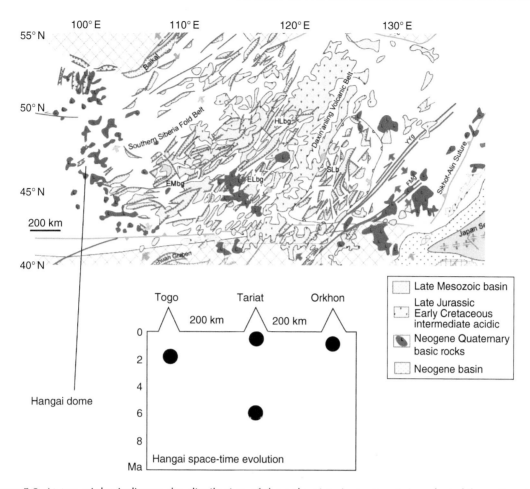

Figure 5.3 At top, violet indicates the distribution of the volcanism in eastern Asia, adapted from *Ren et al.* [2002]. Distribution of the western volcanic regions is distinctively patchy and regular. A scale (200 km) has been added to map. The inset refers to the study of the Hangai dome [*Hunt et al., 2012*], where the spacing of ~ 200 km between volcanic units has been detailed. Note the absence of age progression, which would be expected if the delamination hypothesis were applied to this system.

much greater than the wavelength induced by the surface tension), then instabilities emerge at different wavelengths, depending on very small initial perturbations of the boundary between the fluids [*Pozrikidis*, 1990].

Lister and Kerr [1989] have shown that the most unstable wavelength and the corresponding growth rate of a buoyant horizontal layer of a very low viscosity fluid are both inversely proportional to the cube root of the viscosity of the buoyant fluid. In a more recent analysis, *Lister et al.* [2011] found that the dimensionless wavelength emerging from an unstable horizontal cylinder is $l = 2\pi\lambda / k$, where k is the solution of $k^2 / \ln (2 / k) = 16\lambda$. At the first order, this is solved by $k \sim \lambda^{1/2}$, which is different from $k \sim \lambda^{1/3}$ relative to the 2D layer. In particular, specifically a cylindrical domain in which $\lambda = 0.1$, the wavelength of the instability is about $l = 10$ times its thickness, and becomes only 40 times for $\lambda = 0.001$. A horizontal layer instead develops an instability of $l = 11.76$ for $\lambda = 0.1$, and $l = 55$ for $\lambda = 0.001$.

These studies show that at a first order, the wavelengths of the instability of cylinders and horizontal layers are similar (around 10 for $\lambda = 0.1$ and around 50 for $\lambda = 0.001$). Therefore, focusing on modeling the instability of the cylinder case, which is easier to model, is sufficient for extrapolating the results to a horizontal layer. In particular, analytical studies could not account for the role of a heterogeneously distributed set of diapirs within a cylindrical or horizontal domain. Would a cylinder in which only 10% of the domain is filled with low-viscosity diapirs develop an instability? And at which wavelength? Our models can lend some insight on this problem. Here we apply them to the case of central Asian volcanism.

5.4. NUMERICAL SIMULATIONS

We use the boundary element method to study two-phase flow [*Pozrikidis*, 2002; *Morra et al.*, 2011] and understand the pathways, speed, volumes, and distribution at which partially molten rocks can propagate through the upper mantle and reach the Earth's surface. In particular, we focus on the wavelength displayed by the collective motion of a large number of diapirs placed at variable initial positions and show that the wavelength of the volcanism in eastern Asia is compatible with the one emerging from the rise of the layer of randomly distributed low-viscosity and low-density inclusions.

5.4.1. Numerical Method

The simulation of the collective behavior of a large number of diapirs presents formidable challenges due to the complex evolution of their boundaries. Exploiting the elliptic expression of the Stokes flow, which is typically used for modeling the migration of material in the mantle, we employ the Green functions associated to velocity and stress to formulate a Fredholm integral equation that can be solved on the discretized boundary between the diapirs and the rest of the mantle. This strategy has several advantages, among them (a) only the boundary needs be meshed and (b) the integrals can be accelerated using classic tree codes, or even more efficiently with the fast multipole method [*Greengard and Rokhlin*, 1987]. This approach has been already applied to geodynamic problems in *Morra et al.* [2009] and *Morra et al.* [2012].

In our numerical scheme, the surface of each diapir is discretized into linear triangular elements, originating a well-conditioned and dense linear system [*Zhu et al.*, 2006]. We employ the Generalized Minimal Residual Method (GMRES) algorithm to reach convergence, using fast multipole approach (FFM) to build a matrix multiplier operator, without calculating the associated full dense matrix, whose massive calculation scales like N^2, where N is the number of elements.

Forward time evolution is implemented updating the position of the vertices with an explicit first-order time-stepping. When the resulting mesh is too deformed, a remeshing technique based on the elimination of the triangle edges shorter than a prescribed size and splitting of the edges longer than a prescribed size. More details on the method used are given in *Morra et al.* [2007].

5.4.2. Model Setup

We ran two types of simulations. In the first case, we modeled the rise of a "cluster" of diapirs, randomly distributed inside a cube, varying (i) the size of the cube, *A*, (ii) the relative volume occupied by the diapirs (from 10% to 40%), and (iii) the internal viscosity of each diapir, down to two orders of magnitude less than the external viscosity. In the second type of models, we reproduced the evolution of the rise of an initially long "cylinder" of diapirs. To allow modeling of the longest possible cylinder and analyze the emerging

wavelength of the rising instability, we reduced this model to the minimal setup given by a $2 \times 2 \times n$ lattice. In the largest model n = 150 and the position of each diapir and its size has been randomized around the lattice position.

Many strategies exist for randomizing the position and size of particles and/or bubbles in a predefined domain [*Torquato*, 2005]. For the "cluster" models, we have adopted two methods. (1) In the first, 64 (4^3), 125 (5^3), and 216 (6^3) bubbles have been initially placed in lattice configuration and their x, y, z position has been randomized (uniform probability distribution) in the three directions, plus or minus 50% of the lattice spacing. The radius of each bubble has been randomized as well (uniform probability distribution again), from a minimum value to inhibit very small diapirs that would have increased the calculation time without substantially influencing the solution. (2) In the second set of "cluster" models, the position as well as the radius of the bubbles have been randomized within a 4^3 domain. In both cases, the algorithm is based on a reiteration of both position and radius of every couple of overlapping bubbles. This iterative approach has produced a large number of small bubbles for packings above about 25%.

For every "cluster," the average vertical velocity has been extracted from the velocity of the centroid of each boundary element using the representative volumetric as a weighting factor. The speed has been renormalized by the total volume of the diapirs. For every model viscosity, gravity, and the differential density between diapir interior and mantle have been set to one.

Since the goal of the "cylinder" models is to extract the dominant wavelength of the rising instability, we have adopted the largest model that was computationally feasible with our resources, consisting of 600 diapirs organized in the cylindrical $2 \times 2 \times 150$ structure. Also in this case, the position has been randomly perturbed as well as the radius, to inhibit the rise of resonant artificial instabilities. The wavelength has been extracted by the model measuring the distance between the peaks at the last calculated time-step. Also for these models, the background viscosity, gravity, and differential density have been set to one.

5.5. MODEL RESULTS

5.5.1. Cluster Models

The results obtained with the simulations of the "cluster" models are available in Tables 5.1 and 5.2. Table 5.1 outlines the results of the models with three sets of presets number of diapirs (4^3, 5^3, 6^3) whose position has been perturbed from a lattice configuration. Table 5.2 summarizes the outcomes of the models with randomly placed diapirs restricted by a preset compaction (see section 5.4 for more details). The resulting rising speed for the two sets of

Table 5.1 Results of the models with 64, 125 and 216 diapirs organized in a perturbed lattice configuration and varying inner viscosity.

Rising velocity	Inner viscosity	Number of bubbles	Average distance	Minimum radius	Radius perturbation	Size of the cluster	Rising velocity* cluster size	Compaction	Real volume	Real compaction
0.054	0.01	64	6	1.5	1.0	26.5	1.431	0.155061	1910	0.102635
0.043	0.01	125	6	1.5	1.0	32.5	1.3975	0.155061	3780	0.14
0.036	0.01	216	6	1.5	1.0	38.5	1.386	0.155061	6610	0.141675
0.054	0.05	64	6	1.5	1.0	26.5	1.431	0.155061	2090	0.151186
0.043	0.05	125	6	1.5	1.0	32.5	1.3975	0.155061	3900	0.144444
0.036	0.05	216	6	1.5	1.0	38.5	1.386	0.155061	6650	0.142532
0.053	0.2	64	6	1.5	1.0	26.5	1.4045	0.155061	2010	0.145399
0.042	0.2	125	6	1.5	1.0	32.5	1.365	0.155061	4050	0.15
0.035	0.2	216	6	1.5	1.0	38.5	1.3475	0.155061	6630	0.142103
0.054	0.8	64	6	1.5	1.0	26.5	1.431	0.155061	2070	0.149739
0.042	0.8	125	6	1.5	1.0	32.5	1.365	0.155061	3690	0.136666
0.035	0.8	216	6	1.5	1.0	38.5	1.3475	0.155061	6770	0.145104
0.045	0.01	64	6	2.0	1.0	27	1.215	0.3	3080	0.222800
0.039	0.01	125	6	2.0	1.0	33	1.287	0.3	6520	0.241481
0.034	0.01	216	6	2.0	1.0	39	1.326	0.3	10600	0.227194
0.047	0.05	64	6	2.0	1.0	27	1.269	0.3	3420	0.247395
0.039	0.05	125	6	2.0	1.0	33	1.287	0.3	6380	0.236296
0.034	0.05	216	6	2.0	1.0	39	1.326	0.3	10650	0.228266
0.047	0.2	64	6	2.0	1.0	27	1.269	0.3	3350	0.242332
0.040	0.2	125	6	2.0	1.0	33	1.32	0.3	6390	0.236666
0.034	0.2	216	6	2.0	1.0	39	1.326	0.3	10840	0.232338
0.050	0.8	64	6	2.0	1.0	27	1.35	0.3	3320	0.240162
0.040	0.8	125	6	2.0	1.0	33	1.32	0.3	6420	0.237777
0.034	0.8	216	6	2.0	1.0	39	1.326	0.3	10360	0.222050

Table 5.2 Outcomes of the models with randomly placed diapirs with compaction varying between 0.1 and 0.4.

Rising velocity	Inner viscosity	Average distance	Minimum radius	Radius perturbation	Size of the cluster	Rising velocity* cluster size	Compaction	Real volume	Real compaction
0.058	0.05	6	2.0	2.0	28	1.624	0.1	1380	0.0998263
0.059	0.5	6	2.0	2.0	28	1.652	0.1	1360	0.0983796
0.045	0.05	6	2.0	2.0	28	1.26	0.4	5430	0.3927951
0.045	0.5	6	2.0	2.0	28	1.26	0.4	5450	0.3942418
0.048	0.5	6	2.0	2.0	28	1.344	0.3	4070	0.2944155
0.046	0.05	6	2.0	2.0	28	1.288	0.3	4130	0.2987557
0.047	0.5	6	2.0	2.0	28	1.316	0.35	4750	0.3436053
0.046	0.05	6	2.0	2.0	28	1.288	0.35	4740	0.3428819
0.050	0.05	6	2.0	2.0	28	1.4	0.2	2700	0.1953125
0.052	0.5	6	2.0	2.0	28	1.456	0.2	2850	0.2061631
0.053	0.5	6	2.0	2.0	28	1.484	0.15	2100	0.1519097
0.048	0.05	6	2.0	2.0	28	1.344	0.25	3410	0.2466724
0.051	0.05	6	2.0	2.0	28	1.428	0.15	2130	0.1540798
0.049	0.5	6	2.0	2.0	28	1.372	0.25	3400	0.2459490
0.054	0.5	6	2.0	2.0	28	1.512	0.125	1780	0.1287615
0.0565	0.05	6	2.0	2.0	28	1.582	0.125	1690	0.1222511

models are displayed in Figures 5.4 and 5.5, the first showing rising velocity vs. cluster size, the second rising velocity (renormalized by cluster size) vs. compaction.

Figure 5.6 shows several snapshots from four examples of the rise of a cluster of diapirs. As shown in works on gravity-driven two-phase flow [*Manga and Stone*, 1995], the diapirs at the bottom of the flow elongate and form the shape of a tail, while the diapirs on top flatten and assume a broad, round shape. The viscosity ratio between the interior and exterior of each diapir, λ, has

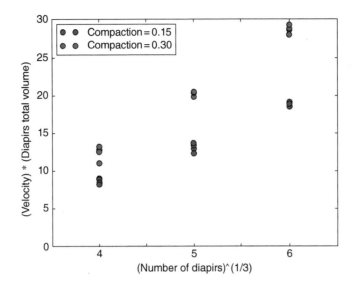

Figure 5.4 Rising velocities for a cluster of 4×4, 5×5 and 6×6, with concentrations of weakened (partially molten rocks) of 20% and 40%, and differential density diapir/exterior $\Delta_\rho = 1$. Different points of the same color refer to viscosity ratios, $\Delta_\mu = 0.01$, 0.05, 0.2, and 0.8 (relative to $\mu_{Mantle} = 1$), indicating that the role of the inner viscosity of the diapirs is minimal compared to the one of compaction and cluster size. As predicted by Stokes law, larger clusters rise faster. Greater compaction implies a larger buoyancy force and shows larger rising speed. As predicted by *Guazzelli and Hinch* [2010], the rising speed is less than proportional due to the velocity backflow.

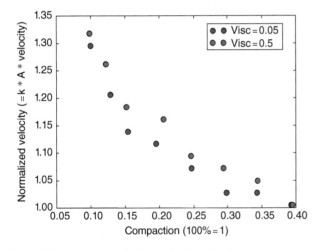

Figure 5.5 Renormalized rising velocities for clusters of compactions between 0.1 (10%) and 0.4 (40%). Renormalization is done multiplying the velocity by the length of the edge of the cube that initially confines the cluster, dividing it by the total mass of the diapirs and multiplying it by the geometry factor k, described in the main text. Setting gravity and viscosity to one, the renormalized velocity tends to one for large volume ratios.

We show that while the overall dynamics of the cluster is little influenced by the interior viscosity of the diapirs, the compaction, that is the relative volume occupied by the diapirs in the volume [*Bercovici et al.*, 2001], is the main factor determining the rising speed of the cluster. For low compactions (e.g. 10%), each diapir in the cluster is more mobile and the cluster dynamics more chaotic, while a high compaction causes an evolution into spherical cluster, similar to the morphology of an isoviscous plume [*Griffiths and Campbell*, 1990; *Ribe et al.*, 2007].

Since viscosity, gravity and differential density are set to one, the Stokes drag

$$F = 6\pi\eta RV = 6\pi RV$$

which at equilibrium compensates the gravity force

$$F = (4/3)\pi R^3 \Delta\rho g = (4/3)\pi R^3$$

where R is the radius of the sphere, η is the fluid viscosity, $\Delta\rho$ is the differential density, g is gravity, V is the velocity, λ is the viscosity ratio, A is the edge length, l is the wavelength of the instability (also L can be used, maybe it is more readable), k is a constant determined in the text, F is the buoyancy force, c is the compaction, and n is the cube root of the number of diapirs in a cluster. The asymptotic velocity is therefore $V = (2/9)*R^2$, and the characteristic timescale is $T = R/V = 9/2R$ [e.g., *Morra et al.*, 2010].

For a nonrigid sphere characterized by a ratio between interior and exterior viscosity $\lambda = \eta/\eta_0$, the renormalized, asymptotic speed for a sphere is [e.g., *Pozrikidis*, 1990]

only a modest effect on the overall dynamic. The results for $\lambda = 0.05$ differ from the ones that have diapirs one order of magnitude more viscous ($\lambda = 0.5$) mainly because the morphology of the tails in the former is more flexible and undulates while accommodating the flow, while the more viscous diapirs assume a straighter shape, reflecting a slower adaptation to the transient flow.

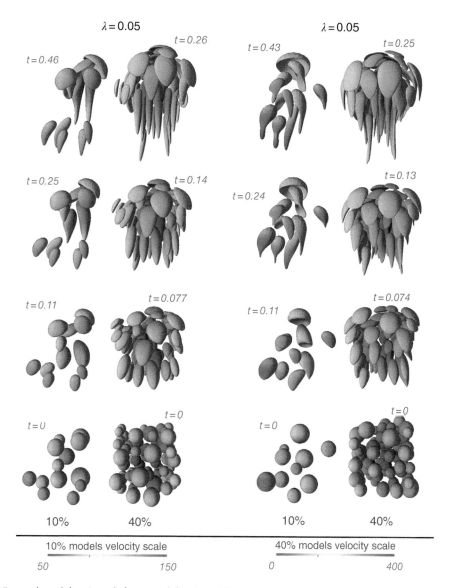

Figure 5.6 Examples of the rise of clusters of diapirs with varying viscosity ratios ($\lambda = 0.5$ and $\lambda = 0.05$) between interior and exterior of the diapir. If the internal viscosity of the diapir is low, its deformation is more pronounced and it rises slightly faster. Greater clusters rise faster as predicted by the Stokes law, but the speed increase is not exactly proportional to the size and mass of the cluster since higher concentrations induce a speed decrease, in agreement with the classic work of *Batchelor* [1972]. "t" is the non-dimensional time.

$$V = (2/3) R^2 (1+\lambda)/(1+3\lambda)$$

whose prefactor varies from 2/9 for infinite λ (rigid sphere) to 2/3 for $\lambda = 0$ (negligible interior viscosity), crossing the isoviscous case, $\lambda = 1$, in which case the prefactor is 1/3.

For a cluster of a bubbles with the average size r randomly placed in a cube of edge length A, the asymptotic velocity requires a coefficient k to be expressed in function of A. Assuming that the drag is the same of a sphere with the same volume, for example

$$A^3 = (4/3)\pi R^3$$

one can write

$$6\pi R V = k A V$$

Given the compaction c, the buoyancy force can be written as

$$F = c A^3 g \Delta \rho = c A^3$$

from which we obtain a reference velocity

$$V = c A^2 / k$$

that we used to renormalized the speed in Figure 5.6.

Figure 5.5 shows the relationship between rising speed and cluster size, this last indicated by n, where the cluster is formed by n^3 diapirs, each randomly displaced from a regular spacing and with randomized radiuses. The different overlapping points refer to distinct λ, ranging from 0.01 to 0.8. The first observation is that the interior viscosity of the diapir plays a minor role for V. A second result is that the rising velocity is at a first order proportional to compaction, as predicted by the above relationship, however, large clusters display a slower speed, which we interpret as due to back-flow to a large cluster, in agreement with *Batchelor* [1972]. Finally, the results are compatible at the first order with a quadratic dependency from n, as predicted by the fact that n is proportional to A.

Figure 5.6 shows the renormalized velocity $V_R = kV/cA^2$ in function of the compaction c for the second set of models in which the position and the radius of each diapir has been completely randomized. This approach requires re-randomizing the position and the size of each couple of overlapping diapirs until a fit is found. Large compactions imply a larger number of smaller diapirs, to allow sufficient filling of the space. In the figure, one can see a weak but clear dependency of the rising velocity from λ, since the cluster of less-viscous diapirs rises slower than the highly viscous one. The stronger feature, however, is the smooth decay, almost exponential, of V_R vs. c, toward the asymptotic solution $V_R = 1$. This unpredicted result allows one to precisely estimate the rising velocity of a cluster of diapirs, given compaction, buoyancy, and exterior viscosity. For low concentrations, greater velocities can be reached, but also a greater dependency on the initial conditions has been found, triggering a larger variability of the cluster speed. This is in agreement with the model outcomes shown in Figure 5.4, where one observes that a smaller number of diapirs can collectively cooperate, if aligned, or rise slower following parallel trajectories. The decay is also in agreement with results obtained for the deposition of suspensions, where the deposition velocity for very low compaction decays closer to 10%, as we also find [e.g., the review of *Guazzelli and Hinch*, 2010, Figure 5.7].

5.5.2. Discrete Cylinder Models

For the cylinder models, we tested several 2×2×n models, with n equal to 100 and 150, and with λ equal to 0.1 and 0.01. In Figure 5.7, we show sections from two viewpoints of the last time-step for the model with $n = 150$ and $\lambda = 0.1$. Along with the other cylindrical simulations, it shows the emergence of an instability at a wavelength approximately equal to 10 times the thickness of the cylinder. Although the compaction of the cylinder is not higher than 10%, with well-spaced diapirs, the emerging wavelength is in strong agreement with the one of a homogeneous cylinder, as found by analytical calculations and laboratory tests (see section 5.3).

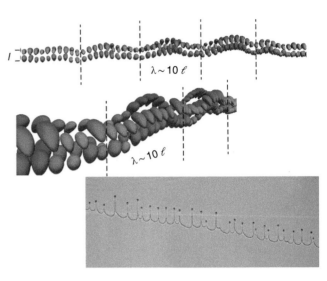

Figure 5.7 At top, see a section of a long cylinder of rising diapirs modeled with the BEM method. One clearly observes that instabilities emerge with a wavelength about 10 times greater than the layer thickness. Below, the figure shows a smaller section of the same model from a different perspective. The inset at bottom is taken from the laboratory experiment by *Lister and Kerr* [1989], that has inspired our simulation. We show that the same instability that they have found for a continuous homogenous cylinder is valid for a discrete distribution of diapirs distributed cylindrically.

Our tests show that the emerging wavelength is independent from both the relative size of each diapir and the distance between the different diapirs. The wavelength is instead proportional to the thickness of the cylinder of diapirs, characterized by a 10x factor. This long instability is perturbed by spurious ones by increasing the scatter between the diapirs and by randomizing their size, however the 10× instability finally always emerges

Similar to the cluster models, we observed that an increase in inner viscosity of each diapir does not increase the wavelength of the instability. This implies that, at least for the low concentrations (<10%) that we tested, the interior viscosity of each diapir cannot be extrapolated from the length emerging instability, which depends only on the cylinder thickness.

The largest run that we tested was on 64 processors for 24 hours. With this computational limit, we could not extrapolate the wavelength for a flat layer, which would have required an order of magnitude more computing power, however, the analytical analysis summarized in section 5.3 tells us that this instability is expected to be very close to the one of a cylindrical. We also ran some models of toroidal structures of diapirs, obtaining the same wavelength instabilities observed for the straight cylinder case. These structures can appear from the evolution of a mature plume head.

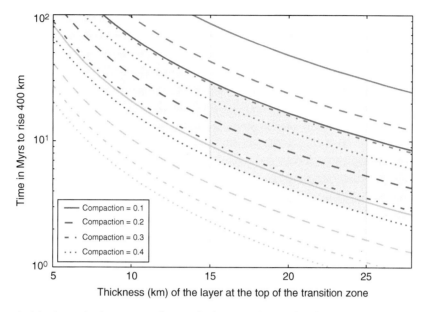

Figure 5.8 Time (in Myrs) required to rise 400 km vs. thickness (in km) of the departure layer of the diapirs. The plot is extrapolated from the results summarized in Figure 5.6, assuming density difference between the diapir and the background $\Delta\rho = 50$ kg/m³ and viscosity of the matrix 10^{18} Pas. The area shaded in blue indicates the average values between a minimum and maximum realistic diapir size. The four types of lines (continuous, segmented, dot segmented, and dots) indicate different values of concentration ($c = 0.1, 0.2, 0.3, 0.4$). The yellow shaded area indicates the model in which the unstable layer on top of the transition zone is around 20 km thick. We assumed that the maximum allowed time for preventing thermal equilibration is about 10 Myrs.

5.6. DISCUSSION

Both the cluster and cylindrical models illustrated in sections 5.4 and 5.5 show that the collective motion of diapirs of partially molten mantle depends on the initial compaction of the diapirs. If the diapirs initially form an isolated assembly, they will rise with a speed that depends almost linearly on the compaction and for constant compaction, quadratically to the linear extension of the cluster (Figure 5.4). If many diapirs are distributed within a cylindrical or horizontal flat domain, even with low concentration ($c < 0.1$) they will rise forming instabilities characterized by wavelength about equal to 10 times the cylinder (or layer) thickness. Our models therefore predict that the initially cylindrical (or flat layer) domains will develop into clusters [see the picture of the experiment in Figure 5.7; *Lister et al.*, 2011] and will manifest at the surface as regularly spaced igneous zones at distances about 10 times the thickness of the layer from which they originate.

These results agree with a model in which partial melts bearing hydrated rocks and carbonates migrate from the stagnant slab caused by the subduction of the Pacific plate during the Cenozoic [*Li and van der Hilst*, 2010]. Our model predicts that hydrous and carbon-bearing rocks will pile at the top of the transition zone, just below the 410 km discontinuity [*Richard and Iwamori*, 2010].

5.6.1. Implications for the Subducted Carbonates in the Transition Zone

When the total H_2O content of the system exceeds the hydrous storage capacity in the minerals of the rock, partial melting occurs [*Karato*, 2006]. In systems with CO_2, melting is determined by carbonate stability and presence of alkali, but it is almost insensitive to CO_2 content [*Litasov et al.*, 2013]. Not only dehydration, but also decarbonation-dehydration melting and freezing can induce partial melts at the conditions of the transition zone, implying that subducted carbonates rather than water are responsible for the upwellings.

The low-density carbonated melt must propagate through the mantle either via percolation or diapir ascent. Percolation can be very complex and its persistence depends on a wide range of parameters, possibly related to porosity waves [*Spiegelmann*, 1993; *Connolly and Podladchikov*, 1998]. However, in the mantle below a few hundred kilometers, no natural porosity exists, therefore free CO_2 or carbon-bearing melt phases can appear only at the grain boundary forming partially molten rocks, if allowed by the dihedral angle [*Karato et al.*, 2006]. In this work, we assume that porosity waves play a role only for smaller-scale features, and that partial melts reach the surface mainly through the rise of an entire diapir.

5.6.2. Implications for the Volcanism in Central and Northeastern Asia

Regional studies of volcanism in Mongolia and around the Baikal lake have found that the distribution of volcanic provinces is regular on thousands-of-kilometers-wide regions. Detailed studies of the Hangai dome [*Hunt et al.*, 2012] have detected a regular spacing between the major magmatic fields (Togo, Tariat, and Orkhon), denoting spatial variations in melting beneath the dome in the order of about 200 km (Figure 5.3).

The geochemical analysis of the lavas relative to the Hangai dome does not display correlations between different eruptions. Fractional crystallization is therefore not supported as a viable mechanism for justifying the regular spacing, which must be therefore caused by deeper processes. This is in agreement with the lack of residual phlogopite in the sources of Orkhon and Togo lavas [*Hunt et al.*, 2012] that would instead be present if the source of the partial melts were in the lithosphere.

Although the collision of India and Asia might be related to the large SW-NE geomorphological structures that characterize central and NE Asia (Figure 5.1), the present research does not support the hypothesis that central Asia volcanism would be caused by lithospheric delamination [*Menzies et al.*, 1993; *Xuzhang et al.*, 2008], associated with the India-Asia collision [*Davies and Bunge*, 2006]. Other Asian volcanic zones (e.g., Gobi Altai) are chemically and isotopically similar to the Hangai dome and suggest a common origin, incompatible with the delamination hypothesis [*Zorin et al.*, 2006; *Barry et al.*, 2007]. *Johnson et al.* [2005] proposed that mantle plumes might be responsible for the magmatism near the Baikal rift.

Our results support the hypotheses that these upwellings emerge from a ~20-km-thick layer in the upper mantle. There small (1 km and less) sparse diapirs of partially molten rock would slowly form due to migration from the stagnant Pacific slab [*Richard and Iwamori*, 2010], just below the 410-km phase transition. Following this line of argument, we can explain the geochemical differentiation in the same volcanic unit in Mongolia [*Hunt et al.*, 2012] by the rise of a cluster of diapirs, instead of a single one. Furthermore, our models find that diapirs do not merge, but only interact mechanically without mixing.

The quadratic dependence of the rising speed vs. cluster size for a given concentration demonstrates that while a single small diapir (100 m–1 km) cannot rise sufficiently fast to overcome equilibration, the collective emerging velocity can become greater by several orders of magnitude. Figure 5.8 shows three families of plots of the time required for the collective diapirs to rise 400 km, vs. the thickness of the layer from which we predict they arise, at the top of the transition zone. We used here differential density between partial melts and a mantle of 50 kg/m^3 and background

viscosity of 10^{18} Pas. It is important to observe that the present model predicts a linear relationship between rising speed and differential density and inverse relationship between rising speed and background viscosity.

Looking at the three families of models in Figure 5.8: in one the instability rises vertically from the unstable layer, in one it envelops an area of mantle above, and in the third, a more probable intermediate case appears that we shaded in blue. For each family of models, four types of lines are shown (continuous, segmented, dot segmented, and dots), which refer to concentrations (0.1, 0.2, 0.3, 0.4). The yellow-shaded area indicates thickness values between 15 and 25 km. Our results show that to rise from the transition zone in less than 10 Myrs, the unstable layer must be about 20 km thick, shown by the overlapping blue and yellow region. A thinner layer would produce much smaller diapirs, which would equilibrate before reaching the surface. The same result could be however reached with a much smaller mantle viscosity, although the tested value, 10^{18} Pas, is already very low.

We therefore propose that the volcanic A, B, and C regions might have been similarly formed from upwelling from the transition zone, as confirmed by high electrical conducting below the entire region [*Kelbert et al.*, 2009], associated with the presence of volatiles. However, regions B and C, which are above the lying slab, might have been formed as a consequence of fluid release from slab bending at the transition zone [e.g., *Faccenda*, 2014]. Why volcanic regions B and C and separated is not clear from our model. The splitting could be due to a dynamic effect during the rise of the melts [e.g., *Kincaid et al.*, 2013].

Region A is located just beyond the edge of the lying slab. As shown in *Faccenna et al.* [2010], volcanism commonly appears near slab edges when they are lying in the transition zone. We therefore propose that the volcanism in region A emerges from a volatile rich transition zone where fluids propagate from stagnant slab [e.g., *Richard and Iwamori*, 2010]. Our calculation shows that clusters of partially molten diapirs will rise if the volatile-rich region is about 20 km thick. The location of the volcanism is likely also related to the strength and thickness of the continental lithosphere, which explains their emergence in that proximity to the existing Mongol-Okhotsk suture zone [*Wang et al.*, 2006].

5.6.3. Implications on C-O-H Circulation in the Upper Mantle

There has been much speculation on the quantity and nature of carbon reservoirs in the deep Earth, because of their involvement in the evolution of life at the Earth's surface and as a possible origin for life inside planetary interiors [*Hazen and Schiffries*, 2013]. The amount of carbon present in the Earth's interior is

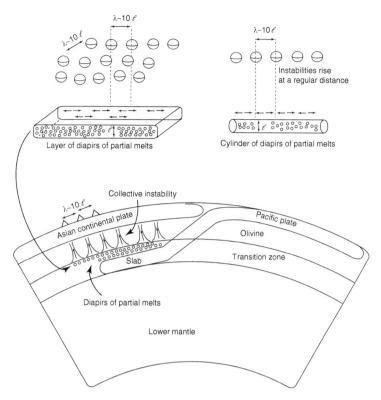

Figure 5.9 Geodynamic scenario of the dynamics at the transition zone. Fluids from hydrous-bearing and carbonbearing rocks rise from the slab to the top of the transition layer. There they diffuse and are advected by the sliding slab below. In the transition zone, the fluid-rich and possibly partially molten 100 m–1 km sized diapirs collectively form upwellings, spaced about 10 times the thickness of the layer. If the layer is at least ~15–20 km thick, the diapirs will rise and reach the surface (see Figure 5.8).

highly uncertain, as it is not known the amount left after Earth's formation as well as many of the characteristics of the present evolution of the carbon cycle in the deep Earth [*Walter et al.*, 2011].

While part of the carbon emerges through volcanic eruption and degassing, it also partly reaches the transition zone and even lower-mantle depths [*Marty et al.*, 2013], as proven by the analysis of natural diamonds found in Kimberlities [*Shirey et al.*, 2013]. At such extreme conditions, recent experiments have shown that CO_2 assumes both fluid and solid forms [*Dasgupta*, 2013] and decomposes in carbon and oxygen at Pressure-Temperature (P-T) mantle conditions corresponding to about 1000 km depth [*Oganov et al.*, 2013; *Shatskiy et al.*, 2013]. *Litasov et al.* [2013] have proposed that diapirs could emerge from subducted carbonate deposits within the subducting oceanic crust at transition zone depths. Our models show how even small amounts of carbonates spaced within the subducted crust could collectively form large-scale diapirs and therefore reach the necessary rising speed to win equilibration and reach the surface (Figure 5.8).

Observation of the weak S-wave velocity anomaly beneath U.S. East Coast combined with geodynamic insight has suggested that water hydrated plumes might rise from the transition zone [*Van der Lee et al.*, 2008]. Also, the analysis of the electrical conductivity of wadsleyite in the transition zone [*Huang et al.*, 2005; *Manthilake et al.*, 2009] accords well with the presence of highly hydrated minerals in the transition zone, although anomalous conductivity could be caused by carbon-rich materials [*Dasgupta and Hirshmann*, 2010].

Mantle rocks lower their melting point by several hundred degrees if water is added. Incompatible components enhance this effect, making melting likely to happen under a broad range of conditions [*Karato*, 2006]. Geochemical observation of water in basalts is compatible with the electrical conductivity of a water-rich (~0.1 wt. %) mantle transition zone and a less wet asthenosphere [*Karato*, 2011]. A 20-km thick layer below 410 km would therefore be in agreement with field observation, numerical models, and geochemical and geophysical observation. Our models show that such a region might also be characterized by patches of partially molten rocks distributed in a 20-km layer, close to a pudding model [*Rüpke et al.*, 2006]. The patchy regions might also develop from the migration of water-rich eclogitic crustal layers [*Karato*, 2011]. These delaminated thin crustal layers increase electrical conductivity, explaining observation.

Several other observations can also lend support to the presence of melting in the mantle. Wetting properties of carbonatite systems at deep-mantle conditions influence pressure solution creep when interacting with olivine. New carbon-bearing Mg-Fe compounds even at lower mantle depths have been recently found by mineral physics experiments [*Boulard et al.*, 2011], suggesting the presence of stable hosts for carbon at depth. Mantle melting is also a viable mechanism for sequestering highly incompatible elements (e.g., hydrogen, helium, argon) and account for several geochemical paradoxes [*Karato et al.*, 2006]. From the geodynamic point of view, a partially molten layer above stagnant slabs favors the horizontal sliding of the slab reducing the mantle viscous drag. This is a viable mechanism for the accommodating the slab shortening due to the fast subduction of the Pacific plate.

5.7. CONCLUSIONS

Intracontinental Cenozoic volcanism in central and northeastern Asia is divided in regularly spaced provinces spread over several regions with distances of around 1000 km long with a breadth of hundred of kilometers (Figure 5.1, regions A, B, and C). From the available maps of volcanism in Mongolia and near the Baikal lake, one can estimate a spacing between volcanic cones around 200 km [*Hunt et al.*, 2012]. Geochemical analyses of the melts from different provinces show similar composition, thus suggesting an analogue formation mechanism in the entire region [*Barry et al.*, 2007].

Mantle tomography identifies a slab lying above the 660-km discontinuity, related to the subduction of the Pacific plate [*Lei and Zhao*, 2005], partially overlapping the volcanically active regions (Figure 5.2). The Big Mantle Wedge hypothesis offers an interpretation at least for the volcanism in regions B and C, which could be caused by upwellings from the Pacific slab when it bends at the transition zone [*Zhao et al.*, 2011]. Volcanism in region A, however, requires a more complex formation mechanism since it does not exactly overlap the stagnant slab. Because mantle tomography is more uncertain below central Asia [*Li and van der Hilst*, 2010], we cannot reconstruct the precise morphology of the flat slab.

We model the formation and the rise of clusters of small diapirs (~1 km of diameter and less), randomly distributed on top of the transition zone, whose existence is suggested by the petrological model of *Bercovici and Karato* [2003], geophysical observables [*Karato et al.*, 2006], but whose thickness and detailed characteristics are unclear [*Karato*, 2011]. We show that the rising speed of a cluster of diapirs is proportional to the square of the size of the cluster and inversely proportional to compaction (the ratio between the volume of partially molten region and the solid mantle). Overall our combined models show that the minimum thickness of the layer where diapir clusters originate is 15–20 km, in order to rise sufficiently rapidly (Figure 5.8), and that a 20-km-thick layer would generate volcanism with 200 km spacing, since the natural wavelength of the instability of such a layer is 10 times greater than the layer thickness.

As illustrated in Figure 5.9 we therefore propose that most of the central and northeastern Asian volcanism originates from an approximately 20-km-thick layer of water-rich and possibly carbon-rich casually distributed diapirs, formed above the stagnant slab and below to the 410 km olivine-wadsleyite phase transition. This is supported by high electrical conductivity observed at transition zone depths below the region where zones A, B, and C are located [*Kelbert et al.*, 2009]. Regions B and C are above the bending of the slab at the transition zone, suggesting that a greater fluid release related to their deformation [*Faccenda*, 2014] might have triggered them. Splitting into B and C regions could also be due to the dynamics of the upwelling flow [e.g., *Kincaid et al.*, 2013].

We propose that the volcanism in region A arises from the combined effect of several mechanisms: (1) the hydration of the transition zone due to long-term slab subduction and stagnation [*Richard and Iwamori*, 2010], which formed the volatile-rich reservoir; (2) the instability of the 20-km-thick hydrated layer near the slab edge [e.g., *Faccenna et al.*, 2010], and (3) penetration of the partially molten diapirs through a weakened continental plate in proximity to the existing Mongol-Okhotsk suture zone [*Wang et al.*, 2006].

ACKNOWLEDGMENT

D. Yuen is supported by a grant from the geochemistry program of the National Science Foundation. G. Morra is supported by the pFund LEQSF-EPS(2014)-PFUND-385 and Research Competitive Subprogram LEQSF(2014-17)-RD-A-14 from the Board of Regents of Louisiana.

BIBLIOGRAPHY

Barry, T. L., A. V. Ivanov, S. V. Rasskazov, E. I. Demonterova, T. J. Dunai, G. R. Davies, and D. Harrison (2007), Helium isotopes provide no evidence for deep mantle involvement in widespread Cenozoic volcanism across Central Asia, *Lithos*, *95*, 415–424.

Batchelor, G. K. (1972), Sedimentation in a dilute dispersion of spheres, *J. Fluid Mech.*, *52*(02), 245–268.

Bercovici, D., and S.-I. Karato (2003), Whole-mantle convection and the transition-zone water filter, *Nature*, *425*(6953), 39–44.

Bercovici, D., Y. Ricard, and G. Schubert (2001), A two-phase model for compaction and damage: 1. General theory, *J. Geophys. Res.*, doi:10.1029/2000JB900430.

Biot, M. A. (1961), Theory of folding of stratified viscoelastic media and its implications in tectonics and orogenesis, *Bull. Geol. Soc. Amer.*, *72*, 1595–1620.

Bonatti, E. (1985), Punctiform initiation of seafloor spreading in the Red Sea during transition from a continental to an oceanic rift, *Nature*, *316*(6023), 33–37.

Boulard, E., A. Gloter, A. Corgne, D. Antonangeli, A.-L., Auzende, J.-P. Perrillat, and G. Fiquet (2011), New host for carbon in the deep Earth, *Proc. National Acad. Sci.*, *108*(13), 5184–5187.

Brown, M. (2010), The spatial and temporal patterning of the deep crust and implications for the process of melt extraction, *Phil. Trans. R. Soc. A*, *368*(1910), 11–51.

Connolly, J. A. D., and Y. Y. Podladchikov (1998), Compaction-driven fluid flow in viscoelastic rock, *Geodinamica Acta*, *11*(2–3), 55–84.

Corti, G., M. Bonini, S. Conticelli, F. Innocenti, P. Manetti, and D. Sokoutis (2003), Analogue modelling of continental extension: A review focused on the relations between the patterns of deformation and the presence of magma, *Earth-Sci. Rev.*, *63*(3–4), 169–247,

Dasgupta, R. (2013), Ingassing, storage, and outgassing of terrestrial carbon through geologic time, *Rev. Mineral Geochem.*, *75*, 183–229.

Dasgupta, R., and M. M. Hirschmann (2010), The deep carbon cycle and melting in Earth's interior, *Earth Planet. Sci. Lett.*, *298*(1), 1–13.

Davies, I. H., and H. P. Bunge (2006), Are splash plumes the origin of minor hotspots? *Geology*, *34*(5), 349–352.

Faccenda, M. (2014), Water in the slab: A trilogy, *Tectonophysics*, doi:10.1016/ j.tecto.2013.12.020.

Faccenna, C., T. W. Becker, S. Lallemand, Y. Lagabrielle, F. Funiciello, and C. Piromallo (2010), Subduction-triggered magmatic pulses: A new class of plumes? *Earth Planet. Sci. Lett.*, *299* (1), 54–68.

Gorczyk, W., and K. Vogt (2013), Tectonics and melting in intra-continental settings, *Gondwana Res.*, doi:10.1016/j.gr.2013.09.021.

Gorczyk, W., B. Hobbs, K. Gessner, and T. Gerya (2013), Intracratonic geodynamics, *Gondwana Res.*, *24*, 838–848.

Greengard, L., and V. Rokhlin (1987), A fast algorithm for particle simulations, *J. Computational Phys.*, *A*, *73*, 325–348.

Griffiths, R. W., and I. H. Campbell (1990), Stirring and structure in mantle starting plumes, *Earth Planet. Sci. Lett.*, *99*(1–2), 66–78

Guazzelli, É., and J. Hinch (2011), Fluctuations and instability in sedimentation, *Annual Rev. of Fluid Mech.*, *43*, 97–116.

Hazen, R. M., and C. M. Schiffries (2013), Why deep carbon? *Rev. Mineral. Geochem.*, *75*, 1–6.

Huang, X., Y. Xu, and S. Karato (2005), Water content in the transition zone from electrical conductivity of wadsleyite and ringwoodite, *Nature*, *434*(7034), 746–749.

Hunt, A. C., I. J. Parkinson, N. B. W. Harris, T. L. Barry, N. W. Rogers, and M. Yondon (2012), Cenozoic volcanism on the Hangai dome, Central Mongolia: Geochemical evidence for changing melt sources and implications for mechanisms of melting, *J. Petrol.*, *53*(9), 1913–1942.

Ionov, D. (2002), Mantle structure and rifting processes in the Baikal-Mongolia region: Geophysical data and evidence from xenoliths in volcanic rocks, *Tectonophysics*, *351*(1), 41–60.

Ivanov, A. V., and K. D. Litasov (2013), The deep water cycle and flood basalt volcanism, *International Geol. Rev.*, *56*(1), 1–14.

Johnson, J. S., S. A. Gibson, R. N. Thompson, and G. M. Nowell (2005), Volcanism in the Vitim volcanic field, Siberia: Geochemical evidence for a mantle plume beneath the Baikal rift zone, *J. Petrol.*, *46*(7), 1309–1344.

Karato S. I. (2006), Influence of hydrogen-related defects on the electrical conductivity and plastic deformation of mantle minerals: A critical review, *Geophys. Monograph, AGU.*, 168, 113.

Karato, S. I. (2011), Water distribution across the mantle transition zone and its implications for global material circulation, *Earth Planet. Sci. Lett.*, *301*, 413–423.

Karato, S. I., D. Bercovici, G. Leahy, G. Richard, and Z. Jing (2006), The transition-zone water filter model for global material circulation: Where do we stand? *Earth's Deep Water Cycle, Geophys. Monograph Series*, *168*.

Kelbert, A., A. Schultz, and G. Egbert (2009), Global electromagnetic induction constraints on transition-zone water content variations, *Nature*, *460*, 1003–1006.

Kincaid, C., K. A. Druken, R. W. Griffiths, and D. R. Stegman (2013), Bifurcation of the Yellowstone plume driven by subduction-induced mantle flow, *Nature Geosci.*, *6*(5), 395–399.

Kuritani, T., J. -I. Kimura, E. Ohtani, H. Miyamoto, and K. Furuyama (2013), Transition zone origin of potassic basalts from Wudalianchi volcano, northeast China, *Lithos*, *156–159*, 1–12.

Lebedev, S., T. Meier, and R. D. van der Hilst (2006), Asthenospheric flow and origin of volcanism in the Baikal Rift area, *Earth Planet. Sci. Lett.*, *249*(3–4), 415–424.

Lei, J., and D. Zhao (2005), P-wave tomography and origin of the Changbai intraplate volcano in Northeast Asia, *Tectonophysics*, *397*(3), 281–295.

Li, C., and R. D. van der Hilst (2010), Structure of the upper mantle and transition zone beneath Southeast Asia from traveltime tomography, *J. Geophys. Res.*, doi:10.1029/ 2009JB006882.

Lister, J. R. (1989), Selective withdrawal from a viscous two-layer system, *Fluid Mech.*, *198*, 231–254.

Lister, J. R., and R. C. Kerr (1989), The effect of geometry on the gravitational instability of a buoyant region of viscous fluid, *Fluid Mech.*, *202*, 577–594

Lister, J. R., C. R. Kerr, N. J. Russell, and A. Crosby (2011), Rayleigh-Taylor instability of an inclined buoyant viscous cylinder, *J. Fluid Mech.*, *671*, 313–338.

Litasov, K. D., A. Shatskiy, E. Ohtani (2013), Earth's mantle melting in the presence of c-o-h-bearing fluid, in S. Karato (ed.), *Physics and Chemistry of the Deep Earth*.

Litasov, K. D., A. Shatskiy, E. Ohtani, and G. M. Yaxley (2012), Solidus of alkaline carbonatite in the deep mantle, *Geology*, *41*(1), 79–82.

Liu, J., J. Han, and W. S. Fyfe (2001), Cenozoic episodic volcanism and continental rifting in northeast China and possible link to Japan Sea development as revealed from K-Ar geochronology, *Tectonophysics*, *339*(3), 385–401.

Manga, M., and H. A. Stone (1995), Collective hydrodynamics of deformable drops and bubbles in dilute low Reynolds number suspensions, *J. Fluid Mech.*, *300*, 231–263.

Manthilake, M. A. G. M., T. Matsuzaki, T. Yoshino, S. Yamashita, E. Ito, and T. Katsura (2009), Electrical conductivity of wadsleyite as a function of temperature and water content, *Phys. of the Earth and Planetary Interiors*, *174*, 10–18.

Marsh, B. D. (1979), Island arc development: Some observations, experiments, and speculations, *J. Geol.*, 687–713.

Marty, B., et al. (2013), Primordial origins of Earth's carbon, *Rev. Mineral. Geochem.*, *75*, 149–181.

Menzies, M. A., W. Fan, and M. Zhang (1993), Palaeozoic and Cenozoic lithoprobes and the loss of >120 km of Archaean lithosphere, Sino- Korean craton, China, *Geological Society, London, Special Publications*.

Miyashiro, A. (1986), Hot regions and the origin of marginal basins in the western Pacific, *Tectonophysics*, *122*(3), 195–216.

Morra, G., D. A. Yuen, L. Boschi, P. Chatelain, P. Tackley, and P. Koumoutsakos (2010), The fate of the slabs interacting with a smooth viscosity discontinuity in the mid lower mantle, *Phys. of the Earth and Planetary Interiors*, *180*, 271–282.

Morra, G., L. Quevedo, and R. D. Müller (2012), Spherical dynamic models of top-down tectonics, *Geochem. Geophys. Geosys.*, *13*(3), Q03005.

Morra, G., L. Quevedo, D. Yuen, and P. Chatelain (2011), Ascent of bubbles in magma conduits using boundary elements and particles, *Procedia Computer Science*, *4*, 1554–1562.

Morra, G., P. Chatelain, P. Tackley, and P. Koumoutsakos (2007), Large scale three-dimensional boundary element simulation of subduction, *Comp.l Sci.- ICCS 2007*, 1122–1129.

Morra, G., P. Chatelain, P. Tackley, and P. Koumoutsakos (2009), Earth curvature effects on subduction morphology: Modeling subduction in a spherical setting, *Acta Geotechnica*, *4*(2), 95–105.

Oganov, A. R., R. J. Hemley, R. M. Hazen, and A. P. Jones (2013), Structure, bonding, and mineralogy of carbon at extreme conditions, *Rev. Mineral. Geochem.*, *75*, 47–77.

Peng, Z., R. Zartman, K. Futa, and D. G. Chen (1986), Pb-, Sr- and Nd-isotopic systematics and chemical characteristics of Cenozoic basalts, eastern China, *Chemical Geology: Isotope Geoscience Section*, *59*, 3–33.

Petit, C., J. Déverchère, E. Calais, V. Sankov, and D. Fairhead (2002), Deep structure and mechanical behavior of the lithosphere in the Hangai-Hövsgö region, Mongolia: New constraints from gravity modeling, *Earth Planet. Sci. Lett.*, *197*(3), 133–149.

Pozrikidis, C. (1990), The instability of a moving viscous drop, *J. Fluid Mech.*, doi:10.1017/S0022112090001203.

Pozrikidis, C. (2002), *Boundary Element Method*, Chapman and Hall/CRC, Boca Raton.

Ren, Y., E. Stutzmann, R. D. van der Hilst, and J. Besse (2007), Understanding seismic heterogeneities in the lower mantle beneath the Americas from seismic tomography and plate tectonic history, *J. Geophys. Res.*, *112*, B01302, doi:10.1029/2005JB004154.

Ren, J., K. Tamaki, S. Li, and Z. Junxia (2002), Late Mesozoic and Cenozoic rifting and its dynamic setting in Eastern China and adjacent areas, *Tectonophysics*, *344*, 175–205, doi:10.1016/S0040-1951(01)00271-2.

Ribe, N., A. Davaille, and U. Christensen (2007), Fluid Dynamics of Mantle Plumes, in J. R. Ritter and U. Christensen (eds.), *Mantle Plumes*, *1*, 1–48.

Richard, G. C., and H. Iwamori (2010), Stagnant slab, wet plumes and Cenozoic volcanism in East Asia, *Phys. of the Earth and Planetary Interiors*, *183*(1), 280–287.

Rüpke, L. H., J. P. Morgan, and J. E. Dixon (2006), Implications of subduction rehydration for Earth's deep water cycle, In: Jacobsen, S. D., and S.v.d. Lee (Eds.), Earth's Deep Water Cycle, American Geophysical Union, Washington, DC, pp. 263–276.

Shatskiy, A., K. D. Litasov, Y. M. Borzdov, T. Katsura, D. Yamazaki, and E. Ohtani (2013), Silicate diffusion in alkali-carbonatite and hydrous melts at 16.5 and 24 GPa: Implication for the melt transport by dissolution-precipitation in the transition zone and uppermost lower mantle, *Phys. of the Earth and Planetary Interiors*, *225*, 1–11.

Shen, X., Z. Huilan, and K. Hitoshi (2008), Mapping the upper mantle discontinuities beneath China with teleseismic receiver functions, *Earth Planets and Space (EPS)*, *60*(7), 713.

Shirey, S. B., P. Cartigny, D. J. Frost, S. Keshav, F. Nestola, P. Nimis, D. G. Pearson, N. V. Sobolev, and Walter, M. J. (2013), Diamonds and the geology of mantle carbon, *Rev Mineral Geochem*, *75*, 355–421.

Spiegelman, M. (1993), Physics of melt extraction: Theory, implications and applications, *Phil. Trans. R. Soc. Lond.*, *342*(1663), 23–41.

Tang, Y., M. Obayashi, F. Niu, S. P. Grand, Y. J. Chen, H. Kawakatsu, S. Tanaka, J. Ning, J. F. Ni (2014), Changbaishan volcanism in northeast China linked to subduction-induced mantle upwelling, *Nature Geosciences*, *7*, 470–475.

Torquato, S. (2005), *Random Heterogeneous Materials: Microstructure and Macroscopic Properties*, Springer-Verlag, New York.

Van der Lee, S., K. Regenauer-Lieb, and D. A. Yuen (2008), The role of water in connecting past and future episodes of subduction, *Earth Planet. Sci. Lett.*, *273*, 15–27.

Walter, M. J., S. C. Kohn, D. Araujo, G. P. Bulanova, C. B. Smith, E. Gaillou, S. B. Shirey (2011), Deep mantle cycling of oceanic crust: Evidence from diamonds and their mineral inclusions, *Science*, *334*(6052), 54–57.

Wang, F., X. H. Zhou, L. C. Zhang, J. F. Ying, Y. T. Zhang, F. Y. Wu, and R. X. Zhu, (2006), Late Mesozoic volcanism in the Great Xing'an Range (NE China): Timing and implications for the dynamic setting of NE Asia.

Whitehead, J. A., H. J. B. Dick, H. Schouten (1984), A mechanism for magmatic accretion under spreading centres, *Nature*, *312*(5990), 146–148.

Windley, B. F., and M. B. Allen (1993), Mongolian plateau: Evidence for a late Cenozoic mantle plume under central Asia, *Geology*, *21*(4), 295–298.

Yarmolyuk, V. V., E. A. Kudryashova, A. M. Kozlovsky, and V. M. Savatenkov (2007), Late Cretaceous-Early Cenozoic volcanism of Southern Mongolia: A trace of the South Khangai mantle hot spot, *J. Volcan. Seis.*, *1*(1), 1–27.

Yin, A. (2010), Cenozoic tectonic evolution of Asia: A preliminary synthesis, *Tectonophysics*, *488*(1), 293–325.

Zhao, D., S. Yu, and E. Ohtani (2011), East Asia: Seismotectonics, magmatism and mantle dynamics, *J. Asian Earth Sci.*, *40*, 689–709.

Zhao, D., Y. Tian, J. Lei, L. Liu, S. Zheng (2009), Seismic image and origin of the Changbai intraplate volcano in East Asia: Role of big mantle wedge above the stagnant Pacific slab, *Physics of the Earth and Planetary Interiors*, *173*(3), 197–206.

Zhou, X., and R. L. Armstrong (1982), Cenozoic volcanic rocks of eastern China: Secular and geographic trends in chemistry and strontium isotopic composition, *Earth Planet. Sci. Lett.*, *58*(3), 301–329.

Zhu, G., A. A. Mammoli, and H. Power (2006), A 3-D indirect boundary element method for bounded creeping flow of drops, *Engineering Analysis with Boundary Element*, *30*, 856–868.

Zhu, G., T. Gerya, and D. A. Yuen (2011), Melt evolution above a spontaneously retreating subducting slab in a three-dimensional model, *J. Earth Sci.*, *22*(2), 137–142.

Zorin, Yu. A., E. Kh. Turutanov, V. M. Kozhevnikov, S. V. Rasskazov, and A. I. Ivanov (2006), Cenozoic upper mantle plumes in east Siberia and central Mongolia and subduction of the Pacific plate, *Doklady Akademii Nauk*, *409*, 217–221.

6

Influence of Variable Thermal Expansivity and Conductivity on Deep Subduction

Nicola Tosi[1], Petra Maierová[2], and David A. Yuen[3]

ABSTRACT

Thickening of slabs subducted in the lower mantle is a robust feature from seismic tomography. Numerical models show clearly that, under suitable circumstances, deeply subducted plates experience buckling instabilities near the transition zone. These can lead to the formation of large-scale folds, which ultimately cause the lower-mantle portion of slabs to be significantly thicker than the upper-mantle part. Such models, however, generally rely on the assumption of constant thermodynamic properties. Here we present subduction simulations that include pressure-, temperature-, and phase-dependent thermal expansivity (α) and conductivity (k), and demonstrate that these parameters exert a dramatic effect on the dynamics of lower-mantle slabs. The decrease of α with pressure causes slabs to lose buoyancy during their descent. This strongly enhances buckling and reduces sinking speed, thereby promoting thermal diffusion and a significant spreading of the slab thermal anomalies in the lower mantle, which is also facilitated by the increase of k with pressure. On the one hand, the temperature dependence of k slightly influences the thermal structure of the slabs, but not their large-scale dynamics. On the other hand, the temperature dependence of α strongly influences surface and sinking velocities, rendering subduction remarkably faster than in cases with constant expansivity.

6.1. INTRODUCTION

Seismic tomographic models clearly indicate that present-day plate tectonics is characterized by a variety of complex styles of lithospheric subduction [e.g., *Li et al.*, 2008]. Slabs can stagnate near the transition zone, being flattened horizontally at various depths [*Fukao et al.*, 2009] as in the case of Japan [*Huang and Zhao*, 2006], the

Mediterranean region [*Piromallo and Morelli*, 2003], or western North America [*Schmid et al.*, 2002]. Following stagnation or because of continental collision, slab break-off can occur. Indeed, detached portions of subducted plates associated with old collision zones have been imaged with great detail throughout the lower mantle in several regions [*van der Meer et al.*, 2010]. Yet another important mode of subduction is the continuous foundering of the oceanic lithosphere, or at least parts of it, into the deep mantle, as in the case of the subduction zones of Tonga-Kermadec [*van der Hilst*, 1995], Central America [*van der Hilst et al.*, 1997], or Java-Sumatra [*Pesicek et al.*, 2008]. Tomographic images of deeply subducted lithosphere also demonstrate that fast seismic anomalies in the lower mantle are significantly broader than in the upper mantle, thus indicating that slabs tend to thicken upon subduction below the 660-km discontinuity [*Li et al.*, 2008; *Wortel and Spakman*, 2000].

[1]*Department of Astronomy and Astrophysics, Technische Universität Berlin, Germany; Department of Planetary Physics, German Aerospace Center (DLR), Berlin, Germany*

[2]*Center for Lithospheric Research, Czech Geological Survey, Prague, Czech Republic*

[3]*School of Environment Studies, China University of Geosciences, Wuhan, China; Minnesota Supercomputing Institute and Department of Earth Sciences, University of Minnesota, Minneapolis, Minnesota, USA*

Subduction Dynamics: From Mantle Flow to Mega Disasters, Geophysical Monograph 211, First Edition.
Edited by Gabriele Morra, David A. Yuen, Scott D. King, Sang-Mook Lee, and Seth Stein.
© 2016 American Geophysical Union. Published 2016 by John Wiley & Sons, Inc.

Slab thickening often has been attributed to the formation of buckling instabilities that arise as the lithosphere subducting through the upper mantle approaches the more viscous lower mantle. This process resembles the way periodic folds form when a sheet of viscous fluid is poured onto a rigid surface [Ribe, 2003]. The dynamics of deep subduction and slab thickening has been the subject of numerous studies, most of which are based on numerical simulations. Gaherty and Hager [1994] used a simple model of a two-layer, compositionally stratified slab sinking vertically through the mantle. They were the first to demonstrate that a viscosity increase of a factor of 30 to 100 across the upper-lower mantle interface causes slabs to thicken by up to a factor of 5 because of pure shear occurring above the viscosity discontinuity, and folding occurring below it. They also reported that compositional effects associated with the presence of a layer of negatively buoyant ecologite overlying the mantle lithosphere are of secondary importance when compared to thermal effects, thus suggesting that the large-scale dynamics of deep subduction is largely controlled by thermal buoyancy. As clearly evidenced by both analogue [Guillou-Frottier et al., 1995] and numerical experiments [Christensen, 1996], the presence of a viscosity contrast is essential for the formation of buckling instabilities. In addition, these studies showed that the latter can also be promoted by trench rollback, which in turn is responsible for the generation of stagnant slabs [e.g., Enns et al., 2005; Billen, 2010; Yoshioka and Naganoda, 2010].

More recently, Ribe et al. [2007] reanalyzed the problem of slab buckling showing that the amount of thickening inferred from tomographic images of the Australian plate subducting beneath Java, and of the Cocos plate subducting beneath Central America is well reproduced by the scaling laws for periodic buckling derived by Ribe [2003] using a numerical model based on thin-sheet theory. More sophisticated simulations of lower mantle subduction have been performed by Běhounková and Čížková [2008] who used a 2D Cartesian setting to model the subduction of an oceanic plate accounting for the effects of the two major phase transitions at 410-km and 660-km depth, and for linear, nonlinear, and Peierl's creep via a mixed rheological formulation. They confirmed that a viscosity jump across the 660-km interface is necessary for subducting slabs to undergo buckling instabilities. However, they also observed that their formation requires both the consideration of the additional negative buoyancy provided by the upward deflection of the 410-km boundary, and of a relatively low-yield stress (100 MPa) in order for the plate to be more easily deformable. Similar calculations have been conducted by Lee and King [2011], who concluded that slab buckling becomes a general feature of the dynamics of plates subducting in the lower mantle when trench rollback is relatively slow and

the viscosity jump across the 660-km boundary relatively large (at least a factor of 30).

All the above studies relied on the assumption of constant thermodynamic parameters. The coefficients of thermal expansion (a) and conduction (k), however, vary considerably both with pressure and temperature. Taking into account their variability becomes particularly important when modeling processes, such as deep subduction, that involve large-scale deformations occurring throughout the entire mantle depth. While considering changes of thermal expansivity and conductivity upon compression is relatively common in global models of mantle convection [e.g., Leitch et al., 1999; Hansen et al., 1993; Zhong et al., 2006, Foley and Becker, 2009; Nakagawa et al., 2010; Tan et al., 2011; Tosi et al., 2010; Tosi and Yuen, 2011; Miyauchi and Kameyama, 2013], the study of the combined effects of their pressure and temperature dependence has received much less attention. Hauck et al. [1999] analyzed the effects of pressure-, temperature- and phase-dependent thermal conductivity (based on the model of Hofmeister [1999]) on the temperature distribution of the subducted lithosphere in the upper mantle. Using simple steady-state models of thermal transport, Hauck et al. [1999] pointed out that discontinuities in conductivity occurring across phase changes affect the thermal field within the slab, with possible implications for the transformation of metastable olivine to spinel. Recently, Maierová et al. [2012] showed numerical models of kinematically prescribed subduction supplemented by a conductivity distribution based on a more sophisticated mineralogy. They reported that although the use of specific compositions can significantly affect the slab temperature distribution, the accompanying effects on thermal buoyancy are minor, thus indicating that variable conductivity plays a secondary role in controlling the large-scale dynamics of subducted plates.

In the framework of fully dynamic models, Schmeling et al. [2003] and Ghias and Jarvis [2008] presented simple simulations of isoviscous thermal convection in 2D Cartesian and 2D cylindrical geometry, respectively, which accounted for the pressure and temperature dependence of a based on the experimental data of Fei [1995] for olivine only. In particular, both studies highlighted the importance of the temperature dependence of a in causing a significant increase of surface heat flow and surface velocity with respect to models with constant or only pressure-dependent expansivity. Recently, Tosi et al. [2013] presented models accounting for the pressure, temperature, and phase dependence of both expansivity and conductivity in a more general context of global thermal convection with a viscoplastic rheology and multiple phase transitions. When using variable a and k, they observed in this study an "increased propensity to local layering, which favors slab stagnation in the transition zone and subsequent thickening in the lower mantle".

Here we extend the work of *Tosi et al.* [2013] by investigating in detail the influence of pressure-, temperature-, and phase-dependent thermal expansivity and conductivity on the dynamics of subduction in the lower mantle. Using two-dimensional, purely thermal models of oceanic subduction in plane layer geometry, we focus in particular on the conditions that promote the formation of buckling instabilities and the accompanying slab thickening in relation to the role of the major phase transitions and of the upper-lower mantle viscosity contrast.

6.2. METHOD AND MODEL

6.2.1. Governing Equations

We have solved the conservation equations of mass, linear momentum, and thermal energy with two phase transitions in a 2D rectangular box with an aspect ratio of four using the extended Boussinesq approximation [e.g., *Christensen and Yuen*, 1985]. In dimensionless form, these read

$$\nabla \cdot \boldsymbol{u} = 0 \qquad (6.1)$$

$$-\nabla p + \nabla \cdot \tau = Ra\left(\alpha T - \sum_{i=1}^{2} \frac{Rb_i}{Ra}\Gamma_i\right)e_z \qquad (6.2)$$

$$\frac{DT}{Dt} = \nabla \cdot \left(k\nabla T\right) + Diu_z\left(T + T_s\right) + \frac{Di}{Ra}\Phi$$
$$+ \sum_{i=1}^{2} Di\frac{Rb_i}{Ra}\frac{D\Gamma_i}{Dt}\gamma_i\left(T + T_s\right), \qquad (6.3)$$

where $\boldsymbol{u}$ is the velocity vector and u_z its vertical component (chosen positively downward), p the dynamic pressure, $\tau \equiv \eta(\nabla \boldsymbol{u} + \nabla \boldsymbol{u}^t)$ the stress tensor, η the viscosity, T the temperature and T_s its surface value, t the time, e_z the unit vector in the vertical direction, α the thermal expansivity, D/Dt the material time derivative, k the thermal conductivity, $\Phi \equiv \tau : \nabla \boldsymbol{u}$ the viscous dissipation, Γ_i and γ_i ($i = 1, 2$) the phase function and Clapeyron slope of the i-th phase transition, Ra and Rb_i the thermal and phase Rayleigh numbers, and Di the dissipation number. The continuity equation (6.1) describes the conservation of mass for an incompressible fluid. In the momentum equation (6.2), where an infinite Prandtl number is assumed, the first and second terms on the right-hand side account for buoyancy forces due to temperature differences and deflections of phase boundaries, respectively.

We considered two phase transitions: at 410-km depth from olivine to spinel ($i = 1$) and at 660-km depth from spinel to perovskite and periclase ($i = 2$). The effect of the i-th phase transition is taken into account through the traditional approach of *Christensen and Yuen* [1985] using a phase function:

$$\Gamma_i = \frac{1}{2}\left(1 + \tanh\left(\frac{z - z_i(T)}{w}\right)\right), \qquad (6.4)$$

where z is the depth, w is the width of the phase transition, and $z_i(T)$ is the temperature-dependent depth at which the transition occurs, i.e.

$$z_i(T) = z_i^0 + \gamma_i\left(T - T_i^0\right), \qquad (6.5)$$

where z_i^0 and T_i^0 are the reference depth and reference temperature of the i-th phase transition. In the thermal energy equation (6.3), heat advection on the left-hand side is balanced by heat diffusion, adiabatic heating/cooling, viscous dissipation, and latent heat release/absorption due to phase changes on the right-hand side. Definition and values of all relevant parameters and variables are reported in Table 6.1.

To solve equations (6.1)–(6.3), we used our finite-volume code YACC (Yet Another Convection Code, https://bitbucket.org/7nic9/yacc-yet-another-convection-code), which has been extensively benchmarked for non-Boussinesq thermal convection [*King et al.*, 2010], for thermal convection with viscoplastic rheology [*Tosi et al.*, 2015], and employed in several other studies [*e.g., Tosi et al.*, 2010; *Tosi et al.*, 2013].

6.2.2. Model Setup

We modeled the subduction of an oceanic plate throughout the entire mantle depth. Figure 6.1a shows a schematic view of the model setup. We considered a Cartesian box with a depth of 2890 km and width of 11,560 km. Sidewalls are reflective and free-slip. The bottom boundary is isothermal and free-slip. The top boundary is isothermal; a no-slip condition is prescribed on the right half to simulate a static overriding plate, and either kinematic or free-slip conditions are prescribed on the left half. The former are used to initiate subduction by setting a constant surface velocity of 5 cm/yr over a time interval of 4 Myr. Afterward, the slab sinks under its own weight by assuming a free-slip upper surface of the subducting plate. The trench is treated as a fixed weak zone of low viscosity extending down to a depth of 150 km, with a dip-angle of 30° and a width of 20 km. The subducting plate is assumed to form at a ridge located at top-left corner of the domain and to be 120 Myr old at the trench; its initial temperature distribution is obtained from the half-space

Table 6.1 Model parameters.

Symbol	Description	Value / Definition
Various parameters		
ρ_0	Reference density	3300 kg m^{-3}
α_0	Reference thermal expansivity	3×10^{-5} K^{-1}
k_0	Reference thermal conductivity	2.5 W m^{-1} K^{-1}
η_0	Reference viscosity	2.7×10^{20} Pa s
T_0	Reference temperature for η_0	1600 K
P_0	Reference pressure for η_0	3×10^6 Pa
T_s	Surface temperature	273 K
T_p	Potential temperature	1623 K
c_p	Heat capacity	1200 J kg^{-1} K^{-1}
g	Gravity acceleration	9.8 m s^{-2}
D	Length scale	2890 km
ΔT	Temperature scale	1390 K
R	Gas constant	8.314 J K^{-1} mol^{-1}
Phase transitions parameters		
w	Phase transitions width	20 km
z_0^1	Reference depth for transition 1	410 km
T_0^1	Reference temperature for transition 1	1794 K
γ_1	Clapeyron slope for transition 1	3×10^6 Pa K^{-1}
$\Delta \rho_1$	Density jump across transition 1	273 kg m^{-3}
z_0^2	Reference depth for transition 2	660 km
T_0^2	Reference temperature for transition 2	1908 K
γ_2	Clapeyron slope for transition 2	-2.5×10^6 Pa K^{-1}
$\Delta \rho_2$	Density jump across transition 2	342 kg m^{-3}
Rheological parameters for the upper mantle		
A_{df}	Prefactor for diffusion creep	1.92×10^{-10} Pa^{-1} s^{-1}
E_{df}	Activation energy for diffusion creep	3×10^5 J mol^{-1} Pa^{-1}
V_{df}	Activation volume for diffusion creep	6×10^{-6} m^3 mol^{-1}
A_{ds}	Prefactor for dislocation creep	2.42×10^{-16} Pa^{-n} s^{-1}
E_{ds}	Activation energy for dislocation creep	5.4×10^5 J mol^{-1} Pa^{-1}
V_{ds}	Activation volume for dislocation creep	15×10^{-6} m^3 mol^{-1}
Rheological parameters for the lower mantle		
A_{df}	Prefactor for diffusion creep	3.65×10^{-15} Pa^{-1} s^{-1}
E_{df}	Activation energy for diffusion creep	2.08×10^5 J mol^{-1} Pa^{-1}
V_{df}	Activation volume for diffusion creep	2.5×10^{-6} m^3 mol^{-1}
A_{ds}	Prefactor for dislocation creep	6.63×10^{-32} Pa^{-n} s^{-1}
E_{ds}	Activation energy for dislocation creep	2.85×10^5 J mol^{-1} Pa^{-1}
V_{ds}	Activation volume for dislocation creep	1.37×10^{-6} m^3 mol^{-1}
Rheological parameters (same for the whole mantle)		
n	Stress exponent	3.5
σ_y	Ductile strength	5×10^8 Pa
C	Cohesion	10^6 Pa
φ	Internal friction angle	30°
Nondimensional numbers		
Ra	Thermal Rayleigh number	$\rho_0^2 c_p \alpha_0\, g\, \Delta T\, D^3 / (\eta_0\, k_0)$
Rb_i	Phase Rayleigh number	$\rho_0^2 c_p\, g \Delta \rho_i\, D^3 / (\eta_0\, k_0)$
Di	Dissipation number	$\alpha_0\, g\, D / c_p$

cooling model [*Turcotte and Schubert*, 2002]. This is also applied to the overriding plate, which has a uniform thickness corresponding to a 120 Myr old lithosphere.

In the shallow mantle, the initial temperature distribution consists of a conductive profile prescribed from the surface down to the base of the subducting and overriding plates, below which an adiabatic profile $T_a(z)$ is assumed that satisfies the equation

$$\frac{dT_a}{dz} = \frac{\alpha g T}{C_p}, \tag{6.6}$$

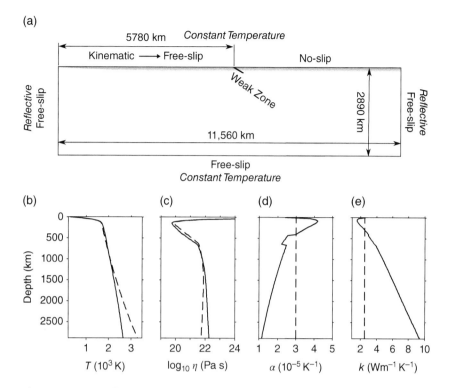

Figure 6.1 (a) Schematic view of the model domain, (b) conductive upper thermal boundary layer combined with an adiabatic temperature profile assuming constant (dashed line) and variable (solid line) thermal expansivity, (c) viscosity profile resulting from diffusion and dislocation creep computed along the temperature distribution of panel b, (d) constant thermal expansivity (dashed line) and thermal expansivity profile computed along the temperature distribution of panel b (solid line), (e) constant thermal conductivity (dashed line) and thermal conductivity profile computed along the temperature distribution of panel b (solid line).

where g is the acceleration of gravity and C_p the heat capacity. When the thermal expansivity α is constant, the solution to equation (6.6) is simply

$$T_a = T_p \exp\left(\frac{\alpha g z}{C_p}\right), \qquad (6.7)$$

where T_p is the potential temperature (dashed line in Figure 6.1b). Depending on the model, the thermal expansivity can be a function of temperature and/or depth (see equation (6.12)). In these cases, the adiabatic profile was obtained by integrating equation (6.6) numerically (solid line below the thermal boundary layer in Figure 6.1b).

We discretized the domain with a structured grid with variable resolution. This varies from a maximum of ~28 km near the sidewalls to a minimum of ~5 km in the trench region, which guarantees that the weak zone is fully resolved. Simulations conducted by doubling the number of grid points in the central part of the domain through which slabs sink did not exhibit significant differences. Furthermore, even though the effective viscosity of the weak zone is known to be important in controlling the coupling between subducting and overriding plate

[e.g., *Androvičová et al.*, 2013], tests performed using different viscosity values showed that our choice of 10^{20} Pa s was sufficiently low to ensure the necessary decoupling.

6.2.3. Transport and Thermodynamic Parameters

We treated the mantle as a viscous fluid with a composite rheology accounting for diffusion and dislocation creep, as well as Byerlee-type plastic yielding. The effective viscosity is thus a function of temperature, depth and strain-rate, and is obtained as the harmonic average of the viscosities corresponding to the individual deformation mechanisms, i.e.

$$\eta = \left(\frac{1}{\eta_{df}} + \frac{1}{\eta_{ds}} + \frac{1}{\eta_{pl}}\right)^{-1}. \qquad (6.8)$$

The expressions (in dimensional form) of the diffusion creep viscosity η_{df}, dislocation creep viscosity η_{ds}, and plastic viscosity η_{pl} read respectively:

$$\eta_{df} = A_{df}^{-1} \exp\left(\frac{E_{df} + P V_{df}}{RT}\right) \qquad (6.9)$$

$$\eta_{ds} = A_{ds}^{-1/n} \varepsilon_{II}^{(1-n)/n} \exp\left(\frac{E_{ds} + PV_{ds}}{nRT}\right) \quad (6.10)$$

$$\eta_{pl} = \frac{\min\left(\sigma_y, C + \sin(\varphi)P\right)}{2\varepsilon_{II}}, \quad (6.11)$$

where A_*, E_*, and V_* are prefactor, activation energy and activation volume, respectively, which differ for diffusion and dislocation creep, and for the upper and lower mantle, n is the stress exponent, $P = \rho_0 gz$ the hydrostatic pressure, R the gas constant, and ε_{II} the second invariant of the strain-rate tensor. In equation (6.11), σ_y is the ductile strength, while the term $C + \sin(\varphi)P$ represents the brittle strength, where C is the cohesion, φ the angle of friction, and z the depth. Numerical values of all relevant parameters are reported in Table 6.1. In particular, it should be noted that we employed the same activation parameters as adopted by *Běhounková and Čížková* [2008] who used different values for the upper and lower mantle on the base of the works of *Karato and Wu* [1993] and *Yamazaki and Karato* [1991], respectively. In Figure 6.1c, we plotted two viscosity profiles computed along the temperature profiles of Figure 6.1b assuming a strain rate of 10^{-15} s^{-1}, no viscosity jump at the interface between upper and lower mantle, and accounting for diffusion and dislocation creep. An upper and a lower cutoff were applied when the viscosity exceeded 10^{26} Pa s and became lower than 10^{18} Pa s.

Following *Tosi et al.* [2013], we took into account both the temperature and depth dependence of the coefficients of thermal expansion and conduction. The former is parameterized as

$$\alpha(T,z) = \left(a_0 + a_1 T + a_2 T^{-2}\right)\exp(-a_3 z), \quad (6.12)$$

where T is the absolute temperature in K, z the depth in m, and a_i ($i = 0,..,3$) are phase-dependent coefficients whose numerical values are reported in Table 6.2. The parameterization (12) was derived by fitting results of

Table 6.2 Phase-dependent coefficients for the parameterizations of thermal expansivity (equation (6.12)) and thermal conductivity (equation (6.13)).

Coefficients	Upper mantle	Transition zone	Lower mantle
Thermal expansivity			
a_0 (K^{-1})	3.15×10^{-5}	2.84×10^{-5}	2.68×10^{-5}
a_1 (K^{-2})	1.02×10^{-8}	6.49×10^{-9}	2.77×10^{-9}
a_2 (K)	-0.76	-0.88	-1.21
a_3 (m^{-1})	1.27×10^{-6}	9.16×10^{-7}	3.76×10^{-7}
Thermal conductivity			
c_0 (W m^{-1} K^{-1})	2.47	3.81	3.48
c_1 (W m^{-2} K^{-1})	1.15×10^{-5}	1.19×10^{-5}	5.17×10^{-6}
c_2	0.48	0.56	0.31

first-principle quasiharmonic calculations [*Wentzcovitch et al.*, 2010], which are accessible online through the VLab open-source database compiled at the University of Minnesota (vlab.msi.umn.edu/resources/thermodynamics). For the upper mantle above the olivine-spinel boundary, we used data for forsterite. In contrast to *Tosi et al.* [2013], we did not distinguish wadsleyite from ringwoodite as they have very similar expansivities. We thus employed wadsleyite data for the whole transition zone. The thermal expansivity in the lower mantle was calculated assuming an assemblage consisting of 80% Mg-perovskite and 20% MgO-periclase. The solid line in Figure 6.1d shows the profile of α corresponding to the adiabatic temperature distribution of panel b (solid line) in comparison with the value of 3×10^{-5} K^{-1} (dashed line) that we used in simulations with constant expansivity.

The thermal conductivity was calculated from the following parameterization [*Tosi et al.*, 2013]

$$k(T,z) = \left(c_0 + c_1 z\right)\left(\frac{300}{T}\right)^{c_2}, \quad (6.13)$$

where the phase-dependent coefficients c_i ($i = 0,..,2$) are also reported in Table 6.2. Equation (6.13) fits thermal diffusivity measurements of upper-mantle phases by *Xu et al.* [2004]. As far as the lower mantle is concerned, it fits instead the experimental data by *Manthilake et al.* [2011], which take into account 3 mol% $FeSiO_3$ in perovskite and 20 mol% FeO in periclase. Similar to the thermal expansivity, we used the conductivity of wadsleyite for the entire transition zone. The solid line in Figure 6.1e shows the profile of k corresponding to the adiabatic temperature distribution of panel b (solid line) in comparison to the value of 2.5 W m^{-1} K^{-1} (dashed line) that we used in simulations with constant conductivity. It should be noted also that the coefficients appearing in equations (6.12) and (6.13) have been derived from calculations (in the case of α) and measurements (in the case of k) performed at given temperature and pressure, and not temperature and depth. As discussed by *Tosi et al.* [2013], the pressure was converted to depth by assuming depth-dependent density and gravity acceleration according to the Preliminary Reference Earth Model [*Dziewonski and Anderson*, 1981].

In many respects, the setup adopted here is similar to those employed by *Běhounková and Čížková* [2008] and *Lee and King* [2011], the major difference being the use, in these two studies, of constant thermodynamic parameters. In addition, contrary to *Běhounková and Čížková* [2008], who employed tracing particles to account for a lubricating low-viscosity crustal layer, we considered a purely thermal, grid-based model as *Lee and King* [2011]. In contrast to the latter study, however, we accounted for non-Boussinesq terms in the thermal energy equation (6.3) and for the role of phase transitions in the buoyancy term of equation (6.2).

6.3. RESULTS

6.3.1. Subduction with Constant Thermal Expansivity and Conductivity

We start describing results from subduction simulations in which both the thermal expansivity and conductivity were held constant. We focused in particular on the role of phase transitions and of the lower- to upper-mantle viscosity contrast ($\Delta\eta_{LU}$ hereafter) in influencing the morphology and time scale of subduction. Table 6.3 contains a list of all the models we ran, with their names and characteristics. Figure 6.2 shows, in a 2000×2000-km-wide region surrounding the slab, temperature anomalies for nine models with $\Delta\eta_{LU}=1$ (panel a), 10 (panels b–e), and 30 (panels f–i), and different combinations of the two-phase transitions in which these are considered or neglected. If no additional viscosity contrast between upper and lower mantle is imposed (panel a), even considering phase transitions, the slab quickly sinks through the mantle, it conserves its thickness, and does not experience buckling instability. When raising the value of $\Delta\eta_{LU}$ to 10 or 30 (panels b and f), the combined effects of viscosity increase and both phase transitions induce buckling and thickening. However, as already noted by *Běhounková and Čížková* [2008], when the exothermic transition at 410-km depth is neglected (panels c and g), the slab sinks straight through the mantle, despite the presence of the endothermic phase transition at 660-km depth, which tends to obstruct mass exchange between upper and lower mantle. The fact that the olivine-spinel phase change is actually responsible for causing buckling instability is also evident when looking at the subsequent panels where either the transition at 660 km (panels d and h) or both (panels e and i) are neglected.

The shapes of the subducted plates obtained with $\Delta\eta_{LU}=30$ closely reflect those obtained with $\Delta\eta_{LU}=10$, although an important difference in the timescale is present. The three-fold increase of $\Delta\eta_{LU}$ causes the timescale of subduction to be ~2.8 times longer, which enhances the importance of thermal diffusion over advection. This affects the temperature of the slabs, which tend to be warmer with the larger value of $\Delta\eta_{LU}$. Because of the same effect, the slabs with higher $\Delta\eta_{LU}$ tend to be thicker. It is worth noting that the increase in thickness is visible throughout their entire length, not only in the lower mantle.

6.3.2. Subduction with Pressure- and Temperature-Dependent Thermal Expansivity and Conductivity

We repeated the numerical experiments described in the previous section including the effects of both the pressure and temperature dependence of thermal expansivity and conductivity. The results are shown in Figure 6.3. If no

viscosity contrast between upper and lower mantle is present, the subduction style is similar to that observed with constant thermodynamic properties (compare Figures 6.2a and 6.3a). However, as soon as a viscosity jump of 10 or 30 is taken into account (panels b–i), the effects of variable α and k become dramatic. All cases in which $\Delta\eta_{LU}=10$ (panels b–e) are clearly characterized by buckling instabilities (see also Figure 6.4), no matter whether or not the two phase transitions are taken into account. The reduction of α with depth causes the slab to lose buoyancy during its descent. This enhances buckling and reduces the sinking speed, thereby promoting thermal diffusion over advection and inducing a significant spreading of the slab thermal anomaly in the lower mantle, which is also facilitated by the increase of thermal conductivity with pressure. With $\Delta\eta_{LU}=30$, these effects are even more pronounced. For example, when both phase transitions are taken into account, the slab thickness in the lower mantle (identified here through the -200 K contour line of temperature anomaly) can become as high as ~600 km (panel f), although it must be noted that in this case subduction becomes extremely slow, with the slab needing up to 400 Myr to reach the mid-lower mantle.

In the simulations with constant thermal expansivity and conductivity, the presence of phase transitions (particularly of the exothermic one at 410 km depth) was pivotal in causing the slab to buckle. On the contrary, when α and k vary with pressure and temperature, phase transitions exert a smaller influence: when these are neglected, buckling instabilities still occur and the dynamics of the slab in the lower mantle is hardly distinguishable from the case when they are taken into account (compare, e.g., Figures 6.3b and 6.3e).

We can gain a more quantitative understanding of the effects described above and of the differences between models with constant and variable α and k by looking at Figure 6.4. Here, for the same models shown in Figures 6.2 and 6.3, we plotted the time series of the plate velocity as measured 100 km left of the trench at a depth of 20 km (panels a and d), and the velocity (panels b and e) and temperature (panel c and f) as a function of depth of a tracing particle, which, at the beginning of the free-slip phase, was positioned also 100 km left of the trench at 20-km depth. All simulations are initially characterized by an abrupt increase of the plate velocity. As clearly shown by the way black-blue and red-green lines tend to group, such increase is larger when the exothermic phase transition at 410-km depth is taken into account. In general, models that incorporate this phase change tend to exhibit a higher sinking velocity both at the surface (panels a and d) and throughout the mantle (panels b and e). Contrary to cases with constant thermodynamic properties (panel a), when variable expansivity and conductivity are considered, buckling instabilities have a

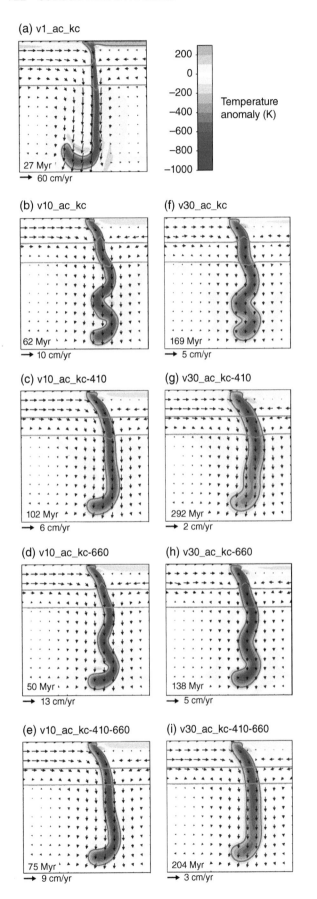

(a) v1_ac_kc

27 Myr
→ 60 cm/yr

Temperature anomaly (K)

200
0
−200
−400
−600
−800
−1000

(b) v10_ac_kc

62 Myr
→ 10 cm/yr

(c) v10_ac_kc-410

102 Myr
→ 6 cm/yr

(d) v10_ac_kc-660

50 Myr
→ 13 cm/yr

(e) v10_ac_kc-410-660

75 Myr
→ 9 cm/yr

(f) v30_ac_kc

169 Myr
→ 5 cm/yr

(g) v30_ac_kc-410

292 Myr
→ 2 cm/yr

(h) v30_ac_kc-660

138 Myr
→ 5 cm/yr

(i) v30_ac_kc-410-660

204 Myr
→ 3 cm/yr

clear expression in the time series of the plate velocity (panel d), in which the slab folding frequencies can be easily recognized (panels a and d). The latter are also affected by the 410-km transition whose presence tends not only to increase subduction speed but also to promote frequent slab folding at the base of the upper mantle. A comparison of solid black with solid blue lines, and of solid red with solid green lines in Figure 6.4d illustrates more quantitatively the effects on buckling exerted by the 660-km phase transition in the presence and absence of the 410-km transition, respectively. In both cases, the endothermic phase change tends to lengthen the buckling period. While in the absence of the 410-km transition (red and green lines) it is difficult to identify a well-defined periodicity, in its presence the buckling period increases from ~13 Myr to ~22 Myr depending on whether the 660-km transition is neglected or taken into account. Because of the pressure dependence of the thermal expansivity, models with variable α and k are characterized by a terminal subduction velocity in the lower mantle smaller than 2 cm/yr (panel e) and systematically lower than that observed in models with constant properties (up to 5 cm/yr as seen in panel b). As a consequence of the lower sinking speed, the temperature within the slab, as measured by a tracing particle, tends to be considerably hotter (about 200 K) when expansivity and conductivity are allowed to vary (panel f) in comparison to cases in which they are kept just constant (panel c).

6.3.3. Reference Model

In order to investigate in more detail the characteristics of deep subduction in the presence of variable α and k, we considered the model v10_azT_kzT (see Table 6.3) as a reference. Details of this model are shown in Figure 6.5, which presents the distribution of different fields after 121 Myr of evolution. In the upper mantle above the transition zone, the slab retains a high viscosity with respect to the background mantle (Figure 6.5c), with a peak difference as high as six orders of magnitude [e.g., *Billen and Hirth*, 2007]. As the slab progressively

Figure 6.2 Temperature anomalies for models with constant thermal expansivity and conductivity, different viscosity contrasts between lower and upper mantle ($\Delta\eta_{LU}$ = 1 in panel a, 10 in panels b–e, and 30 in panels f–i), and different combinations of phase transitions (both at 410- and 660-km depth in panels a, b, and f, only at 660 km in panels c and g, and only at 410 km in panels d and h). Arrows indicate flow directions and magnitude. Gray lines denote phase boundaries, and red lines refer to the −200 K contour of temperature anomaly. The size of the domain shown is 2000 × 2000 km. Model names and their features are listed in Table 6.3.

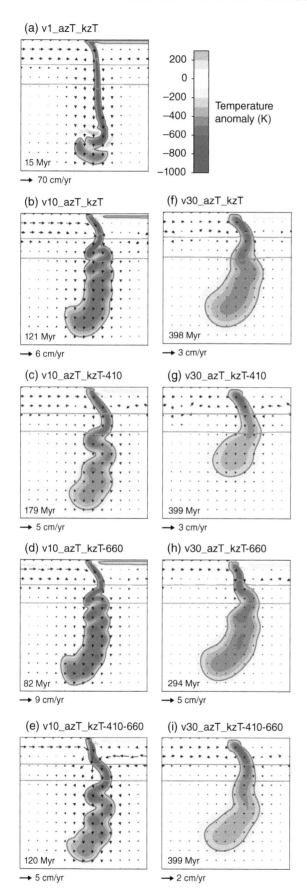

(a) v1_azT_kzT

Temperature anomaly (K)

200
0
-200
-400
-600
-800
-1000

15 Myr

→ 70 cm/yr

(b) v10_azT_kzT

121 Myr

→ 6 cm/yr

(c) v10_azT_kzT-410

179 Myr

→ 5 cm/yr

(d) v10_azT_kzT-660

82 Myr

→ 9 cm/yr

(e) v10_azT_kzT-410-660

120 Myr

→ 5 cm/yr

(f) v30_azT_kzT

398 Myr

→ 3 cm/yr

(g) v30_azT_kzT-410

399 Myr

→ 3 cm/yr

(h) v30_azT_kzT-660

294 Myr

→ 5 cm/yr

(i) v30_azT_kzT-410-660

399 Myr

→ 2 cm/yr

warms and spreads while sinking through the transition zone and in the lower mantle, only a thin highly viscous core is preserved, while the bulk of the thickened plate becomes weaker, with a viscosity only up to about two orders of magnitude higher than the surrounding mantle, in agreement with the requirement of limited strength of lower-mantle slabs imposed by analyses of the gravity field over subduction zones [e.g., *Moresi and Gurnis*, 1996; *Tosi et al.*, 2009]. In the upper mantle, plastic yielding dominates within the cold core of the slab. Dislocation creep acts instead around the slab in the upper mantle where temperature is relatively high but also shear stresses are significant [see e.g., *Billen and Hirth*, 2005]. A similar situation occurs in the deep lower mantle within the slab and in its immediate surroundings, while diffusion creep plays an important role below the 660-km discontinuity and away from the slab where stresses are low. The thermal expansivity and conductivity respectively decrease and increase with depth throughout the largest part of the mantle. Nevertheless, they also exhibit significant variations due to temperature. These are mostly evident in the upper mantle below the lithosphere and within the subducting slab, particularly above the 410-km discontinuity where the plate is coldest, and at the base of the transition zone, where cold material tends to accumulate as a consequence of buckling (Figure 6.5e and f).

The formation of buckling instabilities differs substantially from the case with constant expansivity and conductivity. Figure 6.6 shows the time evolution of subduction when α and k are constant (model v10_ac_kc, upper panels) and depend on depth and temperature (reference model v10_azT_kzT, bottom panels). In the latter case, subduction initially proceeds about twice as fast until the slab tip reaches the upper-lower mantle interface (see Section 3.4 and Figure 6.8). Afterward, the slab buckles there as if it were poured onto a rigid surface before starting to sink in the lower mantle as a thick folded block. On the contrary, in model v10_ac_kc, subducted material does never accumulate at 660-km depth. Although buckling instabilities are evident when looking at the bent shape of the slab tip, these do not lead to the formation of a thick blob as in the case of model v10_azT_kzT. On the one hand, buckling continues steadily when using variable α and k, and the folded slab keeps sinking through the lower mantle with a low velocity caused by the reduction of α with depth. On the other hand, when using constant α and k, the folding amplitude tends to diminish as the slab sinks to greater depths, with the negative

Figure 6.3 Same as Figure 6.2 but for models with pressure- and temperature-dependent thermal expansivity and conductivity prescribed according to equations (6.12) and (6.13). Model names and their features are listed in Table 6.3.

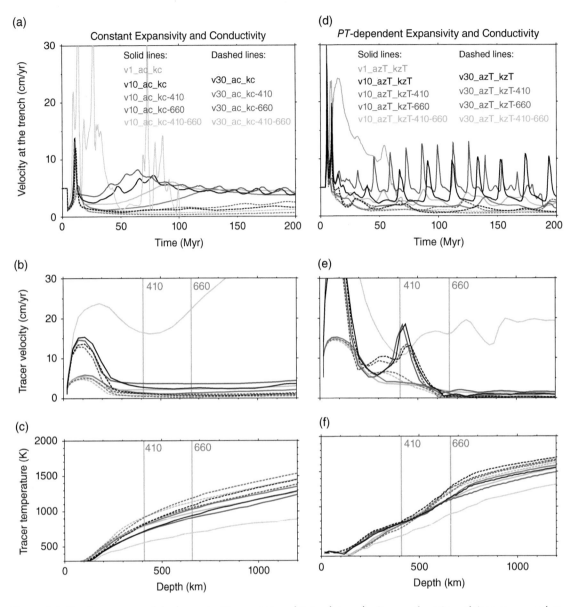

Figure 6.4 For the same models shown in Figures 6.2 and 6.3, plate velocity as a function of time measured at a point located 100 km left of the trench and at a depth of 20 km (a and d), velocity (b and e), and temperature (c and f) as a function of depth of a tracing particle located, at the beginning of the free-slip phase, 100 km left of the trench and 20-km depth.

buoyancy of the lower mantle part of the slab effectively acting to unfold the upper mantle portion.

6.3.4. Separate Roles of Pressure- and Temperature-Dependent Thermal Expansivity and Conductivity

We investigated the separate roles of the pressure and temperature dependence of expansivity and conductivity by comparing different simulation runs against the reference model v10_azT_kzT described in the previous

section. Results of this analysis are shown in Figure 6.7 where we plotted snapshots of the –600 K (thick lines) and –200 K (thin lines) contour lines of the temperature anomaly. The snapshots are taken at different times when the slab tip, defined here as the maximum depth of the –200 K temperature anomaly, reached 1800 km.

The structure of the slabs in the lower mantle is mostly affected by the pressure dependence of the thermal expansivity. In all models in which the latter is taken into account (panels a, d, e, and f), the slab shape

Table 6.3 Model list.

Model name	410 km	660 km	$\Delta\eta_{LU}$	$\alpha(z)$	$\alpha(T)$	$k(z)$	$k(T)$
v1_ac_kc	Yes	Yes	1	No	No	No	No
v10_ac_kc	Yes	Yes	10	No	No	No	No
v30_ac_kc	Yes	Yes	30	No	No	No	No
v10_ac_kc-410	No	Yes	10	No	No	No	No
v30_ac_kc-410	No	Yes	30	No	No	No	No
v10_ac_kc-660	Yes	No	10	No	No	No	No
v30_ac_kc-660	Yes	No	30	No	No	No	No
v10_ac_kc-410-660	No	No	10	No	No	No	No
v30_ac_kc-410-660	No	No	30	No	No	No	No
v1_azT_kzT	Yes	Yes	1	Yes	Yes	Yes	Yes
v10_azT_kzT	Yes	Yes	10	Yes	Yes	Yes	Yes
v30_azT_kzT	Yes	Yes	30	Yes	Yes	Yes	Yes
v10_azT_kzT-410	No	Yes	10	Yes	Yes	Yes	Yes
v30_azT_kzT-410	No	Yes	30	Yes	Yes	Yes	Yes
v10_azT_kzT-660	Yes	No	10	Yes	Yes	Yes	Yes
v30_azT_kzT-660	Yes	No	30	Yes	Yes	Yes	Yes
v10_azT_kzT-410-660	No	No	10	Yes	Yes	Yes	Yes
v30_azT_kzT-410-660	No	No	30	Yes	Yes	Yes	Yes
v10_ac_kzT	Yes	Yes	10	No	No	Yes	Yes
v10_azT_kc	Yes	Yes	10	Yes	Yes	No	No
v10_az_kzT	Yes	Yes	10	Yes	No	Yes	Yes
v10_azT_kz	Yes	Yes	10	Yes	No	Yes	No

Note: 410 km and 660 km indicate whether the corresponding phase transitions are considered or not, $\Delta\eta_{LU}$ refers to the imposed viscosity jump between upper and lower mantle, $\alpha(z)$, $\alpha(T)$, $k(z)$, and $k(T)$ indicate whether the pressure and temperature dependence of the thermal expansivity and conductivity are considered or not.

and thickness are remarkably similar. On the contrary, when α is constant (panels b and c), slabs in the lower mantle, as measured by their temperature anomaly, are not significantly thicker than in the upper mantle, despite the evident formation of buckling instabilities. Quite remarkable is the effect of the temperature dependence of α. When the latter is neglected and only the pressure dependence is considered, the slab morphology does not change appreciably (panel e), but what does change dramatically is the time the slab needs to reach the same depth: 121 Myr when α depends on pressure and temperature compared with 166 Myr when it depends on pressure only. The reason of this behavior is to be sought for in an increase of the plate velocity, which we observed when using temperature-dependent expansivity. As shown in Figure 6.8, the high mantle temperature at the ridge and below the plate determines an increase of the thermal expansivity over a wide region. This in turn generates thermal buoyancy, providing an additional push forcing for the plate, which subducts with a greater velocity (see also Figure 6.9a). Note that this behavior was previously recognized by *Ghias and Jarvis* [2008] in the framework of isoviscous thermal convection models.

A comparison of the reference model (panel a) against v10_azT_kc (panel d), in which k was held constant, suggests that variable thermal conductivity does not strongly impact the way slabs fold and thicken in the lower mantle, but it does influence the timescale of the process, rendering subduction slower. The slab tip reaches a depth of 1800 km after 97 Myr with constant k, while it needs 121 Myr when k depends on pressure and temperature. Responsibility for this is not the decrease of k with temperature but only its increase with pressure. Indeed, model v10_azT_kz, in which k depends solely on pressure (panel f), is essentially indistinguishable from the reference model.

As we did with Figure 6.4, we can describe more quantitatively the effects discussed above looking at Figure 6.9. In all cases, after the slab starts subducting, a rapid increase of the plate velocity is observed, which is mostly evident when α is let to vary (black, red, blue, and green lines in panels a and b). As shown in panel a, the variability of the thermal expansivity is also responsible for clear periodic slab folding occurring at the bottom of the transition zone, manifested by oscillations of plate velocity at the trench. In particular, the temperature dependence of α strongly influences not only the magnitude of

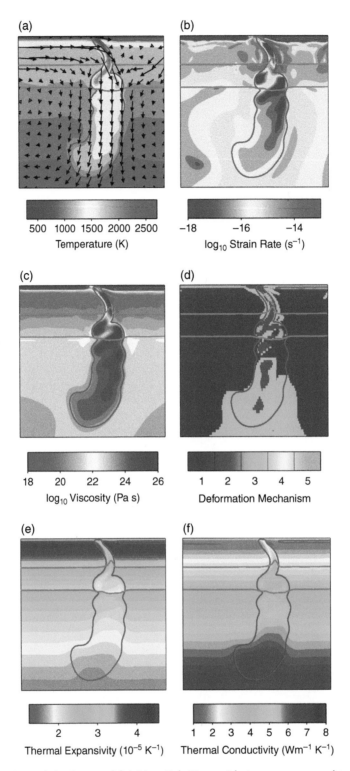

Figure 6.5 Representative subduction model (v10_azT_kzT) considering pressure- and temperature-dependent thermal expansivity and conductivity, both phase transitions at 410 and 660 km depth, and a viscosity contrast of 10 between lower and upper mantle: (a) temperature, (b) second invariant of the strain rate tensor, (c) viscosity, (d) deformation mechanism (1: diffusion creep, 2: plastic yielding, 3: dislocation creep, 4: lower viscosity limit, 5: upper viscosity limit), (e) thermal expansivity, and (f) thermal conductivity.

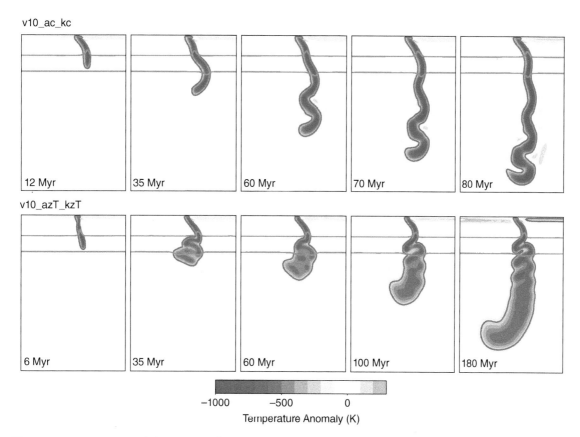

Figure 6.6 Comparison of the time evolution of slab temperature anomalies obtained with constant (model v10_ac_kc, upper panels) and variable thermal expansivity and conductivity (model v10_azT_kzT, bottom panels). The size of the domain shown is 2000×2890 km.

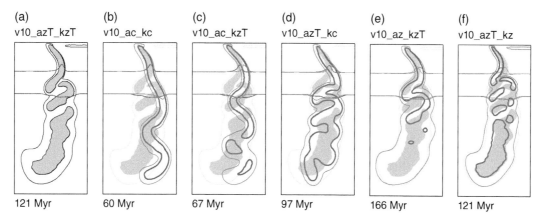

Figure 6.7 Comparison of the reference model v10_azT_kzT (panel a and panels b–g in background) with models (b) v10_ac_kc, (c) v10_ac_kzT, (d) v10_azT_kc, (e) v10_az_kzT, and (f) v10_azT_kz. Thick and thin lines denote the –600 and –200 K contours of temperature anomaly, respectively. Black/gray lines and gray-shaded areas refer to contours of temperature anomaly and phase boundaries of model v10_azT_kzT, and red lines refer to the other models. The time reported below each panel is the time needed by the slabs to reach the same depth. The size of the domain shown is 1000×2000 km.

(a)

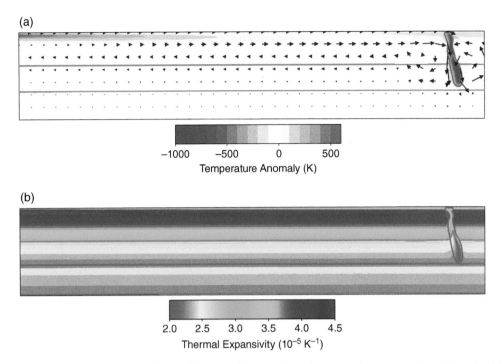

Temperature Anomaly (K)

(b)

Thermal Expansivity (10^{-5} K^{-1})

Figure 6.8 (a) Temperature anomaly and (b) thermal expansivity for the reference model v10_azT_kzT after 6 Myr. Note the high values of expansivity in the vicinity of the hot ridge. The size of the domain shown is 5500 × 1000 km.

the surface plate velocity as previously noted (Figure 6.8), but also the folding period, which roughly doubles when $\alpha(T)$ is neglected (compare blue and green lines in Figure 6.9a). Finally, the effects of the thermal conductivity become evident when looking at Figure 6.9c, which clearly shows that the temperature of the slab core when k varies with pressure and/or temperature (black, magenta, blue, and green lines) tends to be systematically higher with respect to cases in which k is constant (gray and red lines).

6.4. DISCUSSION AND CONCLUSIONS

We used numerical models to investigate the dynamics of slabs subducted in the lower mantle in dependence of three important factors that strongly affect their behavior, namely (1) the viscosity contrast between lower and upper mantle, (2) the two major phase transitions at 410- and 660-km depth, and, for the first time, (3) the pressure and temperature dependence of the coefficients of thermal expansion and conduction.

In absence of a viscosity discontinuity between upper and lower mantle, neither the hampering effect on radial mass flux exerted by the endothermic transition at 660-km depth nor the reduction of thermal buoyancy caused by the decrease of the thermal expansivity upon compression is sufficient to obstruct the downward motion of slabs. These tend instead to cross the transition zone

vertically, maintaining their shape and undergoing little deformation (Figures 6.2a and 6.3a). In order for buckling instabilities to occur, a viscosity increase $\Delta\eta_{LU}$ across the upper-lower mantle interface is therefore required (see, e.g., Figures 6.2b and 6.3b for models with $\Delta\eta_{LU} = 10$ and Figures 6.2f and 6.3f for models with $\Delta\eta_{LU} = 30$). In fact, such increase (although not its precise magnitude) has long been predicted by inversions of geophysical and geodetic data based on models of postglacial rebound and mantle flow [e.g., *Forte and Mitrovica*, 1996; *Mitrovica and Forte*, 2004; *Tosi et al.*, 2005; *Steinberger and Calderwood*, 2006]. In this context, it should be mentioned that some of these models also predict a systematic (and possible large) increase of viscosity with depth below the transition zone, culminating with a peak at a depth of around 2000 km [*Ricard and Wuming*, 1991] or deeper, above the bottom thermal boundary layer [*Calderwood and Steinberger*, 2006]. Although this additional increase is important in controlling the behavior of slabs (or their remnants) as they approach the core-mantle boundary [*Morra et al.*, 2010; *Čížková et al.*, 2012], neglecting it should not influence the way slabs thicken because of buckling as they sink through the transition zone and the shallow lower mantle.

As long as the thermal expansivity is held constant, not only a viscosity increase across the 660-km discontinuity is necessary to observe buckling instabilities, but also buoyancy effects associated with phase transitions need

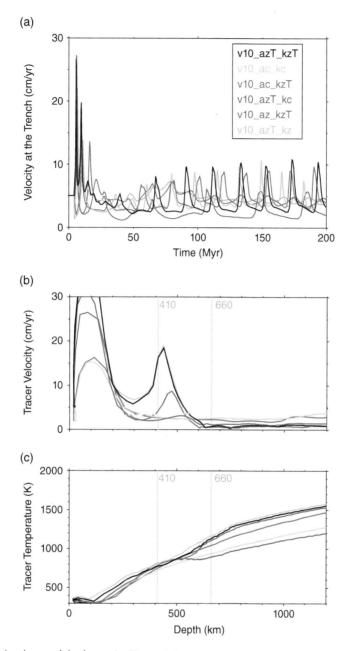

Figure 6.9 As Figure 6.4 but for the models shown in Figure 6.7.

to be taken into account (compare, e.g., Figures 6.2b and 6.2e). However, in contrast to the intuitive idea that the endothermic phase transition from spinel to perovskite and periclase is responsible for this behavior, it is actually the presence of the exothermic transition from olivine to spinel that promotes the formation of large-scale folds in the lower mantle. As it can be seen in Figure 6.4a comparing black and red lines, the initial peak in the plate velocity measured at the trench is due to the presence of the olivine-spinel transition. Such a velocity maximum, which is also observed and is actually more significant when considering variable thermodynamic properties

(Figure 6.4d), occurs as the slab tip approaches the transition zone. The 410-km boundary is then deflected upward and creates an additional localized source of negative buoyancy for the slab, which accelerates and sinks at a faster rate. As shown by *Ribe* [2003], a higher sinking velocity also implies larger folding amplitudes. As a consequence, buckling instabilities are more prominent when the 410-km phase transition is taken into account. From a qualitative point of view, the presence of the 660-km transition exerts a minor effect on the dynamic behavior of slabs in the lower mantle. Nevertheless, it does influence sinking velocities, which are reduced by as much as 60% in comparison to cases in which it is neglected (e.g., compare black and blue lines in Figure 6.4b). Furthermore, it affects the buckling period increasing it by as much as ~70% when the effects of the 410-km phase change are also taken into account (compare solid black and blue lines in Figure 6.4d).

When using constant thermodynamic properties (along with the conditions described above), the subduction style is characterized by the formation of buckling instabilities that do not lead to an actual accumulation of subducted material capable to account for the apparent thickening of slabs recognized in tomographic models. This behavior is readily explained by two factors. On the one hand, we limited ourselves to consider relatively small viscosity contrasts between upper and lower mantle (i.e., $\Delta\eta_{LU} = 10$ or 30), while larger values would promote piling of slab folds near the bottom of the transition zone [e.g., *Lee and King*, 2011]. On the other hand, it is well known that the choice of the ductile strength also plays an important role in controlling the way slabs deform in the transition zone [e.g., *Billen and Hirth*, 2007; *Di Giuseppe et al.*, 2008]. For the same values of $\Delta\eta_{LU}$, the choice of a smaller ductile strength than that adopted here of 0.5 GPa would facilitate deformation and lead to lower mantle slabs significantly thicker than those we observed [*Běhounková and Čížková*, 2008]. Nevertheless, particular care needs to be taken in choosing this parameter. Indeed, by carrying out additional simulations (not shown here), when using $\sigma_y = 0.1$ GPa, we observed frequent slab break-offs, leading to a dripping mode with which subduction of the lithosphere (or of its detached parts) into the lower mantle is not possible. With $\sigma_y = 1$ GPa, on the other hand, we found that slabs tend to thicken in the lower mantle in a way very similar to our reference model ($\sigma_y = 0.5$ GPa), although with a slightly reduced folding frequency.

The viscosity contrast $\Delta\eta_{LU}$ and the presence of phase transitions assume very different roles in models in which variable thermal expansivity is considered. The differences with respect to simulations with constant α are very significant, both from a qualitative and a quantitative point of view. A viscosity contrast of 10 between upper and lower mantle is sufficient to cause the accumulation of folded slabs at the bottom of the transition zone. The thermal anomalies observed when these structures sink through the lower mantle are two to four times broader than the corresponding shallow mantle anomalies (Figure 6.3b–e). These effects become even more dramatic when $\Delta\eta_{LU} = 30$. In this case, the decrease of thermal buoyancy with depth combined with the high viscosity of the lower mantle greatly reduce the sinking velocity of slabs, which tend to stagnate for a sufficiently long time for thermal equilibration to be achieved before they reach the core-mantle boundary (Figure 6.3f–i).

With our models we are certainly not able to put a tight constraint on the viscosity of the lower mantle. However, the requirement of a viscosity increase larger than a factor of 30 to develop periodic slab buckling as recently proposed by *Lee and King* [2011] is likely to be biased because of the lack of consideration of the important role played by variable thermal expansivity. Recent numerical simulations of *Čížková et al.* [2012] show that $\Delta\eta_{LU} = 7$, a nearly isoviscous lower mantle and a 30% decrease of α across the entire mantle yield an optimal sinking speed of slab remnants that closely match the prediction of *van der Meer et al.* [2010] based on seismic tomography and plates reconstructions. Our reference model (Section 3.3) agrees well with this picture and yields a lower mantle slab velocity of ~1 cm/yr (black line in Figure 6.9b), which is close to the value of 1.2 cm/yr proposed by *van der Meer et al.* [2010] for slab remnants. Nevertheless, care needs to be taken when making this comparison: while the speed inferred by *van der Meer et al.* [2010] refers to disconnected blobs freely sinking in the lower mantle, in our case slabs are still attached to the surface plate, which can act as a coherent stress guide and affect in turn the velocity field.

Despite its relative complexity, our reference model also exhibits a buckling amplitude – the ratio between total width of the first fold to the initial thickness of the sinking slab – of about 5, in agreement with the prediction of the scaling laws of *Ribe* [2003].

While largely responsible for the above-described behaviors is the pressure dependence of the thermal expansivity, another interesting finding regards the role of its temperature dependence. Although the consideration of the latter exerts a relatively small effect on the final shape slabs assume in the lower mantle (Figure 6.7e), it does influence to a great extent the surface plate velocity. The decrease of α within the core of cold slabs, which should reduce their sinking speed, is counteracted by an increase of the thermal buoyancy available near the hot ridge zone (Figure 6.8). The net effect is that the initial subduction phase through the uppermost mantle and top part of the transition zone becomes remarkably faster (compare black and blue lines in Figures 6.9b). As a consequence, the slab-folding frequency almost doubles with respect to cases in which α does not depend on temperature (Figure 6.9a). Once slabs have piled up at the bottom of the transition zone, their terminal sinking velocity is no longer affected by the temperature dependence of the expansivity but only by its pressure dependence and by the viscosity of the lower mantle (Figure 6.9b).

The effects of variable conductivity are less evident but also important, particularly as far as the pressure increase of this parameter is concerned. As expected, higher values of k at depth promote heat diffusion over transport, which renders subduction slower than in cases with constant k (Figure 6.7d). Although the temperature dependence is known to be important in controlling the evolution of the thermal structure of the lithosphere both as this grows from midocean ridges [*McKenzie et al.*, 2005) and as it subducts in the upper mantle [*Emmerson and McKenzie*, 2007; *Maierová et al.*, 2012], it does not influence significantly the long-term, large-scale dynamics captured by our models (Figure 6.7f). Nevertheless, as already demonstrated in the above-cited studies, we confirm that models in which the decrease of k with temperature is taken into account lead to cooler slab interiors than models in which this is neglected (compare green and black lines in Figure 6.9c).

Our models also have a number of limitations that will have to be overcome by future and more comprehensive simulations. An important feature they lack is a moving trench. A strong retrograde trench motion is likely to cause slabs to lay flat above the 660-km discontinuity [e.g., *Čížková and Bina*, 2013] and to reduce buckling [e.g., *Lee and King*, 2011], which, as we showed, is a prominent characteristic of slabs entering the lower mantle with a steep angle. The use of periodic rather than reflective conditions at lateral boundaries [e.g., *Enns et al.*, 2005] as well as the use of a free upper surface and/or of a weak crustal layer overlying the subducting plate are also known to affect the motion of the trench and the way slabs interact with the lower mantle [*Quinquis et al.*, 2011]. In a 3D setting, the presence of toroidal motion, intrinsically absent in our 2D configurations, can also limit the emergence of slab folding [*Di Giuseppe et al.*, 2008]. Whether in the above cases the magnitude of the effects induced by variable thermal expansivity and conductivity will remain the same is difficult to predict and will have to be assessed using more general modeling setups.

The coefficients of thermal expansion and conduction are the two key thermodynamic parameters of the mantle that control the way heat is transported in the interior by advection and diffusion. The influence of the pressure dependence of both, and of the temperature dependence of the thermal expansivity on the dynamics of deep subduction is large. Under the assumptions of our 2D models, the effects on the morphology of lower-mantle slabs and on their temporal evolution are similar in importance to those of the mantle viscosity and more relevant than those exerted by the two major phase transformations at 410- and 660-km depth. We suggest, therefore, that global-scale subduction models should routinely account for the pressure and temperature variations of these two parameters.

ACKNOWLEDGMENT

We thank Masanori Kameyama and an anonymous reviewer for their thoughtful comments that helped improve an earlier version of this manuscript. We also thank Hana Čížková for fruitful discussions. N. Tosi acknowledges support from the Helmholtz Association (grant VH-NG-1017), P. Maierová from the Czech Ministry of Education, Youth and Sports (grant number LK11202), and D. A. Yuen from the CMG and geochemistry programs of the National Science Foundation.

REFERENCES

Androvičová, A., H. Čížková, and, A. van den Berg (2013), The effects of rheological decoupling on slab deformation in the Earth's upper mantle, *Stud. Geophys. Geod.*, *57*(3), 460–481, doi:10.1007/s11200-012-0259-7.

Běhounková, M., and H. Čížková (2008), Long-wavelength character of subducted slabs in the lower mantle, *Earth Planet. Sci. Lett.*, *275*, 43–53, doi:10.1016/j.epsl.2008.07.059.

Billen, M. I. (2010), Slab dynamics in the transition zone, *Phys. Earth Planet. Inter.*, *183*(1), 296–308.

Billen, M. I., and G. Hirth (2005), Newtonian versus non-Newtonian upper mantle viscosity: Implications for subduction initiation, *Geophys. Res. Lett.*, *32*, doi:10.1029/2005GL023457.

Billen, M. I., and G. Hirth (2007), Rheologic controls on slab dynamics, *Geochem. Geophys. Geosyst.*, *8*(Q08012), doi:101029/2007GC001597.

Christensen, U. R. (1996), The influence of trench migration on slab penetration into the lower mantle, *Earth Planet. Sci. Lett.*, *140*, 27–39.

Christensen, U. R., and D. A. Yuen (1985), Layered convection induced by phase transitions, *J. Geophys. Res.*, *90*(B12), 10291–10300.

Čížková, H., A. van den Berg, W. Spakman, and C. Matyska (2012), The viscosity of Earth's lower mantle inferred from sinking speed of subducted lithosphere, *Phys. Earth Planet. Inter.*, *200–201*, 56–62, doi:10.1016/j.pepi.2012.02.010.

Čížková, H., and C. R. Bina (2013), Effects of mantle and subduction-interface rheologies on slab stagnation and trench rollback, 379, 95–103, doi: 10.1016/j.epsl.2013.08.011.

Di Giuseppe, E., J. van Hunen, F. Funiciello, C. Faccenna, and D. Giardini (2008), Slab stiffness control of trench motion: Insights from numerical models, *Geochem., Geophys., Geosys.*, *9*(2), doi:10.1029/2007GC001776.

Dziewonski, A. M., and D. L. Anderson (1981), Preliminary reference Earth model, *Phys. Earth Planet. Inter.*, *25*, 297–356.

Emmerson, B., and D. McKenzie (2007), Thermal structure and seismicity of subducting lithosphere, *Phys. Earth Planet Inter.*, *163*(1), 191–208.

Enns, A., T. W. Becker, and H. Schmeling (2005), The dynamics of subduction and trench migration for viscosity stratification, *Geophys. J. Int.*, *160*(2), 761–775, doi:10.1111/j.1365-246X.2005.02519.x.

Foley, B. J., and T. W. Becker (2009), Generation of plate-like behavior and mantle heterogeneity from a spherical, visco-plastic convection model, *Geochem, Geophys. Geosyst.*, *10*(8), doi:10.1029/2009GC002378.

Forte, A. M., and J. X. Mitrovica (1996), New inferences of mantle viscosity from joint inversion of long-wavelength mantle convection and postglacial rebound data, *Geophys. Res. Lett.*, *23*(10), 1147–1150.

Fukao, Y., M. Obayashi, T. Nakakuki, H. Utada, D. Suetsugu, T. Irifune, E. Ohtani, S. Yoshioka, H. Shiobara, T. Kanazawa, and K. Hirose (2009), Stagnant slab: A review, *Ann. Rev. Earth. Planet. Sci.*, *37*, 19–46, doi:10.1146/annurev.earth. 36.031207.124224.

Gaherty, J. B., and B. H. Hager (1994), Compositional vs. thermal buoyancy and the evolution of subducted lithosphere, *Geophys. Res. Lett.*, *21*(2), 141–144.

Ghias, S. R., and G. T. Jarvis (2008), Mantle convection models with temperature- and depth-dependent thermal expansivity, *J. Geophys. Res.*, *113*(B08408), doi:10.1029/2007JB005355.

Guillou-Frottier, L., J. Buttles, and P. Olson (1995), Laboratory experiments on the structure of subducted lithosphere, *Earth Planet. Sci. Lett.*, *133*(1), 19–34.

Hansen, U., D. A. Yuen, S. E. Kroening, and T. B. Larsen (1993), Dynamical consequences of depth-dependent thermal expansivity and viscosity on mantle circulations and thermal structure, *Phys. Earth Planet. Inter.*, *77*, 205–223.

Hauck, S. A., R. J. Phillips, and A. Hofmeister (1999), Variable conductivity: Effects on the thermal structure of subducting slabs, *Geophys. Res. Lett.*, *26*(21), 3257–3260.

Hofmeister A. (1999), Mantle values of thermal conductivity and the geotherm from phonon lifetimes, *Science*, *283*, 1699–1706.

Huang, J., and D. Zhao (2006), High-resolution mantle tomography of China and surrounding regions, *J. Geophys. Res.*, *111*(B9), doi:10.1029/2005JB004066.

Karato, S., and P. Wu (1993), Rheology of the upper mantle: a synthesis, *Nature*, *260*, 771–778.

King, S. D., C. Lee, P. van Keken, W. Leng, S. Zhong, E. Tan, N. Tosi, and M. Kameyama (2010), A community benchmark for 2D Cartesian compressible convection in the Earth's mantle, *Geophys. J. Int.*, *180*, 73–87, doi:10.1111/j.1365-246X.2009.04413.x.

Lee, C., and S. D. King (2011), Dynamic buckling of subducting slabs reconciles geological and geophysical observations, *Earth Planet. Sci. Lett.*, *312*(3), 360–370.

Leitch, A. M., D. A. Yuen, and G. Sewell (1991), Mantle convection with internal heating and pressure-dependent thermal expansivity, *Earth Planet. Sci. Lett.*, *102*, 213–232.

Li, C., R. D. van der Hilst, E. R. Engdahl, and S. Burdick (2008), A new global model for P wave speed variations in Earth's mantle, *Geochem. Geophys. Geosyst.*, *9*(Q07003), doi:10.1029/2007GC001806.

Maierová, P., T. Chust, G. Steinle-Neumann, O. Čadek, and H. Čížková (2012), The effect of variable thermal diffusivity on kinematic models of subduction, *J. Geophys. Res.*, *117*(B07202), doi:101029/2011JB00912019.

Manthilake, G. M., N. de Koker, D. J. Frost, and C. A. McCammon (2011), Lattice thermal conductivity of lower mantle minerals and heat flux from Earth's core, *Proc. Natl. Acad. Sci.*, *108*(44), 17901–17904, doi:10.1073/pnas.1110594108.

McKenzie, D., J. Jackson, and K. Priestley (2005), Thermal structure of oceanic and continental lithosphere, *Earth Planet. Sci. Lett.*, *233*(3), 337–349.

Mitrovica, J. X., and A. Forte (2004), A new inference of mantle viscosity based upon joint inversion of convection and glacial isostatic adjustment data, *Earth Planet. Sci. Lett.*, *225*, 177–189.

Miyauchi, A., and M. Kameyama (2013), Influences of the depth-dependence of thermal conductivity and expansivity on thermal convection with temperature-dependent viscosity, *Phys. Earth. Planet. Inter.*, doi:10.1016/j.pepi.2013.08.001.

Moresi, L., and M. Gurnis (1996), Constraints on the lateral strength of slabs from three-dimensional dynamic flow models, *Earth Planet. Sci. Lett.*, *138*, 15–28.

Morra, G., D. A. Yuen, L. Boschi, P. Chatelain, P. Koumoutsakos, and P. J. Tackley (2010), The fate of the slabs interacting with a viscosity hill in the mid-mantle, *Phys. Earth Planet. Inter.*, *180*, 3–4, 271–282.

Nakagawa, T., P. J. Tackley, F. Deschamps, and J. A. D. Connolly (2010), The influence of MORB and harzburgite composition on thermo-chemical mantle convection in a 3-D spherical shell with self-consistently calculated mineral physics, *Earth Planet. Sci. Lett.*, *296*, 403–412, doi:10.1016/j.epsl.2010.05.026.

Pesicek, J. D., C. H. Thurber, S. Widiyantoro, E. R. Engdahl, and H. R. DeShon (2008), Complex slab subduction beneath northern Sumatra, *Geophys. Res. Lett.*, *35*(20), L20303, doi:10.1029/2008GL035262.

Piromallo, C., and A. Morelli (2003), P wave tomography of the mantle under the Alpine-Mediterranean area, *J. Geophys. Res.*, *108*(B2), 2065.

Quinquis, M. E. T., S. J. H. Buiter, and S. Ellis (2011), The role of boundary conditions in numerical models of subduction zone dynamics, *Tectonophys.* *497*, 57–70.

Ribe, N. M. (2003), Periodic folding of viscous sheets, *Phys. Rev. E*, *68*(3), 036305.

Ribe, N. M., E. Stutzmann, Y. Ren, and R. van der Hilst (2007), Buckling instabilities of subducted lithosphere beneath the transition zone, *Earth Planet. Sci. Lett.*, *254*, 173–179.

Ricard, Y., and B. Wuming (1991), Inferring viscosity and the 3-D density structure of the mantle from geoid, topography and plate velocities, *Geophys. J. Int.*, *105*, 561–572.

Schmeling, H., G. Marquart, and T. Ruedas (2003), Pressure- and temperature-dependent thermal expansivity and the effect on mantle convection and surface observables, *Geophys. J. Int.*, *154*, 2224–229, doi:10.1046/j.1365-246X.2003.01949.x.

Schmid, C., S. Goes, S. van der Lee, and D. Giardini (2002), Fate of the Cenozoic Farallon slab from a comparison of kinematic thermal modeling with tomographic images, *Earth Planet. Sci. Lett.*, *204*(1), 17–32.

Steinberger, B., and A. R. Calderwood (2006), Models of large-scale viscous flow in the Earth's mantle with constraints from mineral physics and surface observations, *Geophys. J. Int.*, *167*, 1461–1481.

Tan, E., E. Choi, P. Thoutireddy, M. Gurnis, and M. Aivazis (2011), On the location of plumes and lateral movement of thermochemical structures with high bulk modulus in the 3-D compressible mantle, *Geochem. Geophys. Geosyst.*, *12*(7), doi:10.1029/2011GC003665.

Tosi, N., and D. A. Yuen (2011), Bent-shaped plumes and horizontal channel flow beneath the 660 km discontinuity, *Earth Planet. Sci. Lett.*, *312*, 348–359, doi:10.1016/j.epsl.2011.10.015.

Tosi, N., D. A. Yuen, and O. Čadek (2010), Dynamical consequences in the lower mantle with the post-perovskite phase change and strongly depth-dependent thermodynamic and transport properties, *Earth Planet. Sci. Lett.*, *298*, 229–243, doi:10.1016/j.epsl.2010.08.001.

Tosi N., R. Sabadini, A. M. Marotta, and L. L. A. Vermeersen (2005), Simultaneous inversion for the Earth's mantle viscosity and mass imbalance in Antarctica and Greenland, *J. Geophys. Res. Solid Earth*, *110*, B07402, doi:10.1029/2004JB003236.

Tosi, N., D. A. Yuen, N. de Koker, and R. M. Wentzcovitch (2013), Mantle dynamics with pressure- and temperature-dependent thermal expansivity and conductivity, *Phys. Earth Planet. Inter.*, *217*, 48–58, doi:10.1016/j.pepi.2013.02.004.

Tosi, N., O. Čadek, and Z. Martinec (2009), Subducted slabs and lateral viscosity variations: effects on the long-wavelength geoid, *Geophys. J. Int.*, *179*, 813–826, doi:10.1111/j.1365-246X.2009.04335.x.

Tosi N., C. Stein, L. Noack, C. Hüttig, P. Maierová, H. Samuel, R. Davies, C. Wilson, S. Kramer, C. Thieulot, A. Glerum, M. Fraters, W. Spakman, A. Rozel and P. Tackley (2015), A community benchmark for viscoplastic thermal convection in a 2-D square box, *Geochem. Geophys. Geosyst.*, *16*, doi:10.1002/2015GC005807.

Turcotte, D. L., and G. Schubert (2002), *Geodynamics*, Cambridge University Press, Cambridge.

van der Hilst, R. D. (1995), Complex morphology of subducted lithosphere in the mantle beneath the Tonga trench, *Nature*, *374*, 154–157.

van der Hilst, R. D., S. Widiyantoro, and E. R. Engdahl (1997), Evidence of deep mantle circulation from global tomography, *Nature*, *386*, 578–584.

van der Meer, D. G., W. Spakman, D. J. J. van Hinsbergen, M. L. Amaru, and T. H. Torsvik (2010), Towards absolute plate motions constrained by lower-mantle slab remnants, *Nature Geosci.*, *3*, 36–40, doi:10.1038/ngeo708.

Wentzcovitch, R. M., Z. Wu, and P. Carrier (2010), First principles quasiharmonic thermoelasticity of mantle minerals, *Rev. Min. Geochem.*, *71*, 99–128, doi:10.2138/rmg.2010.71.5.

Wortel, M. J. R., and W. Spakman (2000), Subduction and slab detachment in the Mediterranean-Carpathian region, *Science*, *290*(5498), 1910–1917.

Xu, Y., T. J. Shankland, S. Linhardt, D. C. Rubie, F. Langenhorst, and K. Klasinski (2004), Thermal diffusivity and conductivity of olivine, wadsleyite and ringwoodite to 20 GPa and 1373 K, *Phys. Earth. Planet. Inter.*, *143–144*, 321–336, doi:10.1016/j.pepi.2004.03.005.

Yamazaki, D., and S. I. Karato (2001), Some mineral physics constraints on rheology and geothermal structure of Earth's lower mantle, *Am. Mineral.*, *86*, 358–391.

Yoshioka, S., and A. Naganoda (2010), Effects of trench migration on fall of stagnant slabs into the lower mantle, *Phys. Earth Planet. Inter.*, *183*, 321–329.

Zhong, S. (2006), Constraints on thermochemical convection of the mantle from plume heat flux, plume excess temperature, and upper mantle temperature, *J. Geophys. Res.*, *111*(B04409), doi:10.1029/2005JB003972.

7

Slab-driven Mantle Weakening and Rapid Mantle Flow

M. A. Jadamec

ABSTRACT

Numerical models of subduction are presented to investigate the relative role of a Newtonian versus composite viscosity, maximum yield stress, and the initial slab dip on the upper-mantle viscosity structure and velocity. The results show that using the experimentally derived flow law to define the Newtonian viscosity (diffusion creep deformation mechanism) and the composite viscosity (both diffusion creep and dislocation creep deformation mechanisms) has a first order effect on the viscosity structure and flow velocity in the upper mantle. Models using the composite viscosity formulation produce a zone of subduction-induced mantle weakening that results in reduced viscous support of the slab. The maximum yield stress, which places an upper bound on the slab strength, can also have a significant impact on the viscosity structure and flow rates induced in the upper mantle, with maximum mantle weakening and mantle flow rates occurring in models with a lower maximum yield stress and shallower slab dip. In all cases, the magnitude of induced mantle flow is larger in the models using the composite viscosity formulation. The laterally variable viscosity that develops in the upper mantle leads to lateral variability in coupling of the mantle to the base of the surface plates.

7.1. INTRODUCTION

Continued research into the processes governing subduction has expanded the classical view of two-dimensional corner flow [e.g., *McKenzie*, 1969] to a slab-driven flow that can be quite complex. More recent, three-dimensional models of subduction indicate the mantle flow field in subduction zones is spatially variable, containing poloidal, toroidal, and trench parallel flow [*Buttles and Olson*, 1998; *Hall et al.*, 2000; *Kincaid and Griffiths*, 2003; *Schellart*, 2004; *Funiciello et al.*, 2006; *Piromallo et al.*, 2006; *Royden and Husson*, 2006; *Behn et al.*, 2007; *Schellart et al.*, 2007; *Honda*, 2009; *Giuseppe et al.*, 2008; *Jadamec and Billen*, 2010; *Gerya*, 2011; *MacDougall et al.*, 2014]. Subduction models using an

experimentally derived non-Newtonian rheology also indicate complex three-dimensional flow in subduction zones, and show that upper-mantle flow rates can be an order of magnitude greater than plate motions [*Conder and Wiens*, 2007; *Jadamec and Billen*, 2010; *Stadler et al.*, 2010; *Jadamec and Billen*, 2012; *Alisic et al.*, 2012]. Thus, the three-dimensional models of subduction show the mantle flow field in subduction zones can be both nonparallel to and faster than surface plate motions and imply, therefore, localized decoupling between the mantle and lithosphere in subduction zones. However, the extent of the lateral variability in plate-mantle coupling is not well understood.

Observations of shear wave splitting are an independent constraint that also indicate complex mantle flow in subduction zones [*Russo and Silver*, 1994; *Savage*, 1999; *Karato et al.*, 2008; *Long*, 2013]. Although shear wave splitting can reflect fabric in the lowermost mantle [*Lynner and Long*, 2014], the uppermost mantle [*Long and Wirth*, 2013], the slab [*Faccenda et al.*, 2008; *Healy et al.*, 2009], and the surface plates [*Fuchs*, 1983; *Hicks*

Department of Earth, Environmental, and Planetary Sciences, Brown University, Providence, Rhode Island; Department of Earth and Atmospheric Sciences, University of Houston, Houston, Texas, USA

Subduction Dynamics: From Mantle Flow to Mega Disasters, Geophysical Monograph 211, First Edition.
Edited by Gabriele Morra, David A. Yuen, Scott D. King, Sang-Mook Lee, and Seth Stein.

et al., 2012], and can vary in its relation to mantle flow direction as function of numerous parameters, including temperature [*Kneller et al.*, 2007], A-, C-, or E-type fabric is expected to dominate in the warm mantle wedge and in the subslab asthenosphere (with the fast seismic axis parallel to the mantle flow direction) [*Savage*, 1999; *Karato et al.*, 2008; *Long and Becker*, 2010; *Long*, 2013].

For example, subduction zones with well-resolved anisotropy of mantle wedge origin, and that use local S waves in particular, suggest mantle flow nonparallel to plate motion, such as in South America [*MacDougall et al.*, 2012], Central America [*Abt et al.*, 2010], northern New Zealand [*Audoine et al.*, 2004], northern Tonga [*Smith et al.*, 2001], the Marianas [*Pozgay et al.*, 2007], the northwest Pacific [*Fouch and Fischer*, 1996], and offshore western Alaska [*Yang et al.*, 1995]. Therefore, the observations of seismic anisotropy are another line of evidence that indicate a difference between the direction of mantle flow and the surface plate motion, implying local decoupling of the surface plates from the mantle in subduction zones [*Jadamec and Billen*, 2010; *Long and Wirth*, 2013].

Previous work has identified mechanisms that can preferentially reduce the viscosity of the mantle in subduction zones [*Hirth and Kohlstedt*, 1996; *Billen and Gurnis*, 2001; *Hirth and Kohlstedt*, 2003; *Honda and Saito*, 2003; *Karato*, 2003], and therefore would be expected to facilitate or moderate the mantle wedge weakening, the magnitude of slab-driven flow, mantle-overriding plate (de)coupling, and the associated development of complex seismic anisotropy. For example, rock deformation experiments indicate upper-mantle viscosity can be locally reduced to 10^{18} Pa s due to sensitivity to increased temperature, water content, melt-fraction, and strain-rate [*Hirth and Kohlstedt*, 2003; *Karato*, 2003], factors which are expected to occur in the mantle wedge due to the slab-driven mantle circulation close to the subduction zone and the release of volatiles from the slab [*Katz et al.*, 2003; *Kelley et al.*, 2006; *Cagnioncle et al.*, 2007; *Wade et al.*, 2008; *Syracuse et al.*, 2010; *Lev and Hager*, 2011; *Jadamec and Billen*, 2012]. A decrease in the mantle viscosity surrounding the slab implies local reduced viscous support of the slab and local reduced resistance to subduction [*Jadamec and Billen*, 2010; *Billen and Jadamec*, 2012].

More recently, geographically referenced 3D geodynamic models using an experimentally derived composite non-linear viscosity formulation were able to fit a range of observables, suggesting the strain-rate dependence as an alternate mechanism to increased temperature, water, or melt fraction for reducing the viscosity of the mantle wedge [*Jadamec and Billen*, 2010; *Stadler et al.*, 2010; *Alisic et al.*, 2012; *Jadamec and Billen*, 2012; *Jadamec et al.*, 2013]. In the region of reduced viscosity surrounding the slab, the numerical models predict localized

mantle flow rates than can be an order of magnitude greater than plate motions [*Jadamec and Billen*, 2010; *Stadler et al.*, 2010]. However, these models are 3D and complex, so it is useful to examine the key parameters in a simplified 2D subduction setting.

A series of two-dimensional numerical models of subduction are presented here to investigate the relative role of a Newtonian versus composite viscosity, the maximum yield stress, and the initial slab dip on the upper-mantle viscosity structure and the magnitude of slab-driven mantle velocity. The numerical models shed light on the mechanisms leading to rapid mantle flow and large gradients in mantle flow rates at subduction zones, decoupling of the supraslab and subslab mantle to the base of the surface plates, definition of the base of the lithosphere, implications for length-scales of complex shear wave splitting at subduction zones, and implications for fixing both slab dip and velocity in numerical modeling of subduction.

7.2. METHODS

7.2.1. Model Setup

The numerical models contain an overriding plate, subducting plate, and underlying mantle, which are distinguished on the basis of temperature and an experimentally derived rheologic flow law (Figure 7.1). The interface between the overriding plate and subducting plate is defined by a viscous shear zone. The driving forces in the system are the negative thermal buoyancy of the subducting slab, which generates a slab pull force. No density variations due to composition or phase changes are included. The resisting forces are the viscous stresses in the mantle, the plate boundary shear zone, and within the interior of the slab. No driving velocities applied to the plates, rather the models predict the flow velocities for both the plates and mantle.

7.2.1.1. Model domain

The model domain spans 0° to 45° longitude and 0 to 2500 km in depth (Figure 7.1). The mesh contains 1248 × 480 elements in the longitudinal and radial directions, respectively. The mesh is locally refined in the longitudinal direction with element size increasing from 0.016° to 0.1525° (~1.7 to 16 km) outwards from the trench. In the radial direction, the mesh is locally refined with an element size of 1.4 km in the upper 350 km that increases to 15 km in the lower mantle. A free-slip boundary condition is used on the upper and lower surface of the model, and reflecting boundary conditions are used on the model sidewalls.

The subduction zone is positioned in the model domain so as to minimize the boundary effects. The trench is located at 18° longitude, and the slab extends to a maximum

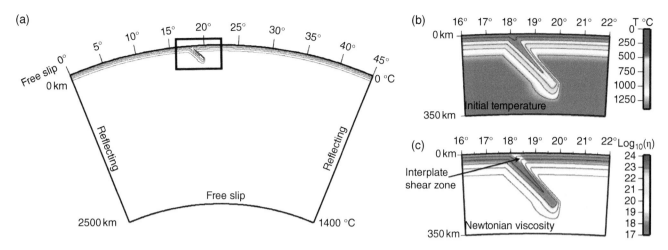

Figure 7.1 Model setup. (a) Model domain. Box inset shows locations of (b), (c). (b) Initial temperature shown in color for box outlined in (a). (c) Newtonian viscosity for box outlined in (a). Thin black lines in (a), (b), (c) are temperature contours at 250°C intervals.

depth of 356 km. Therefore, the subducted slab is located over 1800 km from the left boundary, over 2000 km from the right boundary, and over 2000 km from the model base (Figure 7.1). Each model was run on 48 processors on the Dell Linux cluster, Lonestar, at the Texas Advanced Computing Center. The models were run for up to 300 time steps to examine the subduction evolution. However, the results presented here focus on the instantaneous solution in order to explore the present-day balance of forces and make comparison to the more complex 3D and global instantaneous models that incorporate observed slab geometries.

7.2.1.2. Initial thermal structure

The initial temperature of the overriding plate and subducting plate were calculated and mapped onto the model mesh using SlabGenerator [*Jadamec and Billen*, 2010; *Jadamec et al.*, 2012; *Billen and Jadamec*, 2012; *Jadamec and Billen*, 2012]. A plate-cooling model is used in which the thickness of the thermal lithosphere in the direction perpendicular to the plate surface is a function of the plate age. In these models, the plate age is set to 55 My for both the overriding plate and subducting plate. The plate cooling model has the following infinite series solution:

$$T = T_s + \left(T_m - T_s\right) \left[\frac{y}{y_{Lo}} + \frac{2}{\pi} \sum_{n=1}^{\infty} \frac{1}{n} \exp\left(-\frac{\kappa n^2 \pi^2 t}{y_{Lo}^2}\right) \sin\left(\frac{n\pi y}{y_{Lo}}\right) \right] \quad (7.1)$$

where T_m is the mantle temperature (1400°C), T_s is the temperature at the surface (0°C), κ is the thermal diffusivity (1×10⁻⁶ m²/s), t is the plate age, y is the perpendicular distance to the reference surface, y_{Lo} is the thickness ofxthe lithosphere at large times (125 km), and with the following boundary conditions,

$$T = T_m \;\; at \;\; t = 0, \;\; 0 \le y \le y_{Lo} \quad (7.2)$$

$$T = T_s \;\; at \;\; y = 0, \;\; t > 0 \quad (7.3)$$

$$T = T_m \;\; at \;\; y = y_{Lo}, \;\; t > 0 \quad (7.4)$$

[*Carslaw and Jaeger*, 1959; *Turcotte and Schubert*, 2002]. The series solution is calculated for values of $1 \le n \le 50$.

The boundary condition on the lithospheric plate's top surface, is that the temperature on that surface, T_s, is equal to 0°C where the distance, y, to that surface is equal to 0. Therefore, on the surface of the subducting plate, the temperature will be set to zero, even where the subducting plate is inclined and protrudes into the mantle. This is unrealistic, because the cold surface of the subducting plate will been warmed due to immersion in the mantle. Therefore, a depth-dependent distance correction is added to the distance, y, in equation (7.1) to simulate conductive warming of the inclined slab surface, based on a length-scale diffusion analysis [*Jadamec and Billen*, 2010]. After this correction, the minimum temperature on the slab surface smoothly increases with increasing slab depth, simulating the warming that would occur as the slab is immersed within the mantle. The slab thermal field is then blended into the ambient mantle thermal field using a sigma-shaped smoothing function (Figure 7.1b).

7.2.1.3. Plate boundary shear zone

The overriding plate and subducting plate are separated by a narrow low viscosity layer [*Gurnis and Hager*, 1988; *Kukacka and Matyska*, 2004; *Billen and Hirth*, 2007; *Jadamec and Billen*, 2010]. This plate boundary shear zone is representative of a number of processes operating at the plate interface including the shear heating, variation in grain size, and the presence of water and sediments. The plate boundary shear zone is approximately 25-km thick and 105 km in depth and is fixed in space. A narrow vertical low viscosity layer is also included on the trailing edge of the subducting plate to allow it to decouple from the left model sidewall.

7.2.2. Viscous Flow Modeling

7.2.2.1. Governing equations

The open source finite element code, CitcomCU [*Zhong*, 2006], based on CITCOM [*Moresi and Solomatov*, 1995; *Moresi and Gurnis*, 1996], is used to solve for the viscous flow. CitcomCU solves the Navier-Stokes equation for the velocity, pressure, and temperature assuming an incompressible fluid with a high Prandtl number and assuming no internal heating [*Moresi and Solomatov*, 1995; *Zhong*, 2006], given by the conservation of mass (equation 7.5), momentum (equation 7.6), and energy (equation 7.7),

$$\nabla \cdot \mathbf{u} = 0 \qquad (7.5)$$

$$\nabla \cdot \sigma + \rho_o \alpha \left(T - T_o \right) g \delta_{rr} = 0 \qquad (7.6)$$

$$\frac{\partial T}{\partial t} = -\mathbf{u} \cdot \Delta T + \kappa \nabla^2 T \qquad (7.7)$$

where **u** is the velocity, σ is the stress tensor, ρ_o is the density, α is the coefficient of thermal expansion, g is the acceleration due to gravity, and δ_{rr} is the Kronecker delta, T is temperature, and κ is the thermal diffusivity.

The constitutive relation is defined by

$$\sigma_{ij} = -P\delta_{ij} + \eta_{eff} \dot{\varepsilon}_{ij} \qquad (7.8)$$

where P is the dynamic pressure, η_{eff} is the effective viscosity, and $\dot{\varepsilon}_{ij}$ is the strain-rate tensor. The models are defined and solved in spherical coordinates. The equations of motion are nondimensionalized by the Rayleigh number, Ra, is defined by

$$Ra = \frac{\rho_o \alpha g \Delta T R^3}{\kappa \eta_{ref}} \qquad (7.9)$$

where $\Delta T = T_o - T_{surf}$ and ρ_o, α, g, R, κ, and η_{ref} are as defined in Table 7.1.

7.2.2.2. Rheological implementation

The models use a viscosity formulation previously implemented in CitcomCU [*Jadamec and Billen*, 2010; *Jadamec et al.*, 2012; *Jadamec and Billen*, 2012], and which follows from that implemented in CitcomT [*Billen and Hirth*, 2007]. The viscosity formulation for diffusion

Table 7.1 Dimensionalization parameters.

Parameter	Description	Value
Ra	Rayleigh number	2.34×10^9
g	Acceleration due to gravity, m/s^2	9.8
T_o	Reference temperature, K	1673
T_{surf}	Temperature on top surface, K	273
R	Radius of Earth, m	6371×10^3
η_{ref}	Reference viscosity, Pa·s	1×10^{20}
ρ_o	Reference density, kg/m^3	3300
κ	Thermal diffusivity, m/s^2	1×10^{-6}
α	Thermal expansion coefficient, K^{-1}	2.0×10^{-5}

Table 7.2 Flow law parameters, assuming diffusion and dislocation creep of wet olivine.

Var.	Description	Creep$_{df}$	Creep$_{ds}$
A	Preexponential factor	1.0	9×10^{-20}
n	Stress exponent	1	3.5
d	Grain size, μm (if A in μm)	10×10^3	–
p	Grain size exponent	3	–
C_{OH}	OH concentration, H/10^6 Si	1000	1000
r	Exponent for C_{OH} term	1	1.2
E	Activation energy, kJ/mol	335	480
V	Activation volume, m^3/mol	4×10^{-6}	11×10^{-6}

Source: Values from *Hirth and Kohlstedt* [2003].

creep, η_{df}, and dislocation creep, η_{ds}, is defined by the experimentally determined viscous flow law governing deformation for olivine aggregates [*Hirth and Kohlstedt*, 2003] such that

$$\eta_{df,ds} = \left(\frac{d^p}{A C_{OH}^r} \right)^{\frac{1}{n}} \dot{\varepsilon}^{\frac{1-n}{n}} \exp\left[\frac{E + P_l V}{nR(T + T_{ad})} \right] \qquad (7.10)$$

where P_l is the lithostatic pressure, R is the universal gas constant, T is nonadiabatic temperature, T_{ad} is the adiabatic temperature (with an imposed gradient of 0.3 K/km), and A, n, d, p, C_{OH}, r, E, and V are as defined in Table 7.2, assuming no melt is present [*Hirth and Kohlstedt*, 2003].

For diffusion creep, the strain-rate depends linearly on the stress (n = 1) but nonlinearly on the grain size (p = 3) [*Hirth and Kohlstedt*, 2003]. A grain size of 10 mm is used for the upper mantle giving a background viscosity of 10^{20} Pa s at 250 km. In contrast, for dislocation creep of olivine, the strain-rate dependence is nonlinear (n = 3.5), but there is no grain-size dependence, giving the same background viscosity at 250 km for a strain-rate of 10^{-15} s^{-1} [*Hirth and Kohlstedt*, 2003]. For higher strain-rates, the composite viscosity will lead to lower effective viscosity

Table 7.3 Summary of model parameters and results.

Model	α °	$\sigma_{y_{max}}$ MPa	W_{ss} ° (km)	W_{mw} ° (km)	$\eta_{mw_{min}}$ Pa·s	$\%V_{mw_{max}}$ %
1	30	125	2.68 (298.06)	3.30 (368.19)	17.70	83.16
2	30	250	2.66 (295.55)	2.72 (303.07)	17.90	66.76
3	30	500	1.96 (217.91)	2.16 (240.45)	18.15	24.62
4	30	750	2.10 (232.93)	2.00 (222.92)	18.23	1.53
5	30	1000	2.07 (230.43)	1.98 (220.41)	18.23	0.00
6	45	125	3.78 (420.79)	5.22 (581.08)	17.69	76.84
7	45	250	3.38 (375.70)	4.21 (468.37)	17.84	61.67
8	45	500	3.02 (335.63)	3.31 (368.37)	18.06	18.13
9	45	750	2.90 (323.10)	3.15 (350.65)	18.12	3.36
10	45	1000	2.88 (320.60)	3.11 (345.65)	18.13	0.00
11	60	125	3.40 (378.21)	4.55 (505.94)	18.07	66.42
12	60	250	2.77 (308.07)	3.42 (380.71)	18.26	37.29
13	60	500	2.75 (305.57)	3.15 (350.65)	18.31	27.30
14	60	750	2.48 (275.51)	2.79 (310.58)	18.38	3.81
15	60	1000	2.43 (270.50)	2.72 (303.07)	18.39	0.00

Note: α = initial slab dip; σ_{ymax} = maximum yield stress; W_{ss} = width of region of dynamically reduced viscosity in the subslab mantle for models using the composite viscosity, measured for $\eta < 10^{19}$ Pa·s; W_{mw} = width of dynamically reduced viscosity in the mantle wedge for models using the composite viscosity, measured for $\eta < 10^{19}$ Pa·s; ηmw_{min} = minimum viscosity in the mantle wedge; $\%Vmw_{max}$ = percent change in maximum velocity in the mantle wedge with respect to model using σ_{ymax} = 1000 MPa (positive change indicates velocity increase); Results in table for models using the composite viscosity formulation.

due to most of the deformation being accommodated by the dislocation creep mechanism.

In the upper mantle, a composite viscosity formulation, η_{com}, is applied. For deformation under a fixed stress (driving force) the total strain-rate is the sum of the contributions from deformation by the diffusion (df) and dislocation (ds) creep mechanisms:

$$\dot{\varepsilon}_{com} = \dot{\varepsilon}_{df} + \dot{\varepsilon}_{ds} \qquad (7.11)$$

where $\dot{\varepsilon}_{ij}$ without the subscripts i-j refers to the second invariant of the strain-rate tensor, $\dot{\varepsilon}_{II}$. Substituting $\sigma = 2\eta_{com}\dot{\varepsilon}$, the composite viscosity, η_{com}, has the form

$$\eta_{com} = \frac{\eta_{df}\eta_{ds}}{\eta_{df}+\eta_{ds}} \qquad (7.12)$$

where η_{df} and η_{ds} are according to equation 7.10.

The lower-mantle viscosity is modeled as a Newtonian only viscosity (η_{df}), with a larger effective grain size (70 mm) in order to create a viscosity jump by a factor of 30 from the upper to lower mantle. This simplified viscosity structure for the lower mantle is consistent with the magnitude of the viscosity jump constrained by postglacial rebound [*Mitrovica*, 1996] and models of the long wavelength geoid [*Hager*, 1984], as well as the lack of observations

of seismic anisotropy in the lower mantle and the few experimental constraints on the viscous behavior of perovskite.

Close to the Earth's surface and within the cold core of the subducted slab, the viscosity values determined by equation 7.10 become unrealistically large in that they imply a rock strength much greater than that predicted by laboratory experiments [*Kohlstedt et al.*, 1995]. To allow for plastic yielding where the stresses calculated in the model exceed those predicted from laboratory experiments, the stresses calculated in the model are limited by a depth-dependent yield stress, σ_y. The yield stress increases linearly with depth, assuming a gradient of 15 MPa per km, from 0.1 MPa at the surface to a maximum value of up to 1000 MPa (Table 7.3), with the maximum value based on constraints from experimental observations, low-temperature plasticity, and the dynamics in previous models [*Kohlstedt et al.*, 1995; *Weidner et al.*, 2001; *Hirth*, 2002; *Billen and Hirth*, 2005, 2007]. Taking into account the composite viscosity formulation and the depth-dependent yield stress, the effective viscosity, η_{eff}, is equal to η_{com} if $\sigma_{II} < \sigma_y$, and $\frac{\sigma_y}{\dot{\varepsilon}_{II}}$ if $\sigma_{II} > \sigma_y$, where σ_{II} and $\dot{\varepsilon}_{II}$ are the second invariants of the stress and strain-rate tensors, respectively, and η_{com} is as defined in equation 7.12. At the plate boundary interface, the low-viscosity layer separating the overriding and subducting plates is blended into the viscosity field according to the equation

$$\eta_{wk} = \eta_o 10^{(1-A_{wk})\log_{10}\left(\frac{\eta_{eff}}{\eta_o}\right)} \qquad (7.13)$$

where A_{wk} is a scalar weak zone field, defined using SlabGenerator, assuming a sigma function with values ranging from 0 to 1 [*Jadamec and Billen*, 2010; *Jadamec et al.*, 2012]. Values of 1 correspond to fully weakened regions in the center of the shear zone and values of 0 correspond to unweakened regions. η_{eff} is the effective viscosity as defined above, and η_o is the reference viscosity equal to 1×10^{20} Pa s. The imposed η_{wk} value is an upper bound and will be overwritten if η_{eff} is lower [*Jadamec and Billen*, 2010; *Jadamec et al.*, 2012].

7.3. RESULTS

7.3.1. Newtonian versus Composite Viscosity

For models using the Newtonian viscosity formulation, there are radial variations in the upper-mantle viscosity (Figures 7.2 and 7.4a). The lithosphere viscosity structure is also defined according to the flow law for olivine aggregates, and thus the cold slab is characterized by strong core. For these models, the viscosity contrast between the lithosphere and upper mantle is three to four orders of magnitude and occurs over a distance on the order of 100 km (Figures 7.2 and 7.4a).

For models using the composite viscosity formulation, the upper-mantle viscosity structure differs from that in the Newtonian models in two primary ways. First, lateral viscosity variations occur in the upper mantle in models using the composite viscosity due to the dynamic reduction of viscosity in regions of high strain-rate (Figures 7.3, 7.4b, 7.5, and Table 7.3). This zone of reduced viscosity extends away from the slab surface into the mantle wedge, behind the slab into the subslab region, and beneath the slab tip (Figures 7.3, 7.4b). This reduction occurs in addition to the decrease in the upper-mantle viscosity in the radial direction (Figures 7.4b and 7.6). Second, the models using the composite viscosity show a horizontal layer of reduced viscosity at the base of the lithosphere of the subducting plate, beneath where the subducting plate moves along the model surface. This horizontal layer occurs beneath the overriding plate in the mantle wedge region, but becomes less pronounced beneath the overriding plate far from the subduction zone (Figures 7.3, 7.5, and 7.6). In terms of lithospheric viscosity structure, models using the composite viscosity also show a strong slab core, but the strength of the slab varies along its length with more weakening in the slab hinge (Figures 7.3, 7.4b).

Comparison of the velocity magnitude predicted for the Newtonian versus composite viscosity models, indicates that the predicted velocities are greater in models using the composite viscosity (Figures 7.2, 7.3, and 7.4). For both the Newtonian and the composite viscosity models, the mantle flow rates in the supraslab and subslab mantle increase with decreasing maximum yield stress, σ_{ymax}, resulting in a zone of induced mantle flow surrounding the slab due to torque on the slab as the slab hinge undergoes failure and bending (Figures 7.2 and 7.3). In models with the composite viscosity, there is reduced viscous support of the slab due to the nonlinear weakening of the surrounding mantle (Figure 7.3). For these models, the location of the faster mantle flow rates approximately correlate to the regions of reduced viscosity that also encompasses the slab (Figure 7.3). The remaining sections of the Results will examine the effect of varying the maximum value for the depth-dependent yield stress, σ_{ymax}, and the initial slab dip, α, for models using the composite viscosity formulation.

7.3.2. Yield Stress and Predicted Mantle Viscosity

For all models with the composite viscosity, decreasing the maximum yield strength, σ_{ymax}, leads to a larger zone of reduced viscosity in the supraslab and subslab mantle. The lower the yield stress, the lower the viscosity magnitude that develops in the mantle wedge and subslab regions (Figures 7.3, 7.5, 7.6, and Table 7.3). Figure 7.7 plots the minimum in mantle wedge viscosity, ηmw_{min}, as function of varying σ_{ymax}. Decreasing σ_{ymax} from 1000 MPa to 125 MPa can result in a half an order of magnitude of viscosity reduction in the mantle wedge, for example from $10^{18.2}$ to $10^{17.7}$ Pa s (Figure 7.7). Lowering the maximum yield stress allows for more failure in the slab hinge and therefore less of the ability of the slab to support its own weight, which in turn causes higher strain-rates and a larger reduction in the upper mantle surrounding the slab.

Decreasing the maximum yield stress also affects the spatial dimensions over which the dynamic reduction in upper-mantle viscosity occurs. Lowering the maximum yield stress from 1000 MPa to 125 MPa correlates with a wider region of dynamically reduced mantle viscosity in the subslab direction, W_{ss}, and supraslab direction, W_{mw} (Table 7.3 and Figures 7.3, 7.5, 7.6, 7.8). The width over which the mantle viscosity is reduced to below 10^{19} Pa·s ranges from ~ 218 to 420 km in the subslab mantle and from ~ 223 to 581 km in the supraslab mantle wedge (Table 7.3). Thus, the lateral extent of the viscosity reduction is greater in supraslab direction, leading to a wider mantle wedge (Table 7.3 and Figure 7.8).

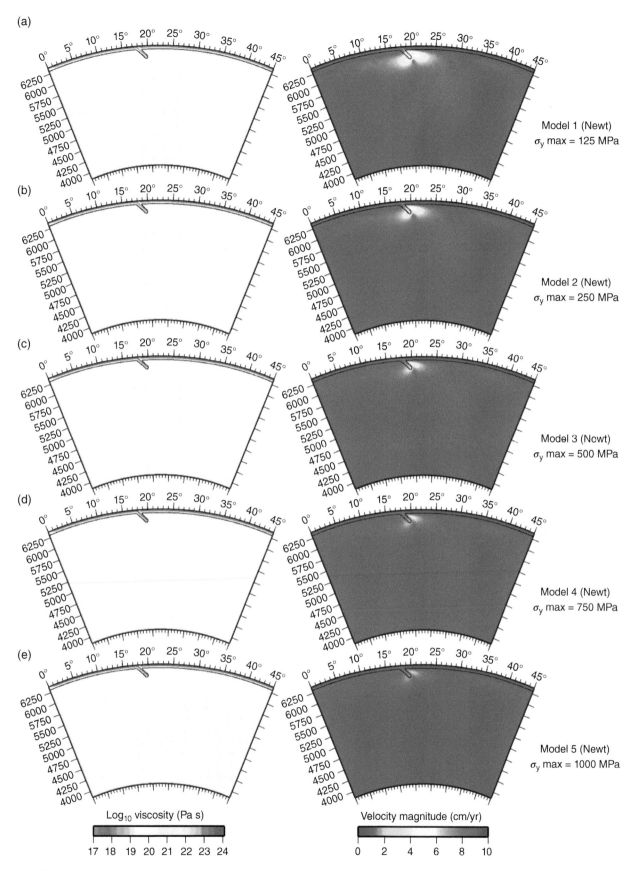

Figure 7.2 Newtonian viscosity (left) and velocity magnitude (right) for models with initial slab dip, α, of 45° and a maximum yield stress, σ_{ymax}, of (a) 125 MPa, (b) 250 MPa, (c) 500 MPa, (d) 750 MPa, and (e) 1000 MPa. Although there is a radial variation in upper-mantle viscosity, there are no lateral viscosity variations in regions of high strain-rate for the Newtonian viscosity. Thin black line marks 1050°C contour.

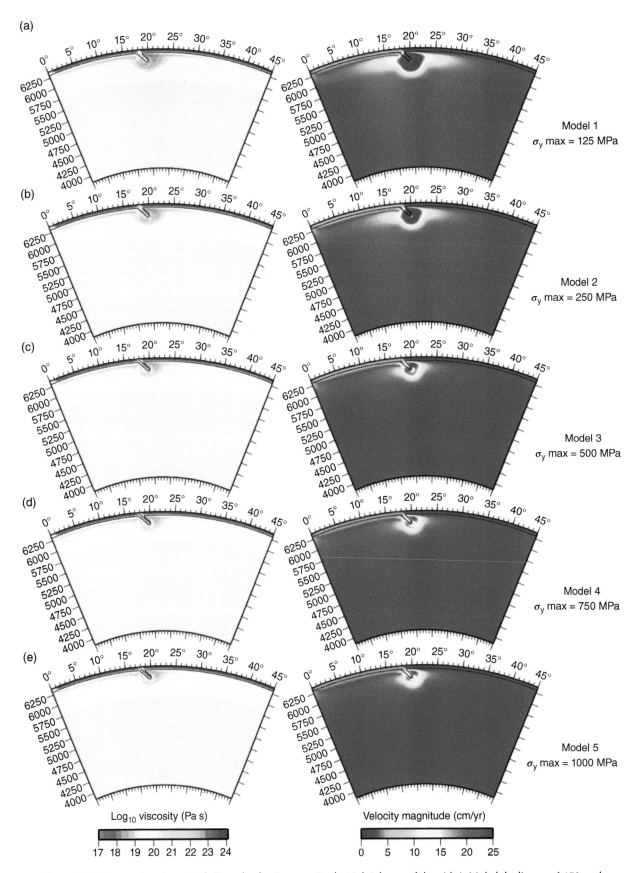

Figure 7.3 Composite viscosity (left) and velocity magnitude (right) for models with initial slab dip, α, of 45° and a maximum yield stress, σ_{ymax}, of (a) 125 MPa, (b) 250 MPa, (c) 500 MPa, (d) 750 MPa, and (e) 1000 MPa. The models show radial variations in upper-mantle viscosity as well as lateral viscosity variations in the mantle in regions of high strain-rate. Decreasing σ_{ymax} leads to weaker slabs than can steepen more easily and to greater mantle weakening and faster mantle flow rates [(a) versus (e), for example]. Velocity magnitude scale bar saturates at 25 cm/yr. Thin black line marks 1050°C contour.

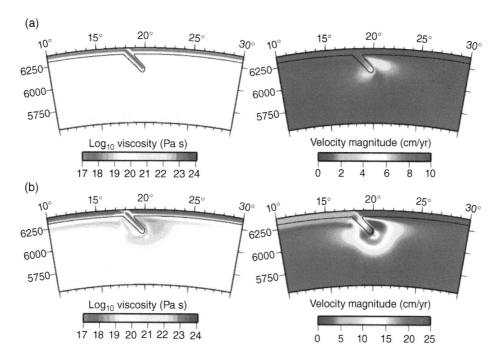

Figure 7.4 Zoomed-in view of subduction area showing viscosity and velocity magnitude for model 3 using (a) Newtonian viscosity and (b) composite viscosity. The viscosity and velocity magnitude plotted over full model domain are shown in Figure 7.2c (corresponding Newtonian model) and Figure 7.3c (corresponding composite viscosity model). Models using the composite viscosity show localized reduction in the upper mantle viscosity as the slab undergoes steepening, leading to lateral viscosity variations in the upper mantle. For the model using the composite viscosity (b) there is also more pronounced weakening in the slab hinge. Thin black line marks 1050°C contour.

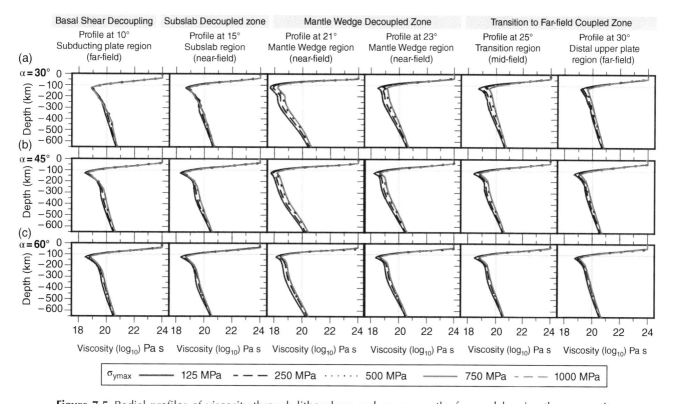

Figure 7.5 Radial profiles of viscosity through lithosphere and upper mantle, for models using the composite viscosity and an initial slab dip, α, of (a) 30°, (b) 45°, and (c) 60°. Profiles at 10° and 15° longitude sample subducting plate lithosphere and underlying mantle. Profiles at 21°, 23°, 25°, and 30° longitude sample the overriding plate lithosphere and underlying mantle, including the mantle wedge. The trench is located at 18° longitude. For reference, horizontal light-gray line marks the 100-km depth interval, and vertical gray line marks 10^{20} Pa·s, which is a viscosity value commonly used for the upper mantle. Profiles show reduction in mantle viscosity with increasing proximity to the driving force of the slab.

7.3.3. Initial Slab Dip and Predicted Mantle Viscosity

The minimum viscosity in the mantle wedge, η mw$_{min}$, as well as the width of the reduced viscosity in the supraslab mantle, W$_{mw}$, and subslab mantle, W$_{ss}$, are also sensitive to the initial dip of the slab (Figures 7.3–7.8, Table 7.3). Models with shallower slab dips are characterized by a larger viscosity reduction (Figure 7.7), but one that can occur over a smaller lateral extent (Figure 7.8). Models with the steeper initial slab dip of 60° tend to have to less of a reduction in viscosity magnitude surrounding the slab (Figure 7.7), but one that may occur over a wider extent (Figure 7.8, models with slab dip 60° versus 30°). For example, the minimum viscosity in the mantle wedge in models with an initial slab dip of 60° is on average stronger than that for models with the shallower slab dip by approximately half an order of magnitude.

Comparison of the results for models with an initial dip of 45° versus 30°, indicates that for the models with the shallower slab dip, those with the initial dip of 45° have the most pronounced viscosity reduction in the supraslab and subslab regions of the mantle (Figures 7.3–7.8, Table 7.3). This result demonstrates the effect of including an overriding plate in the models. Because there is an overriding plate, the model with the initial 30° dip has more of the slab in contact with the overriding plate, leaving less of the slab freely protruding into the mantle. Therefore, there is less driving force in the mantle for models with the initial 30° dip, leading to a smaller viscosity reduction.

7.3.4. Yield Stress and Predicted Mantle Velocity

The mantle velocity surrounding the slab is sensitive to the maximum yield stress, σ_{ymax}, used in the models. Decreasing the maximum yield stress results in a wide zone of increased velocity magnitude in supraslab mantle, below the slab tip, and in the subslab mantle (Figures 7.2 and 7.3 and Table 7.3). The effect is nonlinear, with small changes in velocity magnitude occurring in models with higher maximum yield stresses ($\sigma_{ymax} > 500$ MPa) and larger changes in the mantle flow rates occurring in models with lower yield stresses ($\sigma_{ymax} < 500$ MPa) (Figure 7.3).

For example, decreasing the maximum yield stress leads to a nonlinear increase in maximum velocity in the mantle

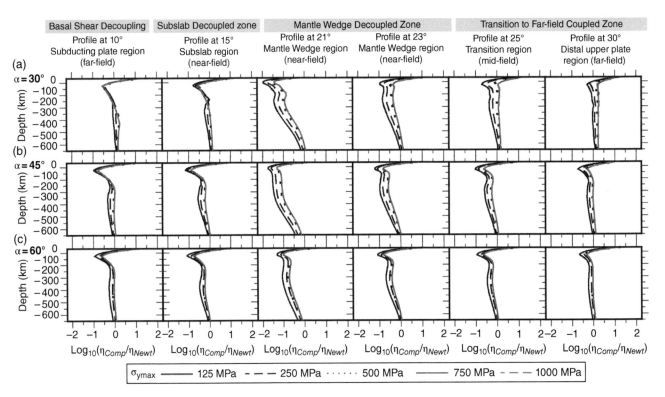

Figure 7.6 Radial profiles showing ratio of the $\log_{10}$ of composite viscosity and $\log_{10}$ of Newtonian viscosity through lithosphere and upper mantle, for models with an initial slab dip, α, of (a) 30°, (b) 45°, and (c) 60°. Profiles at 10° and 15° longitude sample subducting plate lithosphere and underlying mantle. Profiles at 21°, 23°, 25°, and 30° longitude sample the overriding plate lithosphere and underlying mantle, including the mantle wedge. The trench is located at 18° longitude.

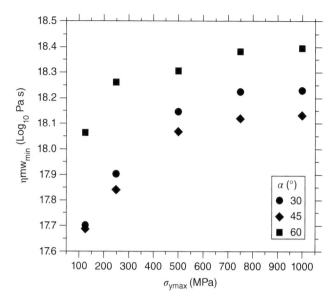

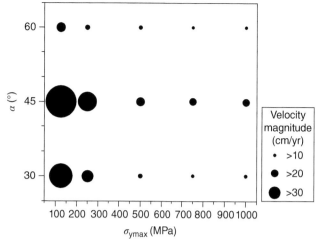

Figure 7.7 Minimum mantle wedge viscosity, ηmw_{min}, as a function of maximum yield stress, σ_{ymax} for models using the composite viscosity. Decreasing the maximum yield stress (weakening the slab) results in smaller mantle wedge viscosities. Initial slab, α, in each model shown with symbols. Minimum values measured from vertical profiles through the mantle wedge located approximately 1.5 degrees east of the code nose of the mantle wedge.

Figure 7.9 Maximum velocity magnitude in the mantle wedge plotted over parameter space of maximum yield stress, σ_{ymax}, and initial slab dip, α, for models using the composite viscosity. Models using a large σ_{ymax} (strong slab) predict slower velocities in the mantle wedge, whereas, models using a smaller σ_{ymax} (weaker slab) predict faster mantle wedge velocities. Models with a smaller dip result in faster induced flow. However, if the slab dip becomes too shallow, less of the slab protrudes into the mantle beneath the overriding plate and thus a smaller magnitude of flow is induced.

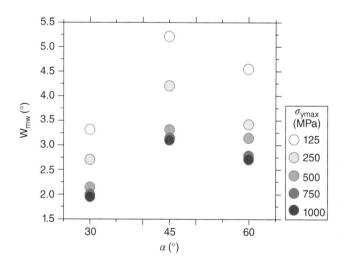

Figure 7.8 Lateral width of dynamically reduced mantle wedge viscosity, W_{mw}, as a function of slab dip, α, for models using the composite viscosity. Points colored according to maximum yield stress, σ_{ymax}, in each model. W_{mw} increases with decreasing σ_{ymax}, leading to a wider zone of reduced mantle wedge viscosity for models with weaker slabs. Distance measured in degrees (°) outward from slab surface over which viscosities are below 10^{19} Pa·s at a depth slice of 175 km.

wedge (Figure 7.9). There is a significant increase in mantle wedge velocity for models with σ_{ymax} < 500 MPa, but this effect is less pronounced for models with maximum yield stress of 750 MPa to 1000 MPa (Figure 7.9). This variation also can be seen in a plot of the percent change in maximum velocity in the mantle wedge, $\%\text{Vmw}_{max}$, for each model set with respect to the model in that set with the maximum yield stress (Figure 7.10). Here, there is a small change in the maximum mantle wedge velocity when decreasing the maximum yield stress from 1000 MPa to 750 MPa (Figure 7.10). However, a large change is observed between 750 MPa and 500 MPa, and this becomes more pronounced going from 500 MPa to 250 MPa (Figure 7.10). In these cases, the weaker slab fails more easily, inducing the faster mantle flow.

7.3.5. Initial Slab Dip and Predicted Mantle Velocity

The initial slab dip also exerts a control on the zone of increased mantle velocity surrounding the slab (Figures 7.9 and 7.10). Models with a shallower initial slab dip result in larger mantle velocities, with the largest velocity magnitudes occurring in models with the initial dip of 45° (Figures 7.9 and 7.10). Models with the steeper initial slab dip of 60° result in a slab-induced mantle flow field of both smaller magnitude and spatial dimension than that for the models with the shallower initial slab

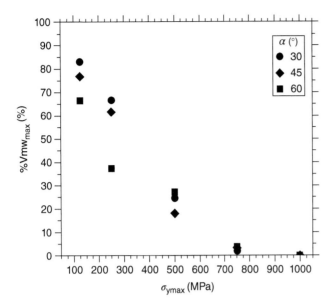

Figure 7.10 Percent change of maximum velocity in the mantle wedge, $\%Vmw_{max}$, as a function of maximum yield stress, σ_{ymax}, for models with a composite viscosity. The percent change is calculated with respect to the model that used $\sigma_{ymax} = 1000$ MPa for each slab dip, α. Therefore, for models using $\sigma_{ymax} = 1000$ MPa, there is 0% change. Lowering σ_{ymax} from 1000 MPa to 125 MPa (weakening the slab) leads to a positive percent change (speed up) in mantle wedge velocity.

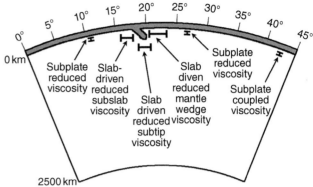

Figure 7.11 Conceptual diagram of subduction driven mantle weakening based on two-dimensional models of subduction with a strain-rate dependent viscosity.

dips. This is because the torque that tends to rotate the slab to the vertical is smaller for the slab with the steeper initial slab dip [*Tovish et al.*, 1978]. In contrast, models with the shallower initial slab dips resulted in greater slab-induced mantle flow due to the tendency for the shallower slabs to steepen. However, if the slab becomes too shallow, less of the slab protrudes into the mantle and thus a smaller magnitude of flow is induced.

7.4. DISCUSSION

A series of two-dimensional models are presented to investigate the relative role of the Newtonian versus composite viscosity, the maximum yield stress, and the initial slab dip on the viscosity structure and flow velocity in the upper mantle for subduction models that include an overriding plate. The model results indicate that a zone of slab-driven mantle deformation develops and encompasses the subducting slab in a two-dimensional subduction setting.

The results show that using the experimentally derived flow law to define the Newtonian viscosity (diffusion creep deformation mechanism) and the composite viscosity (both diffusion creep and dislocation creep deformation mechanisms) has a first-order effect on the viscosity structure and flow velocity in the upper mantle. The

results also demonstrate the maximum value of the depth-dependent yield stress, which places an upper bound on the slab strength, can have a significant impact viscosity structure and flow rates induced in the upper mantle.

The maximum flow velocities occur in models with a lower maximum yield stress and shallower slab dip, whereas, the slowest flow velocities occur in models with the largest maximum yield stress and steeper initial slab dip. However, if the slab becomes too shallow, less of the slab protrudes into the mantle and thus a smaller magnitude of flow is induced. In all cases, the magnitude of induced mantle flow is larger in the models using the composite viscosity formulation.

Figure 7.11 illustrates, in a conceptual 2D diagram of a subduction zone, the extent of the slab-driven deformation in the subslab mantle, subslab tip mantle, and supraslab mantle. A conceptual diagram of how the subduction-induced mantle deformation zone would be manifested in a more complicated three-dimensional subduction setting is provided in Figure 7.12, based on recent 3D numerical models with a composite viscosity formulation [*Jadamec and Billen*, 2010, 2012].

7.4.1. Slab-driven Zone of Upper-mantle Deformation

Models with the composite viscosity are characterized by a zone of dynamically reduced viscosity and increased mantle velocities in the supraslab, subslab, and subslab tip mantle (Figures 7.3, 7.4, and 7.11). The viscosity reduction occurs because the power-law relationship between stress and strain-rate for deformation by dislocation creep in olivine leads to a faster rate of deformation (higher strain-rates) for a given stress, which can be expressed as a reduction in the effective viscosity in models using the composite viscosity [*Hirth and Kohlstedt*, 2003; *Billen and Hirth*, 2007; *Jadamec and Billen*, 2010].

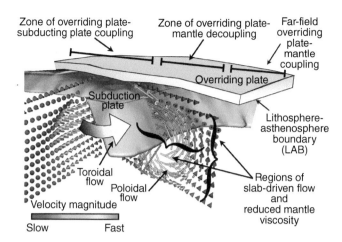

Figure 7.12 Illustration of slab-driven complex mantle flow and weakening from three-dimensional models of subduction with a strain-rate dependent viscosity and a realistic slab geometry (modified from *Jadamec and Billen* [2012]).

Therefore, models using a Newtonian viscosity [*Piromallo et al.*, 2006; *Stegman et al.*, 2006; *Schellart et al.*, 2007] do not exhibit this zone of dynamically reduced viscosity encompassing the slab in regions of high strain-rate (Figures 7.2 and 7.4).

Although models using a Newtonian viscosity do show faster velocities close to the slab during slab steepening or slab rollback [*Kincaid and Griffiths*, 2003; *Schellart*, 2004; *Funiciello et al.*, 2006; *Piromallo et al.*, 2006; *Stegman et al.*, 2006; *Schellart et al.*, 2007], the maximum velocities and the velocity gradients are larger for models using the composite rheology because of the local reduction in mantle viscosity [*Jadamec and Billen*, 2010; *Stadler et al.*, 2010; *Billen and Jadamec*, 2012]. This is consistent with a non-Newtonian viscosity behavior where a power-law dependence between stress and strain-rate is characterized by larger velocity gradients than in a Newtonian viscosity [*Turcotte and Schubert*, 2002], and has also been shown in models of mantle plumes using a nonlinear rheology [*Larsen et al.*, 1999].

The region of weakened mantle and faster flow velocities that develops and encompasses the slab extends on the order of 200 to 600 km from the slab into the supraslab mantle, below the slab tip, and behind the slab in the sub-slab mantle (Figures 7.3, 7.5, 7.11 and Table 7.3). The largest values occur with the lower maximum yield stress in models using the composite viscosity, whereas, the smaller values correspond to models with the higher yield stress (Figure 7.8). The results show that the zone of weakened mantle and increased mantle velocities is not a function of the lateral boundaries being located too close to the slab. In these models, the model sidewall is over 2500 km away from the trench in the supraslab direction, over 1800 km away behind the slab in the subslab direction, and over

2000 km below from the slab tip at the model base (Figures 7.1, 7.3, and 7.11). The models suggest that the slab-driven mantle weakening and localized increased mantle flow phenomenon occurs in most subduction zones, and is primarily modulated by the mantle deformation mechanism [*Karato et al.*, 1986; *Hirth and Kohlstedt*, 2003], slab strength, and slab buoyancy, as suggested in 3D regional [*Jadamec and Billen*, 2010; *Billen and Jadamec*, 2012; *Jadamec and Billen*, 2012] and global models [*Stadler et al.*, 2010; *Alisic et al.*, 2012] (Figures 7.11 and 7.12).

The range of widths predicted for slab-driven mantle flow is consistent with that from previous studies of subduction [*Kincaid and Griffiths*, 2003; *Schellart*, 2004; *Funiciello et al.*, 2006; *Piromallo et al.*, 2006; *Stegman et al.*, 2006; *Schellart et al.*, 2007; *Capitanio and Faccenda*, 2012], although few models have investigated the spatial extent of the dynamically reduced mantle viscosity region using the composite viscosity formulation [*Andrews and Billen*, 2009; *Jadamec and Billen*, 2010; *Alisic et al.*, 2012].

7.4.2. Shear Wave Splitting Observations and Slab-driven Upper-mantle Deformation

A zone of supraslab and subslab mantle deformation has been suggested previously during early interpretations of seismic shear wave splitting observations in the vicinity of subduction zones [*Russo and Silver*, 1994; *Fouch and Fischer*, 1996; *Savage*, 1999; *Fischer et al.*, 2000; *Smith et al.*, 2001; *Pozgay et al.*, 2007]. These studies observed seismic fast axes in the supraslab and subslab mantle commonly oriented nonparallel to the direction of local surface plate motion.

In some cases, the seismic fast axis can be an indicator of mantle flow direction through the rock fabric developed in olivine under the shearing conditions in the upper mantle. Rock deformation experiments indicate that olivine, the dominant constituent in the upper mantle, deforms primarily by dislocation creep under typical upper-mantle conditions [*Jung and Karato*, 2001; *Hirth and Kohlstedt*, 2003; *Karato et al.*, 2008]. This results in a lattice-preferred orientation (LPO) in the olivine aggregates characterized by an A-, B-, C-, D-, or E-type fabric depending predominantly on the pressure, temperature, stress, water content, and melt fraction, that produces anisotropy in the seismic waves [*Jung and Karato*, 2001; *Kaminiski and Ribe*, 2002; *Karato et al.*, 2008; *Long and Silver*, 2008; *Long and Becker*, 2010].

A-, C-, or E-type fabric is expected to dominate in the mantle wedge and asthenosphere (with the seismic fast axis parallel to the mantle flow direction) and B-type fabric in the inner cold nose of the mantle wedge (with the seismic fast axis perpendicular to the mantle flow direction) [*Jung and Karato*, 2001; *Kaminiski and Ribe*, 2002; *Kneller et al.*, 2007; *Karato et al.*, 2008; *Long and Silver*,

2008; *Long and Becker*, 2010]. This suggests therefore that the numerous seismic studies that resolve seismic anisotropy within the mantle wedge and show fast seismic axes nonparallel to plate motion where A-, C-, or E-type fabric are expected to dominate, indicate there is a local mantle flow field surrounding the slab characterized by flow vectors that may commonly be nonparallel to the direction of local surface plate motion. (See the comprehensive shear wave splitting catalog in *Long and Wirth* [2013] and also *Long* [2013].)

Therefore, the geodynamic models, which show localized small-scale circulation close to the subduction zone (e.g., Figure 7.12), independently predict a similar result to the observations of shear wave splitting in the uppermost mantle. Moreover, it is this same dislocation creep deformation mechanism that leads to both (a) the lattice-preferred orientation (LPO) in olivine in the upper mantle leading to the anisotropy in the seismic waves [*Savage*, 1999; *Kaminiski and Ribe*, 2002; *Karato et al.*, 2008; *Long and Silver*, 2009] and (b) the power-law strain-rate dependence in the viscous flow law that leads to the dynamic reduction of the viscosity surrounding the slab in the geodynamic models [*Hirth and Kohlstedt*, 2003; *Billen and Hirth*, 2007; *Jadamec and Billen*, 2010; *Stadler et al.*, 2010; *Jadamec and Billen*, 2012]. Thus, to a first order the slab-driven mantle weakening region in the supraslab and subslab mantle and related zone of localized increased mantle flow velocity are consistent with the seismic observations. Although a quantitative comparison of the geodynamic flow modeling to shear wave splitting is beyond the scope of this paper, previous studies have indicated a consistent result with slab-driven mantle flow correlating to the observations of complex splitting close to subduction zones [*Hall et al.*, 2000; *Kneller and van Keken*, 2007; *Jadamec and Billen*, 2010; *Miller and Becker*, 2012; *Faccenda and Capitanio*, 2013].

7.4.3. Mantle Decoupling at the Base of the Lithosphere

Although studies of postglacial rebound and global plate velocity models indicate the long-term viscosity of the upper mantle is on the order of 10^{20} Pa·s [*Hager*, 1991; *Mitrovica and Forte*, 2004], the models here suggest the upper-mantle viscosity may be several orders of magnitude lower in the mantle wedge (10^{18} to possibly 10^{17} Pa s) (Figures 7.3, 7.5, and 7.7). This is consistent with rock deformation experiments that indicate mantle-wedge viscosity can be reduced to 10^{18} Pa s due to sensitivity to increased temperature, water content, melt-fraction, and strain-rate [*Hirth and Kohlstedt*, 2003; *Karato*, 2003], factors that are expected to occur in the mantle wedge due to the slab-driven mantle circulation close to the subduction

zone and the release of volatiles from the slab [*Billen and Gurnis*, 2001; *Katz et al.*, 2003; *Kelley et al.*, 2006; *Cagnioncle et al.*, 2007; *Wade et al.*, 2008; *Syracuse et al.*, 2010; *Lev and Hager*, 2011; *Jadamec and Billen*, 2012].

The models suggest the strain-rate dependence as an alternate mechanism to increased temperature or melt fraction for reducing mantle wedge viscosity, consistent with the result of previous models using the composite viscosity [*Jadamec and Billen*, 2010; *Stadler et al.*, 2010; *Alisic et al.*, 2012; *Jadamec and Billen*, 2012]. For example, recent 3D geodynamic models of the Alaska subduction zone that incorporate a strain-rate dependent (non-Newtonian viscosity) predict a zone of low upper mantle viscosity and rapid mantle flow close to the slab, that transitions to higher mantle viscosity and less vigorous mantle flow with increasing distance from the driving force of the slab [*Jadamec and Billen*, 2010; *Jadamec and Billen*, 2012]. The incorporation of the strain-rate dependent viscosity into the models provides a better fit to observations of shear wave splitting in the mantle wedge and around the lateral slab edge in the eastern Alaska subduction zone [*Jadamec and Billen*, 2010; *Christensen and Abers*, 2010; *Hanna and Long*, 2012; *Jadamec and Billen*, 2012].

A natural consequence of a laterally variable upper-mantle viscosity, is the lateral variability in coupling between the upper mantle and the base of the lithospheric plates. This suggests lateral variability in tractions along the base of the lithosphere, which can be important in fitting mantle tractions to global and regional models [*Davies*, 1978; *Hager and O'Connell*, 1981; *Lithgow-Bertelloni and Guynn*, 2004; *Becker*, 2006; *Conrad and Lithgow-Bertelloni*, 2006; *Ghosh and Holt*, 2012]. Figures 7.11 and 7.12 provide a conceptual illustration of the lateral variability in coupling of the mantle to the base of the lithosphere, with localized decoupling occurring within several hundred kilometers of the subduction zone.

In addition to the region of reduced viscosity that encompasses the slab in models using the composite viscosity, the models show that a thin layer of low viscosity also develops in the asthenosphere beneath the plates (Figures 7.3, 7.5, 7.6, and 7.11). The layer is more pronounced beneath the moving downgoing plate, where the strain-rates between the plate and uppermost mantle are larger. The thin layer facilitates the surface motion of the subducting plate, as seen in previous global models [*Becker*, 2006; *Alisic et al.*, 2012]. The models show that using the composite viscosity formulation leads to a sharper definition of the rheological base of the lithosphere, which could be important in the interpretation of the lithosphere-asthenosphere boundary from seismic data [*Rychert and Shearer*, 2009; *Fischer et al.*, 2010; *Lekic and Romanowicz*, 2011; *Karato*, 2012].

7.4.4. Implications of Slab Strength on Upper-mantle Flow

Allowing the maximum yield stress to vary from 1000 MPa to 125 MPa results in progressively more pronounced failure in the slab hinge and reduced resistance of the slab-to-slab bending for models using the composite viscosity and the depth-dependent yield stress (Figures 7.3 and 7.4). The failure in the slab hinge has important trade-offs with the dynamic effects of the zone of weakened upper mantle surrounding the slab, where there is both reduced viscous support of the slab and a reduction in the viscous stresses on the slab resisting subduction. These effects are expected to be especially important during the slab steepening phase identified in time-dependent numerical and laboratory models of subduction, and can result in mantle flow rates close to the subduction zone that are an order of magnitude greater than surface plate motions.

Laboratory models with a viscous slab show that the slab dip does not remain constant throughout the life of the subduction zone, but rather the slab undergoes slab steepening during the early free-falling phase of subduction [*Schellart*, 2004; *Funiciello et al.*, 2006]. Numerical models that allow the slab dip to evolve through time show this as well [*Cizkova et al.*, 2002; *Billen and Hirth*, 2007; *Giuseppe et al.*, 2008; *Sharples et al.*, 2014]. During early subduction, as the length of slab protruding into the mantle increases, the slab reaches a point where it can no longer fully support its own weight and it begins to pivot and the slab dip steepens [*Spence*, 1987; *Conrad and Hager*, 1999; *Lallemand et al.*, 2008; *Ribe*, 2010]. The torque that tends to rotate the slab to the vertical is greater for slabs with shallower slab dips [*Tovish et al.*, 1978]. The slab dip then continues to vary as a function of the negative buoyancy of the slab, the internal strength of the slab, the overriding plate properties, the viscous support of the mantle, the dynamic pressure in the mantle, and interaction with the transition zone. Thus, the slab steepening is a natural part of the evolution of a subduction zone, and the slab strength as well as viscous support of the slab can play a large role in modulating the rate and extent of slab steepening and consequently the magnitude of induced mantle flow.

Two key parameters in controlling these effects are the maximum yield stress imposed in the models and the stress exponent in the experimentally derived flow law for olivine. Previous high-resolution regional 3D models using a composite viscosity formulation with a stress exponent of 3.5 for dislocation creep and using a depth dependent yield stress with maximum values of 500 MPa show a better fit to shear wave splitting observations, and were able to reproduce the major regions of active uplift and subsidence in south-central Alaska [*Jadamec and Billen*, 2010, 2012; *Jadamec et al.*, 2013]. Global models with local adaptive mesh refinement allowing for high resolution at plate boundaries found a better fit to a range of global observables with a maximum yield stress of 100 MPa and a stress exponent of 3.0 [*Alisic et al.*, 2012]. This implies there is a trade-off between the stress exponent and the maximum yield stress in fitting the observational constraints.

On Earth, the slab strength likely varies to some extent between subduction zones due to the variations in slab properties and subduction conditions, and may be reflected in part by the range in observed slab dips world wide [*Jarrard*, 1986; *Gudmundsson and Sambridge*, 1998; *Lallemand et al.*, 2005; *Syracuse and Abers*, 2006; *Hayes et al.*, 2012]. Nonetheless, better resolving slab strength as well as the flow law parameters are critical for understanding subduction dynamics [*King*, 2001; *Moresi and Gurnis*, 1996; *Zhong and Gurnis*, 1996; *Conrad and Hager*, 1999; *Kameyama et al.*, 1999; *Cizkova et al.*, 2002; *Billen and Hirth*, 2007; *Giuseppe et al.*, 2008; *Jadamec and Billen*, 2010; *Alisic et al.*, 2012; *Hines and Billen*, 2012]. In particular, in addition to modulating the localized subduction-induced mantle flow rates [*Jadamec and Billen*, 2010; *Stadler et al.*, 2010; *Alisic et al.*, 2012; *Billen and Jadamec*, 2012; *Jadamec and Billen*, 2012], mantle rheology and slab strength have been shown play a key role in slab detachment [*Andrews and Billen*, 2009; *Duretz et al.*, 2011], trench curvature [*Bellahsen et al.*, 2005], subduction initiation [*Regenauer-Lieb et al.*, 2001; *Billen and Hirth*, 2005], and the magnitude of slab pull and plate motions [*Conrad and Lithgow-Bertelloni*, 2002; *Faccenna et al.*, 2007; *Capitanio et al.*, 2009; *Faccenda et al.*, 2009; *Ghosh and Holt*, 2012].

In terms of upper mantle flow rates, previous high-resolution 3D regional and global models that have used the composite viscosity formulation with a depth-dependent yield stress show the magnitude of the slab-induced mantle flow can be an order of magnitude greater than plate motions [*Jadamec and Billen*, 2010; *Stadler et al.*, 2010; *Alisic et al.*, 2012; *Jadamec and Billen*, 2012]. The high-resolution two-dimensional models here illustrate the effect of the composite viscosity formulation and varying the slab strength in a simplified 2D subduction setting. The models show an increase in velocity magnitude with increasing failure in the slab hinge and a decrease in velocity magnitude with less failure in the slab hinge. Thus, in the limit, a very strong viscous slab approaches the fully rigid slab used in corner flow models that fix the slab dip and velocity. This suggests that numerical models that use a non-Newtonian viscosity but that fix both the slab dip and slab velocity (e.g., *Kneller and van Keken* [2008]; *van Keken et al.* [2008]) may underestimate the velocity magnitude in the mantle wedge, as these models would not allow failure in the slab hinge or the ability for the slab to steepen and pivot.

7.4.5. Implications for Modeling Subduction

Connecting rheology to the phenomenology of the subduction process, such as the slab evolution and slab-driven mantle deformation, is an important avenue of research. Although the results presented here are from two-dimensional subduction models, the models were run on 48 processors and require high resolution (1248 × 480 elements, with a resolution of up to 1.4 km). This is because the incorporation of the composite viscosity formulation leads to large viscosity variations over short distances, with the gradients increased in models using a lower yield strength. In addition, the inclusion of an overriding plate requires definition of the plate interface across which there are also large viscosity variations (approximately four orders of magnitude) occurring over short distances (<100 km). Large viscosity variations of this kind pose a challenge for computational codes, and models with complex 3D geometries require substantially greater numbers of elements, increasing the computational demands [Spera et al., 1982; Moresi and Solomatov, 1995; Moresi et al., 1996; Tackley, 1996; Zhong et al., 2000; May and Moresi, 2008; Geenen et al., 2009; Burstedde et al., 2009; Furuichi et al., 2011; Jadamec et al., 2012]. Finding optimal methods for handling complex geometry and complicated, nonlinear rheologies will allow mantle convection codes to simulate a much broader range of models, opening up many new applications and opportunities for comparing the models with regional observations [Moresi and Gurnis, 1996; Billen et al., 2003], such as shear wave splitting in the mantle [Kneller and van Keken, 2007; Jadamec and Billen, 2010; Miller and Becker, 2012], locations of active shortening and subsidence in the overriding plate [Jadamec et al., 2013; Eakin et al., 2014], and the transport of geochemical signatures [Hoernle et al., 2008; Durance et al., 2012; Liu and Stegman, 2012; Lin, 2014].

The models presented here include the numerical formulation for the experimentally derived flow law for olivine aggregates, which causes a dynamic viscosity reduction in regions of high strain-rate. However, it is important to point out that the models do not include mechanical anisotropy in the viscosity, that is the viscosity is not preferentially weakened in a given direction. Mechanical anisotropy in the viscosity may be expected to occur in the olivine during deformation accommodated by dislocation creep, as the fabric development may result in preferentially weakened directions. This has been shown to have an effect on the

development of lithospheric instabilities [Lev and Hager, 2008], the temperature in the mantle wedge [Lev and Hager, 2011], and the sustainability of the plate interface [Sharples et al., 2014].

It is also important to point out that although the experimentally derived flow law for olivine aggregates implemented in this study does contain a parameter for grain size and water content [Hirth and Kohlstedt, 2003], in the models presented here, the grain size and water content are held constant for the upper mantle. However, the role of dynamic grain size variability could be important [Karato et al., 1986; Hall and Parmentier, 2003; Hirth and Kohlstedt, 2003; Faul and Jackson, 2005; Hansen et al., 2011; Wada et al., 2011], as could spatially varying the water content [Karato, 2003; Cagnioncle et al., 2007; Rychert et al., 2008; Le Voci et al., 2014]. In addition, determining the dominant deformation regimes in the upper mantle is still an active area of research, with recent rock deformation experiments suggesting that the crystallographic preferred orientation of olivine due to deformation by diffusion creep could also produce shear wave splitting in the uppermost mantle [Miyazaki et al., 2013].

7.5. CONCLUSIONS

A series of two-dimensional models of subduction are presented to investigate parameters controlling slab-induced mantle weakening and increased mantle flow rates in a simplified tectonic setting. The results show that using the experimentally derived flow law to define a Newtonian viscosity and to define a composite viscosity (including the nonlinear effect of the dislocation creep deformation mechanism) has a first-order effect on the viscosity structure and flow velocity in the upper mantle. Models using the composite viscosity formulation produce a zone of subduction-induced mantle weakening that results in reduced viscous support of the slab. The results demonstrate that the maximum value of the depth-dependent yield stress, which places an upper bound on the slab strength, can also have a significant impact viscosity structure and flow rates induced in the upper mantle, with maximum mantle weakening and mantle flow rates occurring in models with a lower maximum yield stress and shallower slab dip. In all cases, the magnitude of induced mantle flow is larger in the models using the composite viscosity formulation. The models demonstrate that the composite viscosity formulation leads to a laterally variable upper-mantle viscosity, and, consequently, lateral variability in coupling of the mantle to the base of the surface plates. The results are consistent with three-dimensional models of subduction with a composite rheology and show that the rapid mantle flow predicted by these models is not due to the proximity of the model sidewalls.

ACKNOWLEDGMENT

I thank Karen Fischer and Matt Knepley for insightful discussions on the modeling results and manuscript. Thoughtful reviews by Satoru Honda and Weronika Gorczyk greatly improved the manuscript. This work was supported by National Science Foundation grant EAR-1316416 awarded to M. Jadamec. Models were run on Lonestar, at the Texas Advanced Computing Center, through XSEDE allocations TG-EAR120010 and TG-TRA110020. Thank you to Vicki Halberstadt and the staff at XSEDE. CitcomCU was modified in collaboration with Magali Billen with details given in *Jadamec and Billen* [2010], *Jadamec and Billen* [2012], and *Jadamec et al.* [2012], and was sourced from the CitcomCU C code [*Moresi et al.*, 1996; *Zhong*, 2006] hosted by the Computational Infrastructure for Geodynamics (http://geodynamics.org/cig/software/citcomcu/). Figures were made with GMT [*Wessel and Smith*, 1991].

REFERENCES

Abt, D. L., K. M. Fischer, G. A. Abers, M. Protti, V. Gonzalez, and W. Strauch (2010), Constraints from upper mantle anisotropy surrounding the Cocos slab from SK(K)S splitting, *J. Geophys. Res.*, *115*(B06316), doi:10.1029/2009JB006710.

Alisic, L., M. Gurnis, G. Stadler, C. Burstedde, and O. Ghattas (2012), Multiscale dynamics and rheology of mantle flow with plates, *J. Geophys. Res.-Sol Ea* (1978–2012), *117*(B10).

Andrews, E. R., and M. I. Billen (2009), Rheologic controls on the dynamics of slab detachment, *Tectonophysics*, *464*, 60–69, doi:10.1016/j.tecto.2007.09.004.

Audoine, E., M. K. Savage, and K. Gledhill (2004), Anisotropic structure under a back arc spreading region, the Taupo Volcanic Zone, New Zealand, *J. Geophys. Res.-Sol Ea* (1978–2012), *109*(B11).

Becker, T. W. (2006), On the effect of temperature and strain-rate dependent viscosity on global mantle flow, net rotation, and plate-driving forces, *Geophys. J. Intern.*, *167*(2), 943–957.

Behn, M. D., G. Hirth, and P. B. Kelemen (2007), Trench-parallel anisotropy produced by foundering of arc lower crust, *Science*, *317*, 108–111.

Bellahsen, N., C. Faccenna, and F. Funiciello (2005), Dynamics of subduction and plate motion in laboratory experiments: Insights into the "plate tectonics" behavior of the Earth, *J. Geophys. Res.*, *110*(B01401), doi:10.1029/2004JB002999.

Billen, M., and M. A. Jadamec (2012), Origin of localized fast mantle flow velocity in numerical models of subduction, *Geochem. Geophys. Geosys.*, *13*, Q01,016, doi: 10.1029/2011GC003856.

Billen, M. I., and G. Hirth (2005), Newtonian versus non-Newtonian upper mantle viscosity: Implications for subduction initiation, *Geophys. Res. Lett.*, *32*(L19304), doi:10.1029/2005GL023457.

Billen, M. I., and G. Hirth (2007), Rheological controls on slab dynamics, *Geochem. Geophys. Geosys.*, *8*(Q08012), doi:10.1029/2007GC001597.

Billen, M. I., and M. Gurnis (2001), A low viscosity wedge in subduction zones, *Earth and Planetary Science Letters*, *193*, 227–236.

Billen, M. I., M. Gurnis, and M. Simons (2003), Multiscale dynamics of the Tonga-Kermadec subduction zone, *Geophys. J. Intern.*, *153*, 359–388.

Burstedde, C., O. Ghattas, G. Stadler, T. Tu, and L. Wilcox (2009), Parallel scalable adjoint-based adaptive solution of variable-viscosity stokes flow problems, *Comp. Meth.s App. Mechan. Eng.*, *198*(21–26), 1691–1700.

Buttles, J., and P. Olson (1998), A laboratory model of subduction zone anisotropy, *Earth and Planetary Science Letters*, *164*, 245–262.

Cagnioncle, A. M., E. M. Parmentier, and L. T. Elkins-Tanton (2007), Effect of solid flow above a subducting slab on water distribution and melting at convergent plate boundaries, *J. Geophys. Res.*, *112*(B09402).

Capitanio, F., and M. Faccenda (2012), Complex mantle flow around heterogeneous subducting oceanic plates, *Earth Planet. Sci. Lett.*, *353*, 29–37.

Capitanio, F. A., G. Morra, and S. Goes (2009), Dynamics of plate bending at the trench and slab-plate coupling, *Geochem. Geophys. Geosys.*, *10*(4).

Carslaw, H., and J. Jaeger (1959), Conduction of heat in solids, Oxford: Clarendon Press, 2nd ed., 1.

Christensen, D. H., and G. A. Abers (2010), Seismic anisotropy under central Alaska from SKS splitting observations, *J. Geophys. Res.*, *115*(B04315), doi:10.1029/2009JB006712.

Cizkova, H., J. van Hunen, A. P. van den Berg, and N. J. Vlaar (2002), The influence of rheological weakening and yield stress on the interaction of slabs with the 670 km discontinuity, *Earth Planet. Sci. Lett.*, *199*, 447–457.

Conder, J., and D. Wiens (2007), Rapid mantle flow beneath the Tonga volcanic arc, *Earth Planet. Sci. Lett.*, *264*(1-2), 299–307.

Conrad, C. P., and B. H. Hager (1999), Effects of plate bending and fault strength at subduction zones on plate dynamics, *J. Geophys. Res.*, *104*(B8), 17,551–17,571.

Conrad, C. P., and C. Lithgow-Bertelloni (2002), How mantle slabs drive plate tectonics, *Science*, *298*, 207–209.

Conrad, C. P., and C. Lithgow-Bertelloni (2006), Influence of continental roots and asthenosphere on plate-mantle coupling, *Geophys. Res. Lett.*, *33*(5), L05,312.

Davies, G. F. (1978), The roles of boundary friction, basal shear stress and deep mantle convection in plate tectonics, *Geophys. Res. Lett.*, *5*(3), 161–164.

Durance, P. M. J., M. A. Jadamec, T. J. Falloon, and I. A. Nicholls (2012), Magmagenesis within the Hunter Ridge Rift zone resolved from olivine-hosted melt inclusions and geochemical modelling, with insights from geodynamic models, *Austral. J. Earth Sci.*, *59*(6), 913–931.

Duretz, T., T. Gerya, and D. May (2011), Numerical modelling of spontaneous slab breakoff and subsequent topographic response, *Tectonophysics*, *502*(1), 244–256.

Eakin, C. M., C. Lithgow-Bertelloni, and F. M. Davila (2014), Influence of Peruvian flat-subduction dynamics on the evolution of western Amazonia, *Earth Planet. Sci. Lett.*, *404*, 250–260.

Faccenda, M., and F. Capitanio (2013), Seismic anisotropy around subduction zones: Insights from three-dimensional modeling of upper mantle deformation and SKS splitting calculations, *Geochem. Geophys. Geosys. 14*(1), 243–262.

Faccenda, M., L. Burlini, T. V. Gerya, and D. Mainprice (2008), Fault-induced seismic anisotropy by hydration in subducting oceanic plates, *Nature*, *455*(7216), 1097–1100.

Faccenda, M., T. V. Gerya, and L. Burlini (2009), Deep slab hydration induced by bending-related variations in tectonic pressure, *Nature Geosci.*, *656*, 790–793.

Faccenna, C., A. Heuret, F. Funiciello, S. Lallemand, and T. W. Becker (2007), Predicting trench and plate motion from the dynamics of a strong slab, *Earth and, Planetary Science Letters*, *257*, 29–36.

Faul, U., and I. Jackson (2005), The seismological signature of temperature and grain size variations in the upper mantle, *Earth Planet. Sci. Lett.*, *234*(1), 119–134.

Fischer, K. M., E. M. Parmentier, A. R. Stine, and E. Wolf (2000), Modeling anisotropy and plate-driven flow in the Tonga subduction zone back arc, *J. Geophys. Res.*, *105*(B7), 16,181–16,191.

Fischer, K. M., H. A. Ford, D. L. Abt, and C. A. Rychert (2010), The lithosphere-asthenosphere boundary, *Annual Reviews of Earth and Planetary Sciences*, *38*, 549–573.

Fouch, M. J., and K. M. Fischer (1996), Mantle anisotropy beneath northwest pacific subduction zones, *J. Geophys. Res.*, *101*(B7), 15,987–16,002.

Fuchs, K. (1983), Recently formed elastic anisotropy and petrological models for the continental subcrustal lithosphere in southern Germany, *Phys. Earth Planet. Inter.*, *31*(2), 93–118.

Funiciello, F., M. Moroni, C. Piromallo, C. Faccenna, A. Cenedese, and H. A. Bui (2006), Mapping mantle flow during retreating subduction: Laboratory models analyzed by feature tracking, *J. Geophys. Res.*, *111*(B03402), doi:10.1029/2005JB003792.

Furuichi, M., D. May, and P. Tackley (2011), Development of a stokes flow solver robust to large viscosity jumps using a schur complement approach with mixed precision arithmetic, *J. Comp. Phys. 230*(24), 8835–8851.

Geenen, T., M. ur Rehman, S. MacLachlan, G. Segal, C. Vuik, A. van den Berg, and W. Spakman (2009), Scalable robust solvers for unstructured FE geodynamic modeling applications: Solving the stokes equation for models with large localized viscosity contrasts, *Geochem. Geophys. Geosys.*, *10*, Q09,002.

Gerya, T. (2012), Future directions in subduction modeling, *J. Geodynam.*, *2011*.

Ghosh, A., and W. Holt (2012), Plate motions and stresses from global dynamic models, *Science*, *335*(6070), 838–843.

Giuseppe, E. D., J. van Hunen, F. Funiciello, C. Faccenna, and D. Giardini (2008), Slab stiffness control of trench motion: Insights from numerical models, *Geochem. Geophys. Geosys.*, *9*(2), Q02,014, doi:10.1029/2007GC001776.

Gudmundsson, O., and M. Sambridge (1998), A regionalized upper mantle (RUM) seismic model, *J. Geophys. Res.*, *103*(B4), 7121–7136.

Gurnis, M., and B. H. Hager (1988), Controls of the structure of subducted slabs, *Nature*, *335*(22), 317–321.

Hager, B. H. (1984), Subducted slabs and the geoid: constraints on mantle rheology and flow, *J. Geophys. Res.*, *89*(B7), 6003–6015.

Hager, B. H. (1991), Mantle viscosity: A comparison of models from postglacial rebound and from the geoid, plate driving forces, and advected heat flux, in E. B. R. Sabodini and K. Lambeck (eds), *NATO Advanced Research Workshop on Glacial Isostasy, Sea-Level, and Mantle Rheology*, *334*, 493–513, Springer, New York.

Hager, B. H., and R. J. O'Connell (1981), A simple global model of plate dynamics and mantle convection, *J. Geophys. Res.*, *86*, 4843–4867.

Hall, C. E., and E. M. Parmentier (2003), Influence of grain size evolution on convective instability, *Geochem. Geophys. Geosys.*, *4*(3), 27 pp.

Hall, C. E., K. M. Fischer, E. M. Parmentier, and D. K. Blackman (2000), The influence of plate motions on three-dimensional back arc mantle flow and shear wave splitting, *J. Geophys. Res.*, *105*(B12), 28,009–28,033.

Hanna, J., and M. D. Long (2012), SKS splitting beneath Alaska: Regional variability and implications for subduction processes at a slab edge, *Tectonophysics*, *530–531*, 272–285.

Hansen, L., M. Zimmerman, and D. L. Kohlstedt (2011), Grain boundary sliding in San Carlos olivine: Flow law parameters and crystallographic-preferred orientation, *J. Geophys. Res.-Sol Ea (1978–2012)*, *116* (B8).

Hayes, G., D. Wald, and R. Johnson (2012), Slab1. 0: A three-dimensional model of global subduction zone geometries, *J. Geophys. Res.*, *117*(B1), B01,302.

Healy, D., S. M. Reddy, N. E. Timms, E. M. Gray, and A. V. Brovarone (2009), Trench-parallel fast axes of seismic anisotropy due to fluid-filled cracks in subducting slabs, *Earth Planet. Sci. Lett.*, *283*(1), 75–86.

Hicks, S. P., S. E. Nippress, and A. Rietbrock (2012), Sub-slab mantle anisotropy beneath south-central Chile, *Earth Planet. Sci. Lett.*, *357*, 203–213.

Hines, J., and M. Billen (2012), Sensitivity of the short-to intermediate-wavelength geoid to rheologic structure in subduction zones, *J. Geophys. Res.*, *117*(B5), B05, 410.

Hirth, G. (2002), Laboratory constraints on the rheology of the upper mantle, in S. Karato and H. R. Wenk (eds.), *Plastic Deformation of Minerals and Rocks, Reviews in Mineralogy and Geochemistry*, *51*, 97–120.

Hirth, G., and D. Kohlstedt (2003), Rheology of the upper mantle and the mantle wedge: A view from the experimentalists, in J. Eiler (ed.), *Inside the Subduction Factory, Geophysical Monograph*, *138*, 83–105, American Geophysical Union, Washington, DC.

Hirth, G., and D. L. Kohlstedt (1996), Water in the oceanic upper mantle: Implications for rheology, melt extraction and the evolution of the lithosphere, *Earth Planet. Sci. Lett.*, *144*, 93–108.

Hoernle, K., et al. (2008), Arc-parallel flow in the mantle wedge beneath Costa Rica and Nicaragua, *Nature*, *451*, 1094–1098.

Honda, S. (2009), Numerical simulations of mantle flow around slab edges, *Earth Planet.y Sci. Lett.*, *277*(1), 112–122.

Honda, S., and M. Saito (2003), Small-scale convection under the back-arc occurring in the low viscosity wedge, *Earth Planet. Sci. Lett.*, *216*, 703–715.

Jadamec, M. A., and M. I. Billen (2010), Reconciling surface plate motions and rapid three-dimensional flow around a slab edge, *Nature*, *465*, 338–342.

Jadamec, M. A., and M. I. Billen (2012), The role of rheology and slab shape on rapid mantle flow: Three-dimensional numerical models of the Alaska slab edge, *J. Geophys. Res.*, *117*(B02304), doi:10.1029/2011JB008563.

Jadamec, M. A., M. I. Billen, and O. Kreylos (2012), Three-dimensional simulations of geometrically complex subduction with large viscosity variations, in *XSEDE '12 Proceedings of the 1st Conference of the Extreme Science and Engineering Discovery Environment: Bridging from the eXtreme to the Campus and Beyond*, 1–8, Association for Computing Machinery.

Jadamec, M. A., M. I. Billen, and S. M. Roeske (2013), Three-dimensional numerical models of flat slab subduction and the Denali fault driving deformation in south-central Alaska, *Earth Planet. Sci. Lett.*, *376*, 29–42.

Jarrard, R. D. (1986), Relations among subduction parameters, *Rev. Geophys.*, *24*(2), 217–284.

Jung, H., and S.-I. Karato (2001), Water-induced fabric transitions in olivine, *Science*, *293*, 1460–1463.

Kameyama, M., D. A. Yuen, and S.-I. Karato (1999), Thermal-mechanical effects of low-temperature plasticity (the Peierls mechanism) on the deformation of a visco-elastic shear zone, *Earth Planet Sci Lett.*, *168*, 159–172.

Kaminiski, E., and N. M. Ribe (2002), Timescales of the evolution of seismic anisotropy in mantle flow, *Geol. Geochem. Geophys.*, *3*(1), doi:10.1029/2001GC000222.

Karato, S. (2003), Mapping water content in the upper mantle, in J. Eiler (ed.), *Subduction Factory, AGU Monograph*, American Geophysical Union, Washington, DC.

Karato, S. (2012), On the origin of the asthenosphere, *Earth Planet. Sci. Lett.*, *321*, 95–103.

Karato, S., H. Jung, I. Katayama, and P. Skemer (2008), Geodynamic significance of seismic anisotropy of the upper mantle: New insights from laboratory studies, *Ann. Rev. Earth and Planetary Sci.*, *36*(59-95), doi:10.1146/annurev.earth.36.031207.124120.

Karato, S., M. S. Paterson, and J. D. Fitzgerald (1986), Rheology of synthetic olivine aggregates: influence of grain size and water, *J. Geophys. Res.*, *91*, 8151–8176.

Katz, R., M. Spiegelman, and C. Langmuir (2003), A new parameterization of hydrous mantle melting, *Geochem. Geophys. Geosys.*, *4*(9), 1073.

Kelley, K. A., T. Plank, T. L. Grove, E. M. Stolper, S. Newman, and E. Hauri (2006), Mantle melting as a function of water content beneath back-arc basins, *J. Geophys. Res.-Sol Ea (1978–2012)*, *111*(B9).

Kincaid, C., and R. W. Griffiths (2003), Laboratory models of the thermal evolution of the mantle during rollback subduction, *Nature*, *425*, 58–62.

King, S. D. (2001), Subduction zones: observations and geodynamic models, *Phys. Earth Planet. Inter.*, *127*, 9–24.

Kneller, E. A., and P. E. van Keken (2007), Trench-parallel flow and seismic anisotropy in the Mariana and Andean subduction systems, *Nature*, *450*(7173), 1222–1226.

Kneller, E. A., and P. E. van Keken (2008), Effect of three-dimensional slab geometry on deformation in the mantle wedge: Implications for shear wave anisotropy, *Geochem. Geophys. Geosys.*, *9*(Q01003), doi:10.1029/2007GC001677.

Kneller, E. A., P. E. van Keken, I. Katayama, and S. Karato (2007), Stress, strain and B-type olivine fabric in the fore-arc mantle: Sensitivity tests using high-resolution steady-state subduction models, *J. Geophys. Res.*, *112*(B04406), doi: 10.1029/2006JB004544.

Kohlstedt, D. L., B. Evans, and S. J. Mackwell (1995), Strength of the lithosphere: Constraints imposed from laboratory experiments, *J. Geophys. Res.*, *100*(9), 17,587– 17,602.

Kukacka, M., and C. Matyska (2004), Influence of the zone of weakness on dip angle and shear heating of subducted slabs, *Phys. Earth Planet. Inter.*, *141*, 243–252.

Lallemand, S., A. Heuret, and D. Boutelier (2005), On the relationships between slab dip, back-arc stress, upper plate absolute motion, and crustal nature in subduction zones, *Geochem. Geophys. Geosys.*, *6*(9), Q09,006, doi:10.1029/2005GC000917.

Lallemand, S., A. Heuret, C. Faccenna, and F. Funiciello (2008), Subduction dynamics as revealed by trench migration, *Tectonics*, *27*(3), 15 pp.

Larsen, T., D. Yuen, and M. Storey (1999), Ultrafast mantle plumes and implications for flood basalt volcanism in the northern Atlantic region, *Tectonophysics*, *311*(1–4), 31–44.

Lekic, V., and B. Romanowicz (2011), Tectonic regionalization without a priori information: A cluster analysis of upper mantle tomography, *Earth Planet. Sci. Lett.* *308*(1), 151–160.

Lev, E., and B. H. Hager (2008), Rayleigh-Taylor in stabilities with anisotropic lithospheric viscosity, *Geophys. J. Inter.*, *173*(3), 806–814.

Lev, E., and B. H. Hager (2011), Anisotropic viscosity changes subduction zone thermal structure, *Geochem. Geophys. Geosys.*, *12*, Q04009, doi:10.1029/2010GC003382.

Le Voci, G., D. Davies, S. Goes, S. Kramer, and C. Wilson (2014), A systematic 2-D investigation into the mantle wedge's transient flow regime and thermal structure: Complexities arising from a hydrated rheology and thermal buoyancy, *Geochem. Geophys. Geosys.*, *15*(1), 28–51.

Lin, S. C. (2014), Three-dimensional mantle circulations and lateral slab deformation in the southern Chilean subduction zone, *J. Geophys. Res.-Sol Ea*, *119*(4), 3879–3896.

Lithgow-Bertelloni, C., and J. H. Guynn (2004), Origin of the lithospheric stress field, *J. Geophys. Res.-Sol Ea (1978–2012)*, *109*(B1).

Liu, L., and D. Stegman (2012), Origin of Columbia River flood basalt controlled by propagating rupture of the Farallon slab, *Nature*, *482*(7385), 386–389.

Long, M., and E. A. Wirth (2013), Mantle flow in subduction systems: The mantle wedge flow field and implications for wedge processes, J. Geophys. Res., doi: 10.1002/jgrb.50063.

Long, M. D. (2013), Constraints on subduction geodynamics from seismic anisotropy, *Rev. Geophys.*, *51*(1), 76–112.

Long, M. D., and P. G. Silver (2008), The subduction zone flow field from seismic anisotropy: A global view, *Science*, *319*, 315–318.

Long, M. D., and P. G. Silver (2009), Shear wave splitting and mantle anisotropy: Measurements, interpretations, and new directions, *Surv. Geophys.*, *30*, 407–461.

Long, M. D., and T. W. Becker (2010), Mantle dynamics and seismic anisotropy, *Earth Planet. Sci. Lett.*, *297*(3), 341–354.

Lynner, C., and M. D. Long (2014), Lowermost mantle anisotropy and deformation along the boundary of the african llsvp, *Geophys. Res. Lett.*, *41*(10), 3447–3454.

MacDougall, J., K. Fischer, and M. Anderson (2012), Seismic anisotropy above and below the subducting Nazca lithosphere in southern South America, *J. Geophys. Res.-Sol Ea (1978–2012)*, *117*(B12).

MacDougall, J. G., C. Kincaid, S. Szwaja, and K. M. Fischer (2014), The impact of slab dip variations, gaps, and rollback on mantle wedge flow: Insights from fluids experiments, *Geophys. J. Intet.*, ggu053, doi:10.1093/gji/ggu053.

May, D., and L. Moresi (2008), Preconditioned iterative methods for Stokes flow problems arising in computational geodynamics, *Phys. Earth Planet. Inter.*, *171*(1–4), 33–47.

McKenzie, D. P. (1969), Speculations on the consequences and causes of plate motions, *Geophys. J. Inter.*, *18*(1), 1–32.

Miller, M., and T. Becker (2012), Mantle flow deflected by interactions between subducted slabs and cratonic keels, *Nature Geosci.*, *5*(10), 726–730.

Mitrovica, J., and A. Forte (2004), A new inference of mantle viscosity based upon joint inversion of convection and glacial isostatic adjustment data, *Earth Planet. Sci. Lett.*, *225*(1), 177–189.

Mitrovica, J. X. (1996), Haskell [1935] revisited, *J. Geophys. Res.*, *101*(B1), 555– 569, doi:10.1029/95JB03208.

Miyazaki, T., K. Sueyoshi, and T. Hiraga (2013), Olivine crystals align during diffusion creep of earth's upper mantle, *Nature*, *502*(7471), 321–326.

Moresi, L., and M. Gurnis (1996), Constraints on the lateral strength of slabs form three-dimensional dynamic flow models, *Earth Planet. Sci. Lett.*, *138*, 15–28.

Moresi, L., S. Zhong, and M. Gurnis (1996), The accuracy of finite element solutions of Stoke's flow with strongly varying viscosity, *Phys. Earth Planet. Inter.*, *97*, 83–94.

Moresi, L. N., and V. S. Solomatov (1995), Numerical investigation of 2D convection with extremely large viscosity variations, *Physics of Fluids*, *7*(9), 2154–2162.

Piromallo, C., T. W. Becker, F. Funiciello, and C. Faccenna (2006), Three-dimensional instantaneous mantle flow induced by subduction, *Geophys. Res. Lett.*, *33*(L08304), doi:10.1029/2005GL025390.

Pozgay, S. H., D. A. Wiens, J. A. Conder, H. Shiobara, and H. Sugioka (2007), Complex mantle flow in the mariana subduction system: Evidence from shear wave splitting, *Geophys. J. Inter.*, *170*, 371–386.

Regenauer-Lieb, K., D. A. Yuen, and J. Branlund (2001), The initiation of subduction: criticality by addition of water, *Science*, *294*, 578–580.

Ribe, N. (2010), Bending mechanics and mode selection in free subduction: A thin-sheet analysis, *Geophys. J. Inter.*, *180*(2), 559–576.

Royden, L. H., and L. Husson (2006), Trench motion, slab geometry and viscous stresses in subduction systems, *Geophys. J. Inter.*, *167*(2), 881–905.

Russo, R. M., and P. G. Silver (1994), Trench-Parallel flow beneath the Nazca plate from seismic anisotropy, *Science*, *263*(5150), 1105–1111.

Rychert, C. A., and P. M. Shearer (2009), A global view of the lithosphere-asthenosphere boundary, *Science*, *324*(5926), 495–498.

Rychert, C. A., K. M. Fischer, G. A. Abers, T. Plank, E. Syracuse, J. M. Protti, and W. Strauch (2008), Variations in attenuation in the mantle wedge beneath Costa Rica and Nicaragua, *Geochem. Geophys. Geosys.*, *9*(10), Q10S10, doi: 10.1029/2008GC002040.

Savage, M. K. (1999), Seismic anisotropy and mantle deformation: What have we learned from shear wave splitting? *Rev. Geophys.*, *374*(12), 65–106.

Schellart, W. P. (2004), Kinematics of subduction and subduction-induced flow in the upper mantle, *J. Geophys. Res.*, *109* (B07401), doi:10.1029/2004JB002970.

Schellart, W. P., J. Freeman, D. R. Stegman, L. Moresi, and D. May (2007), Evolution and diversity of subduction zones controlled by slab width, *Nature*, *446*, 308–311, doi:10.1038.

Sharples, W., M. A. Jadamec, L. N. Moresi, and F. Capitanio (2014), Overriding plate controls on subduction evolution, *J. Geophys. Res.*, *119*, 6684–6704, doi: 10.1002/2014JB011163.

Smith, G. P., D. A. Wiens, K. M. Fischer, L. M. Dorman, S. C. Webb, and J. A. Hilderbrand (2001), A complex pattern of mantle flow in the Lau Backarc, *Science*, *292*, 713–716.

Spence, W. (1987), Slab pull and the seismotectonics of subducting lithosphere, *Rev. Geophys.*, *25*(1), 55–69.

Spera, F. J., D. A. Yuen, and S. J. Kirschvink (1982), Thermal boundary layer convection in silicic magma chambers: Effects of temperature-dependent rheology and implications for thermogravitational chemical fractionation, *J. Geophys. Res.-Sol Ea (1978–2012)*, *87*(B10), 8755–8767.

Stadler, G., M. Gurnis, C. Burstedde, L. C. Wilcox, L. Alisic, and O. Ghattas (2010), The dynamics of plate tectonics and mantle flow: From local to global scales, *Science*, *329*, 1033–1038.

Stegman, D. R., J. Freeman, W. P. Schellart, L. Moresi, and D. May (2006), Influence of trench width on subduction hinge retreat rates in 3D models of slab rollback, *Geochem. Geophys. Geosys.*, *7*(3), Q03,012, doi:10.1029/2005GC001056.

Syracuse, E., P. van Keken, and G. Abers (2010), The global range of subduction zone thermal models, *Phys. Earth Planet. Inter.*, *183*(1), 73–90.

Syracuse, E. M., and G. A. Abers (2006), Global compilation of variations in slab depth beneath arc volcanoes and implications, *Geochem. Geophys. Geosys.*, *7*(5), Q05,017, doi:10.1029/ 2005GC001045.

Tackley, P. (1996), Effects of strongly variable viscosity on three-dimensional compressible convection in planetary mantles, *J. Geophys. Res.*, *101*(B2), 3311–3332.

Tovish, A., G. Schubert, and B. P. Luyendyk (1978), Mantle flow pressure and the angle of subduction: Non-Newtonian corner flows, *J. Geophys. Res.*, *83*, 5892– 5898.

Turcotte, D. L., and G. Schubert (2002), *Geodynamics*, 2nd ed., Cambridge University Press, New York.

van Keken, P. E., et al. (2008), A community benchmark for subduction zone modeling, *Phys. Earth Planet. Inter.*, *171*(1), 187–197.

Wada, I., M. D. Behn, and J. He (2011), Grain-size distribution in the mantle wedge of subduction zones, *J. Geophys. Res.-Sol Ea (1978–2012)*, *116*(B10).

Wade, J., T. Plank, E. Hauri, K. Kelley, K. Roggensack, and M. Zimmer (2008), Prediction of magmatic water contents via measurement of H2O in clinopyroxene phenocrysts, *Geology*, *36*(10), 799–802.

Weidner, D., J. Chen, Y. Xu, Y. Wu, M. Vaughan, and L. Li (2001), Subduction zone rheology, *Phys. Earth Planet. Inter.*, *127*(1–4), 67–81.

Wessel, P., and W. H. F. Smith (1991), Free software helps map and display data, *EOS Transactions of AGU*, *72*(441), 445–446.

Yang, X., K. M. Fischer, and G. A. Abers (1995), Seismic anisotropy beneath the Shumagin islands segment of the Aleutian-Alaska subduction zone, *J. Geophys. Res.-Sol Ea (1978–2012)*, *100*(B9), 18,165–18,177.

Zhong, S. (2006), Constraints on thermochemical convection of the mantle from plume heat flux, plume excess temperature, and upper mantle temperature, *J. Geophys. Res.*, *111*(B04409), doi:10.1029/2005JB003972.

Zhong, S., and M. Gurnis (1996), Interaction of weak faults and non-Newtonian rheology produces plate tectonics in a 3D model of mantle flow, *Nature*, *383*, 245–247.

Zhong, S., M. T. Zuber, L. Moresi, and M. Gurnis (2000), Role of temperature-dependent viscosity and surface plates in spherical shell models of mantle convection, *J. Geophys. Res.*, *105*(B5), 11,063–11,082.

8

Influence on Earthquake Distributions in Slabs from Bimaterial Shear Heating

Byung-Dal So[1] and David A. Yuen[2]

ABSTRACT

The deviatoric stress regime inside subducting slabs plays an essential role in explaining the distributions of deep focus earthquakes. The relationship between the level of deviatoric stress and aseismic zone (i.e., 300 km to 450 km deep) is not well understood. We suggest that shear heating around the 410 km two-phase coexistence loop (i.e., olivine and spinel) may reduce the deviatoric stress level, and then can reduce the tendency toward transformational faulting. The phase loop makes it possible for a reduction in elastic modulus and consequently act as the bimaterial interface for providing intense shear heating within the loop. When the ratio in elastic moduli between the phase loop and surrounding region is larger than ~2, the stress drop is more than ~0.4 GPa in the phase loop. A localized zone with a low deviatoric stress level represents a new concept in subduction dynamics and may be able to explain the aseismic zone.

8.1. INTRODUCTION

The origin and spatial distribution of deep focus earthquakes (DFEs) within subducting slabs is an important issue in geophysics. In the past, people have proposed various mechanisms for understanding how DFEs occur within deep-subducting slabs where the normal brittle failure is not a preferable regime due to extremely high confining pressure. The most widely discussed mechanisms include: dehydrational embrittlement [e.g., *Meade and Jeanloz*, 1991; *Jung et al.*, 2004], transformational faulting related with the olivine-spinel phase transition [e.g., *Green and Brunley*, 1989; *Kirby et al.*, 1996], and the slab melting by shear heating [e.g., *Ogawa*, 1987; *John et al.*, 2009] around

the preexisting weak zone [*Chen and Wen*, 2015], and the differential volume contraction between the subductin gcrust and mantle lithosphere [*Liu and Zhang*, 2015]. Since slabs extending ahead of the Wadati-Benioff zones are clearly observed using seismic tomography imaging [e.g., *Fukao et al.*, 2001], we may deduce that the DFEs must be continuously active along the subducting slab. However, most subducting slabs have a peculiar spatial distribution related to the depth of seismic activity. The seismically active regions are divided into two different zones, from the top to 300 km deep and from 450 km deep to the 660-km phase boundary. On the other hand, between 300 km and 450 km of depth, no or few DFEs are detected [*Sykes*, 1966; *Isacks et al.*, 1968; *Houston*, 2007]. This pattern raises the question of why a zone of weak seismicity exists between two seismically active zones in spite of continuously increasing pressure and temperature over these depths [*Ji and Salisbury*, 1993; *Frohlich*, 2006; *Myhill*, 2013].

Two possible mechanisms for the lack of DFEs at depths of 300–450 km (hereafter, we use the acronym ASZ for this aseismic zone) have been suggested. One is the lack of hydrous minerals that can be dehydrated

[1]*Research Institute of Natural Sciences, Chungnam National University, Daejeon, South Korea*
[2]*School of Environment Studies, China University of Geosciences, Wuhan, China; Minnesota Supercomputing Institute and Department of Earth Sciences, University of Minnesota, Minneapolis, Minnesota, USA*

Subduction Dynamics: From Mantle Flow to Mega Disasters, Geophysical Monograph 211, First Edition.
Edited by Gabriele Morra, David A. Yuen, Scott D. King, Sang-Mook Lee, and Seth Stein.
© 2016 American Geophysical Union. Published 2016 by John Wiley & Sons, Inc.

under the pressure-temperature condition of the ASZ [*Green et al.*, 2010]. The other is that the olivine-spinel phase transition, which should occur at around the 410-km phase boundary and can cause DEFs through transformational faulting, is delayed due to sluggish kinetics until the depth of ~500 km due to a relatively low temperature in the core of the slab [e.g., *Iidaka and Suetsugu*, 1992]. However, in recent studies, low- (i.e., olivine) and high- (i.e., spinel) pressure polymorphs can coexist around the phase boundary in the subducting slab [*Sung and Burns*, 1976; *Rubie and Ross II*, 1994]. If there is a coexistence of olivine and spinel in the ASZ, this should explain why no DFEs are shown in the ASZ even though the olivine-spinel phase transition is ongoing in the ASZ. Based on experimental studies [e.g., *Schubnel et al.*, 2013], the DEFs by the transformational faulting require large deviatoric stresses (i.e., 1–2 GPa). We may speculate that the deviatoric stress is decreased by additional mechanisms. Shear heating can be the most efficient process for reducing the deviatoric stress [*Peacock et al.*, 1994; *Tackley*, 1998; *Gerya et al.*, 2004]. Through the shear heating, the deviatoric stress around the shear zone can be relaxed by the release of the stored strain-energy as a form of thermal energy [*Schmalholz and Podladchikov*, 2013]. We need now to address which mechanism may induce intense shear heating at this depth inside the slab.

The coexistence of low- and high-pressure polymorph is known as the phase loop [*Jackson and Anderson*, 1970] (see Figure 8.1). Recent experimental mineral physics research on the phase loop, which is associated with the olivine-spinel phase transition, has established that compressional waves are reduced by more than ~40%, corresponding to a ~70% reduction in the effective bulk moduli [*Li and Weidner*, 2008]. Moreover, the elastic shear modulus can also be significantly softened in the phase loop [*Ricard et al.*, 2009], as confirmed in the study by *Li and Weidner* [2008]. This finding has inspired us to recognize that a visible bimaterial interface, which refers to a boundary with a large contrast in elastic moduli, is present in the slab around the 410-km phase boundary. It has been well understood that bimaterial property in shear modulus can trigger shear instability with a preferred propagation orientation [*Ben-Zion and Shi*, 2005; *Ampuero and Ben-Zion*, 2008]. Recently, *So et al.* [2012] argued that larger contrasts in shear moduli between two attached lithospheres generate more energetic shear heating around the interface. In this paper, we numerically investigate how shear heating from the bimaterial instability at the phase loop impacts the thermal structure of the subducting slab. Furthermore, we discuss the dynamical implications for the lack of DFEs in the ASZ.

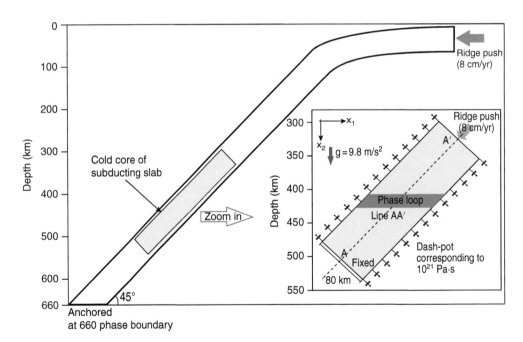

Figure 8.1 The description of the domain for our calculations. We select the cold core of the oceanic lithosphere in a subducting slab with a 45° angle. In the left box, the zoomed domain is shown. The domain (390-km long × 80-km thick) contains the phase loop (25-km thick) with a different shear modulus compared to the surrounding region. The dash-pot elements, which act as an isotactic restoring force from the asthenosphere with a viscosity of 10^{21} Pa·s [*Regenauer-Lieb et al.*, 2001], are set along the right- and left-side boundaries.

8.2. NUMERICAL METHOD AND MODEL SETTINGS

We have performed two-dimensional finite element simulation using ABAQUS code [Hibbitt, Karlsson and Sorensen, Inc., 2009]. Because the focus of this study is on the evolution of the thermal structure around the 410-km phase boundary, the domain of our calculation is the cold core of the entire subducting lithosphere containing the phase loop (see Figure 8.1). The depth range of the domain is 275 km–530 km, and the thickness is 80 km. The phase loop and the rest region are respectively meshed by rectangular elements with different sizes of 0.5×0.5 km^2 and 1×1 km^2. The phase loop has a much finer resolution, because the instability is expected to appear around the loop. The angle of subduction is 45°. The lithosphere has an elastoplastic rheology under the assumption of plane strain. Mass, momentum, and energy are conserved within the numerical scheme during the calculation. Continuity, objective derivative of the stress tensor, and energy conservation are described by the following equations (8.1–8.3):

$$\frac{\partial v_i}{\partial x_i} = 0, \tag{8.1}$$

$$\frac{\partial \sigma_{ij}}{\partial x_j} = -\rho g_i, \tag{8.2}$$

and

$$\rho C_P \frac{DT}{Dt} = \rho C_P \left[\frac{\partial T}{\partial t} + v_i \frac{\partial T}{\partial x_i} \right] = \frac{\partial}{\partial x_j} \left(k \frac{\partial T}{\partial x_j} \right) + \tau_{ij} \dot{\varepsilon}_{ij}^{plastic}. \tag{8.3}$$

x_i, v_i, and g_i represent the spatial coordinate, velocity, and gravitational acceleration in the i-th direction, respectively. The g_1 is zero and g_2 is 9.8 m/s^2. ρ, C_P, and k density, specific heat and thermal conductivity, respectively. The σ_{ij} is the total stress tensor, which is the sum of the deviatoric stress tensor (τ_{ij}) and pressure (P) (see equation 8.4).

$$\sigma_{ij} = -P\delta_{ij} + \tau_{ij} \quad \text{where } \delta_{ij} \text{ is the Kronecker delta.} \tag{8.4}$$

D/Dt is the material derivative. The meaning and value of each variable are shown in Table 8.1. In equation 8.3, the last term on the right-hand side is the shear heating, which refers to the conversion of stored elastic energy into thermal energy. Once thermal energy appears, the system goes into a mode involving highly nonlinear feedback among lithospheric plastic strength, plastic strain-rate deformation and shear heating. The total

Table 8.1 Descriptions of input parameters

Symbol	Values [Unit]	Descriptions
ρ	3300 [kg/m^3]	Density
C_P	900 [J/(kg·K)]	Specific heat
G_{sur}	4×10^{10} [Pa]	Shear modulus outside of the phase loop
σ_{yield}	10-300 [MPa]	Von Mises' yield strength
n	4.48	Power law exponent
A_{litho}	5.5×10^{-25} [Pa^{-n}·s^{-1}]	Prefactor
R	8.314 [J/(K·mol)]	Universal gas constant
E^*	350 [kJ/mol]	Activation energy
V^*	14×10^{-6} [m^3/mol]	Activation volume

Source: Karato and Ogawa [1982].

strain-rate ($\dot{\varepsilon}_{ij}^{total}$) is mathematically assumed to be the sum of the elastic ($\dot{\varepsilon}_{ij}^{elastic}$) and plastic ($\dot{\varepsilon}_{ij}^{plastic}$) strain-rates [e.g., So and Yuen, 2014]:

$$\dot{\varepsilon}_{ij}^{total} = \dot{\varepsilon}_{ij}^{elastic} + \dot{\varepsilon}_{ij}^{plastic} \tag{8.5}$$

where $\dot{\varepsilon}_{ij}^{total} = \frac{1}{2} \left[\frac{\partial v_i}{\partial x_j} + \frac{\partial v_j}{\partial x_i} \right]$ and $\dot{\varepsilon}_{ij}^{elastic} = \frac{1}{2G} \frac{D\tau_{ij}}{Dt}$.

G refers to the elastic shear modulus. For a simple von Mises yielding criterion (see equation 8.6), the plastic strain-rate will begin and follow the plastic creeping law from experimental studies dealing with olivine (see equation 8.7) [Chopra and Paterson, 1981]:

$$J_2 \geq \sigma_{yield} \tag{8.6}$$

and

$$\dot{\varepsilon}_{ij}^{plastic} = A J_2^{n-1} \tau_{ij} \exp\left(-\frac{E^* + PV^*}{RT} \right) \tag{8.7}$$

where $J_2 = \left(\frac{1}{2} \tau_{ij} \tau_{ij} \right)^{1/2}$.

E^* and V^* are the activation energy and activation volume, respectively. E^* can be effectively reduced by the content of water [e.g., Yoshino et al., 2006]. Although the basic value of σ_{yield} is 100 MPa, we will test some cases with different values of σ_{yield}, from 10 MPa to 300 MPa.

Once the slab reaches the 660-km phase boundary related to wadsleyite → perovskite + magnesiowüstite [Ringwood, 1991], the sinking of the slab into the lower mantle is prevented due to the negative Clapeyron slope at the phase boundary [e.g., Schubert et al., 1975; Ito and Takahashi, 1989]. For this reason, the stress regime of the slab is down-dip compression in seismological [e.g., Isacks and Molnar, 1971] and numerical studies [e.g., Stadler et al., 2012]. Following these examples, the bottom and top boundary conditions are fixed and constant compressional rate (i.e., 8 cm/yr), respectively. The dash-pot elements, which act as an isostatic restoring force from the purely viscous asthenosphere with a viscosity of 10^{21} Pa·s [*Regenauer-Lieb et al.*, 2001], are imposed along the two lateral boundaries. We checked that the viscous dissipation by the dash-pot element is negligibly small, compared with the heating in the phase loop. If we adopt a power-law viscosity into the mantle, the viscous dissipation will be stronger [e.g., *Gerya et al.*, 2004].

The formation of the phase loop, which has a lower elastic shear modulus than the surrounding region, is a complicated process because it simultaneously depends on temperature, pressure, and rate of subduction. We can estimate that the phase loop exists at the depth range between 350 km and 450 km with a thickness of 10 km–100 km [*Rubie and Ross II*, 1994]. The cold subducting slab (i.e., old lithosphere and fast subduction) has a thicker and deeper phase loop because the phase transition is kinetically hindered in lower temperature environments. We have predefined a simple and representative phase loop with a 25-km thickness at the depth of 410 km (see Figure 8.1). The fractions of high-pressure polymorph in the phase loop are definitely 0 and 1 at the top and bottom of the loop, respectively [*Rubie and Ross II*, 1994; *So and Yuen*, 2015]. We therefore assumed a quadratic distribution of the shear

modulus with the minimum shear modulus at the center of the loop (see Figure 8.1). The contrast of shear moduli (R_s) is varied from 1 to 5 for each calculation to determine the thermal-mechanical effects of elastic lateral heterogeneity on the thermal state of the slab structure.

Because the domain of interest lies in the cold portion of the slab, the distribution of the initial temperature is simply linear from 400°C at the top to 650°C at the bottom (see Figure 8.1), according to the temperature distribution using variable thermal conductivity derived by *Emmerson and McKenzie* [2007]. In addition, the right and left boundaries are thermally insulated. The heat transport through the interface between the cold core and the hot shell can be retarded because of the contrasts in thermal conductivity due to the large difference in the temperature [e.g., *van den Berg et al.*, 2004; *Maierová et al.*, 2012; *So and Yuen*, 2013].

8.3. RESULTS

First, we will compare the cases of $R_s = 1$ and 3. The case of $R_s = 1$ represents the model where the whole domain has a homogeneous distribution for the shear modulus (i.e., no phase loop and no shear modulus contrast). The case of $R_s = 3$ means that we predefined a quadratic distribution of the shear modulus with a minimum value of 1.33×10^{10} Pa for the loop, and the rest of the domain has a shear

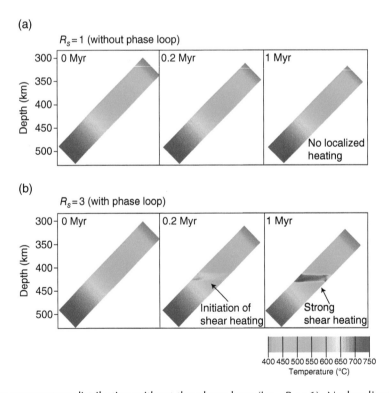

Figure 8.2 (a) The temperature distribution without the phase loop (i.e., $R_s = 1$). No localized shear heating is generated, and no notable high-temperature zone is shown. (b) The temperature distribution with the phase loop (i.e., $R_s = 3$). The heat is narrowly localized around the phase loop.

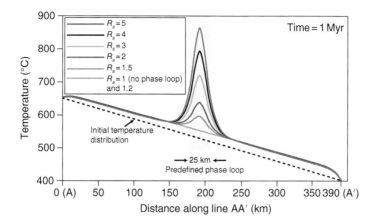

Figure 8.3 Temperature distributions along line AA' with varying R_s. In the case of larger R_s, the temperature elevation is larger than for smaller R_s. To reduce the deviatoric stress around the phase loop, R_s should be larger than 2 with 1 Myr compression.

modulus of 4×10^{10} Pa. Figures 8.2a–b show the thermal structure without and with the phase loop. No localized shear zone is formed in the case of $R_s = 1$ (see Figure 8.2a). Although more compression is applied, the localized shear zone is not shown, rather broadly distributed temperature increases are observed. Plastic yielding also occurs in the elastically homogeneous model. However, shear heating cannot be so narrowly confined, and, therefore, the heating does not significantly impact the state of deviatoric stress in subducting slabs. This is consistent with results from previous studies dealing with shear instability at the bimaterial interface [e.g., *Langer et al.*, 2010]. On the other hand, the model with the phase loop exhibits a clear localized shear zone around the loop (see Figure 8.2b). In the phase loop, the small temperature perturbation is generated after deforming for 0.2 Myr shortening. When more compression is exerted, the heating enhances and is localized around the loop. The temperature at the center of the loop is ~100°C higher than that at the lower end of the domain. We can observe a slight increase in temperature outside the loop. However, most of the heat is focused inside the loop, which is mechanically weakened enough to generate the positive feedback between the strain and intense localized heating.

The decrease in elastic modulus in the phase loop requires further study because the phase loop is a complicated system that contains many types of minerals as well as olivine and wadsleyite. To consider the uncertainty in the decrease of the shear modulus, we adopt many values of R_s (i.e., from 1 to 5) to test various cases. The activation energy, which is an important factor in determining the magnitude of shear heating, is held constant at 350 kJ/mole for all models. Figure 8.3 shows the temperature increases along Line AA' with different values of R_s. Very little shear heating is generated in cases with $R_s = 1$, 1.2, and 1.5 (see the orange and blue lines in Figure 8.3). In the case of $R_s < \sim 1.5$, the temperature distribution is similar to that with no phase loop. When R_s is

greater than 2, the temperature in the phase loop increases by 100–300°C at 1 Myr. We can predict that the deviatoric stress is significantly reduced around the phase loop. This may have an implication for the ASZ between 300-km and 450-km deep, which is revealed by earthquake distribution.

Mechanical strength plays a strong role in seismic activities. When weakening of the mechanical strength strongly interacts with the heat dissipation, the accumulated mechanical energy is not large enough to release large seismicity. For this reason, we investigate the evolution of mechanical strength with various R_s. The average values of stress over the phase loop under different R_s are calculated. We plotted the deformation time versus the areal average stress in Figure 8.4a. The larger R_s is, the more efficient weakening of strength occurs in the phase loop. The difference in strength between $R_s = 1.5$ and $R_s = 5$ is almost 30% at 1 Myr. Figure 8.4b shows the relation between strain and areal average stress over the phase loop. In phase loops with larger R_s, a larger amount of elastic energy is stored in the loop. The gray shaded areas under each curve in Figure 8.4b indicates the accumulated elastic energy before plastic yielding. This implies that phase loops with smaller shear moduli (i.e., larger R_s) can store more elastic energy. The zone with a larger elastic energy density is more likely to emit efficiently energy in the form of intense shear heating [*Regenauer-Lieb et al.*, 2012]. In Figure 8.4a–b, the level of maximum stress in subducting slabs is ~0.6 GPa, and the stress drop during the heat dissipation is ~0.3 GPa. In previous numerical studies, ~1 GPa has been proposed as an average stress level in subducting slabs [e.g., *Čížková et al.*, 2002; *John et al.*, 2009]. The estimated stress drop based on paleo-earthquake data can go up to ~0.6 GPa [*Andersen et al.*, 2008]. On the other hand, a lower stress level of ~0.1 GPa was calculated using seismic source parameters [*Prieto et al.*, 2013]. Therefore, we found that

(a)

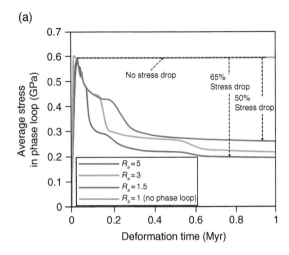

(b)

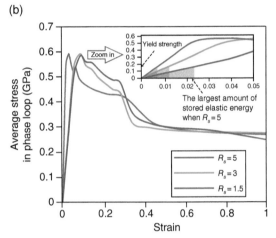

Figure 8.4 (a) Time-varying areal average of stress over the phase loop with different R_s. More efficient weakening occurs in the case of larger R_s than for smaller ones. The weakened phase loop will show weak seismicity. (b) Areal average of stress with strain. Before plastic yielding, larger R_s stores more elastic energy than smaller R_s. This difference in the accumulated elastic energy will influence the thermal structure and the deformation regime after yielding (see gray shaded area in b).

the stress level of the present study lies within the results from previous works. The average values of stress along line AA′ with different R_s are plotted after 1 Myr of deformation (see Figure 8.5). Around the phase loop, the significant drop of stress is shown when large R_s is assigned. On the other hand, the stress drop far from the loop is small. This indicates that the phase loop is not a preferable condition for transformational faulting, which can cause DFEs. In the region near point A, all cases are showing the stress drop, which is caused by the fixed boundary condition.

In order to confirm the relationship between the stored elastic energy density and the amount of dissipation, we plot the time variation of stored elastic energy density in the phase loop. The case of larger R_s can accumulate a

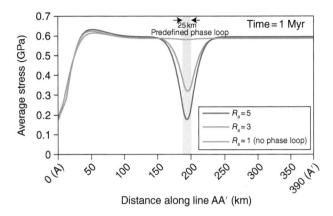

Figure 8.5 The average values of stress along line AA′ with different R_s at 1 Myr deformation. The stress drop around the phase loop is much larger than that far from the loop.

larger amount of elastic energy. After plastic yielding, all phase loops with various values of R_s begin to release the elastic energy (see Figure 8.6a). Apparently, the case with $R_s = 5$ shows the greatest reduction of elastic energy density within the loop. On the other hand, the surrounding part of the phase loop shows a small amount of elastic energy storage. Moreover, the energy release is very weak (see Figure 8.6b). This indicates that the imbalance of elastic energy densities between the phase loop and the surrounding part cause the significantly concentrated shear heating on the phase loop.

We trace the shear strain along line AA′ for varying R_s in Figure 8.7. The deviatoric stress can enhance the differential movement with attendant shear heating at this depth and lead to the stress drop transforming the deformation regime into rapid aseismic ductile creep [*Ide et al.*, 2007]. The intensively localized negative shear strain in the phase loop is shown. Moreover, the larger R_s induces larger shear strain differences. Each curve of shear strains seems to have two portions. One is the localized shear strain in the phase loop. The other is the strain related with a simple buckling by a compressional boundary condition, and has a small finite amplitude. For the case of $R_s = 1$ (see the orange line in Figure 8.7), a buckling instability over a whole domain is not generated because of a homogeneous strength of the domain. We also tested the effect of predefined plastic yield strength. Since the yield strength, as well as the elastic heterogeneity, can decide the amount of stored elastic energy, the yield strength may control long-term and global-scale shear heating dissipation. We fixed the R_s at 3, then varied the yield strength from 10 MPa to 300 MPa. Figure 8.8 shows the temperature distribution along line AA′ (see Figure 8.1) after the deformation of 1 Myr with varying the predefined plastic strength. The cases with unrealistically low (i.e., 10 MPa) yield strength, compared with the value derived from laboratory experiments [*Byerlee*,

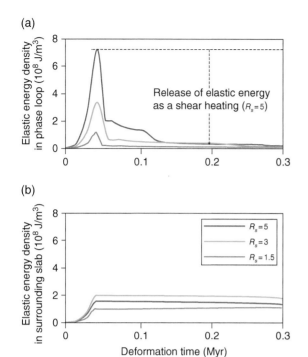

Figure 8.6 (a) The time evolution of elastic energy density in the phase loop. The case of larger R_s accumulates the larger amount of elastic energy. After plastic yielding (see the dotted arrow), the drop of elastic energy is the biggest in the case of the R_s = 5. (b) The time evolution of elastic energy density in the surrounding part. A much smaller amount of elastic energy is stored.

1978], do not show an appropriate thermal structure for the efficient stress drop around the 410-km phase boundary. The cases for 10 MPa show the broadly distributed temperature increase, which is similar with the case of small R_s. The absolute amount of elastic energy in cases with lower yield strength is too low to trigger off the bimaterial instability and promote it to a matured shear zone. The case of 300 MPa yield strength induced the broad and large temperature increase without the localized thermal structure (see the black solid line in Figure 8.8). Due to large yield strength, the outside region of the phase loop accumulates enough elastic strain energy, and then releases large thermal energy, as much as the phase loop does. On the other hand, a localized shear zone is clearly observable in the cases of realistic yield strength (i.e., from 100 to 200 MPa) [*Byerlee*, 1978]. We may deduce that the important factor for efficient shear localization is not only a large enough R_s but also a suitable range of yield strength.

8.4. DISCUSSION

We have attempted to verify our hypothesis that the shear heating from bimaterial interface due to the existence of the phase loop may be able to explain why there is weak seismicity in the depth range between 300 km and 450 km in subducting slabs. We showed that a large deviatoric stress drop occurs in the ASZ (i.e., 300–450 km

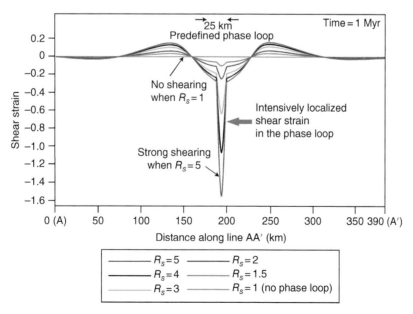

Figure 8.7 The shear strain along line AA′ with varying R_s. When a larger R_s is adopted, the stress difference between the loop and the surrounding areas is greater. This means that the positive feedback with shear heating in cases of larger R_s is stronger. This is a more conducive environment for the deviatoric stress drop, which can restrict the transformational faulting and the subsequent DFEs.

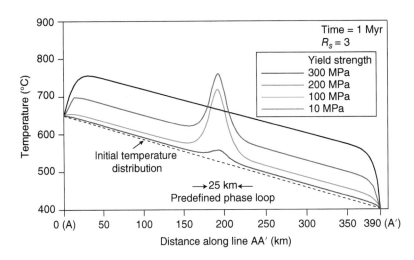

Figure 8.8 Temperature distributions along line AA′ with varying yield strengths. In the case of small yield strength (i.e., 10 and 50 MPa), the temperature distribution does not exhibit well-localized structure due to a small elastic energy accumulation. For the case of unrealistically large yield strength (i.e., 300 MPa), the localized thermal structure does not appear. The only range between 100 and 200 MPa generates localized shear heating, which can selectively reduce the deviatoric stress level.

depth) by the shear heating and strain softening when the shear modulus sufficiently decreases in the phase loop. Many studies have expended much effort to constrain the thermal structure and the stress distribution of the subducting slab. For instance, *Emmerson and McKenzie* [2007] proposed that the low temperature (<~600°C) in the cold core of the subducting slab can persist down to a depth of 660 km because of temperature-dependent thermal conductivity, which can retard heat diffusion from the hot mantle to the cold slab. Even though they were successful in explaining the existence of DFEs at levels deeper than 450 km, their models did not consider how the ASZ is formed at depths between 300 km and 450 km. *Ogawa* [1987] also discussed one of the most important factors for triggering DFEs. He argued that the shear melting of slab materials is followed by seismic slip. However, the mechanism for the ASZ was not taken into account.

Although a couple of hypotheses based on shear heating are frequently adopted for geodynamical situations in Earth's near-surface such as folding instability [e.g., *Hobbs et al.*, 2007] and the initiation of subduction [e.g., *Regenauer-Lieb et al.*, 2001], the shear heating from a bimaterial interface (i.e., phase loop) is first applied in modeling of deep subduction dynamics. It is because the elasticity has been thought to be ignored in a long time-scale geodynamics due to the short timescale physics of viscoelasticity. Some have argued the importance of elasticity in short timescale geodynamical phenomena such as the slip behavior [*Weertman*, 1980] and vibration at the fault plane [*Adams*, 1995]. However, few studies have focused on elastic energy storage in the embedded

zone with lower elastic modulus. Especially, *Kaus and Becker* [2007] argued that the elasticity contrast does not significantly affect Rayleigh-Taylor instability in mantle convection. Their elastic contrast was given for just long wavelength undulation, which is not the suitable contrast for the viscous dissipation by short wavelength localized elastic energy accumulation. On the contrary, we have shown that the short wavelength elastic heterogeneity in the slab can induce a localized shear heating, which can strongly influence the stress level of the slab. However, the problem of our study is that the shear modulus reduction in the phase loop has not been seismically detected but rather has been observed in laboratory experiments [*Li and Weidner*, 2008]. Since the expected depth of phase loop is too deep and its thickness is on the order of tens of kilometers, it is difficult that the loop is seismically detected [*Ricard et al.*, 2009]. With a consideration of the impact of bimaterial shear instability near the Earth's surface such as asymmetric mechanical rupture propagation and the distribution of aftershocks in geophysical [e.g., *Rubin and Gillard*, 2000] and numerical studies [e.g., *Dalguer and Day*, 2009], we should consider the localized bimaterial effect on the slab dynamics. According to recent experimental and semitheoretical studies [*Li and Weidner*, 2008; *Ricard et al.*, 2009], the attenuation of the shear velocity is ~40%, which refers to ~70% decrease in the elastic modulus. In our study, the cases of small R_s (< ~1.5) do not show strong shear heating to cause a large drop in deviatoric stress. On the other hand, for cases of R_s greater than 2, the shear heating is sufficiently large to cause the stress drop. For instance,

So et al. [2012], in their numerical study, found that a critical R_s value (i.e. ~1.7) for triggering an asymmetric shear instability. In seismological observations, the asymmetric features of aftershocks [*Rubin and Gillard*, 2000] and the patterns of damage [*Dor et al.*, 2006] along the San Andreas faults appear at the boundary between the Pacific and North American plates, which is intrinsically a bimaterial interface. The seismic velocity in the stiff Pacific plate is 20% faster than that in the compliant North American plates [e.g., *Eberhart-Phillips and Michael*, 1998; *Allam and Ben-Zion*, 2012], which corresponds to R_s = ~1.5.

We also investigated the elastic energy stored in the phase loop before plastic yielding and the strength of the loop after plastic yielding. *Regenauer-Lieb et al.* [2012] argued that lithospheres with lower elastic moduli, which accumulate larger elastic strain energies before the yielding, can more strongly influence the long-term evolution of thermal and mechanical structures than those with higher elastic moduli. We found that the stored elastic energy in the loop is much larger due to the larger strain associated with greater values of R_s. The loop with a larger amount of stored elastic energy more easily dissipates the energy as a form of shear heating. Another important issue is the contrast in plastic yield strength. Since we fixed the yield strength to be the same for both the phase loop and the surrounding region, the contrast in elastic modulus mainly controls the amount of elastic energy in the loop. If we use different yield strengths for the loop and surrounding region, the stored elastic energy is simultaneously controlled by the contrasts in the yield strength and the elastic modulus. Even though the yield strength of rocks has been thought to be similar regardless of types of rock, a slight difference in the yield strength from type of rock, water contents, and temperature can induce a very contrasting rheological behavior in geological timescale.

In addition to the bimaterial interface caused by the phase loop in the subducting slab, the bimaterial condition is commonplace in natural geomaterials at various spatial scales, such as the grain, glacial, fault, and plate boundary scales [*Lutter et al.*, 2004; *Deichmann et al.*, 2000], and should be considered in studies involving short wavelength shear instability. In spite of the importance and relevance of the elastic heterogeneity, few numerical studies have applied the elastic heterogeneity within the subducting slab and other geological situations. However, we have produced a distinct result by adopting the phase loop in the subducting slab. The complicated system of subducting slabs includes other bimaterial interfaces, and these interfaces have the potential to cause shear heating according to the mechanism we have discussed. We have not considered the time evolution of thermal energy with chemical reactions. For instance, dehydration of hydrous mineral is a strong endothermic reaction. If we implemented the process of dehydration in our study, the temperature increase within the phase loop could be smaller, because the heat energy must be consumed by the dehydration [*Veveakis et al.*, 2010]. Moreover, the enhanced dehydration by shear heating can cause DFEs [*Green et al.*, 2010]. Therefore, we should further investigate the relationship between the shear heating and endothermic dehydration reaction.

8.5. CONCLUSIONS

The ASZ is a region of great interest because it is located between two seismically active zones. With subduction toward the deep mantle, the pressure and temperature of the slab must be elevated; thus, the deformation regime of the deeper parts should be more aseismic. However, the ASZ is more aseismic than the hotter lower part of the slab. Since low- and high-pressure polymorphs (i.e., olivine and spinel, respectively) simultaneously exist, the olivine-spinel phase transition usually occurs around the 410-km phase boundary. This indicates that the DFEs with transformational faulting should occur. However, just a small number of DFE activity is recorded. The necessary condition for effective transformational faulting is a large deviatoric stress. Thus, in this study, we have suggested that additional mechanisms may decrease the deviatoric stress level, and then the ASZ appears around the phase loop by applying recent results regarding bimaterial instability and the 410-km phase loop.

The bimaterial interface due to the phase loop, which is the coexistence zone of lower and higher pressure polymorphs and has a lower shear modulus, results in the storage of dense elastic energy and strong shear heating for the deviatoric stress drop. Models containing a phase loop with R_s > ~2 showed a sufficient stress drop (i.e., ~500 MPa) due to the bimaterial shear heating. This shear heating can induce the ASZ around the phase loop. This explanation may be a new approach for the ASZ inside subducting slabs. Moreover, the pressure-induced amorphous zone also has a low shear modulus [e.g., *Binggeli et al.*, 1994]. Both the phase loop and the amorphized zone can act in concert as a bimaterial interface, thus influencing the thermal-mechanical structure of the slab. Hence, observational constraints of these elastic heterogeneities in the slab should be investigated. Finally, we should apply in the future the elastic contrast to explain the ASZ and buckling inside the subducting slab. Moreover, recent studies suggested [e.g., *Gerya*, 2015; *Schmalholz and Duretz*, 2015; *So and Yuen*, 2015] that the pressure deviation from lithostatic pressure (i.e., tectonic over-pressure and under-pressure) after the plastic yielding can affect the pressure-temperature condition and mechanical behaviors of deforming lithosphere.

ACKNOWLEDGMENT

We acknowledge with thanks our discussions with Shun-ichiro Karato and constructive comments by an anonymous reviewer and Manolis Veveakis. This work was supported by the National Research Foundation of Korea (NRF) grants funded by the Korean government (NRF-2014R1A6A3A04055841) for B.-D. So and by U.S. National Science Foundation grants in the Collaboration of Mathematics and Geosciences (CMG) program and Geochemistry for D. A. Yuen.

REFERENCES

Adams, G. G. (1995), Self-excited oscillations of two elastic half-spaces sliding with a constant coefficient of friction, *J. Appl. Mech.*, *62*, 867–872.

Andersen, T. B., K. Mair, H. Austrheim, Y. Y. Podladchikov, and J. C. Vrijmoed (2008), Stress-release in exhumed intermediate-deep earthquakes determined from ultramafic pseudotachylyte, *Geology*, *36*, 995–998.

Allam, A. A., and Y. Ben-Zion (2012), Seismic velocity structures in the southern California plate-boundary environment from double-difference tomography, *Geophys. J. Int.*, *190*(2), 1181–1196.

Ampuero, J.-P, and Y. Ben-Zion (2008), Cracks, pulses and macroscopic asymmetry of dynamic rupture on a biomaterial interface with velocity-weakening friction, *Geophys. J. Int.*, *173*, 674–692, doi: 10.1111/j.1365-246X.2008.03736.x.

Ben-Zion, Y., and Z. Shi (2005), Dynamic rupture on a material interface with spontaneous generation of plastic strain in the bulk, *Earth Planet. Sci. Lett.*, *236*(1–2), 486–496.

Binggeli, N., N. R. Keskar, and J. R. Chelikowsky (1994), Pressure-induced amorphization, elastic instability, and soft modes in a-quartz, *Phys. Rev. B*, *49*, 3075–3081.

Byerlee, J. D. (1978), Friction of rock, *Pure Appl. Geophys.*, *116*(4–5), 615–626.

Chen, Y., and L. Wen (2015), Global large deep-focus earthquakes: Source process and cascading failure of shear instability as a unified physical mechanism, *Earth and Planet. Sci. Lett.*, *423*, 134–144, doi:10.1016/j.epsl.2015.04.031

Chopra, P. N., and M. S. Paterson (1981), The experimental deformation of dunite, *Tectonophysics*, *78*(1–4), 453–473.

Čížková, H., J. van Hunen, A. P. van den Berg, and N. J. Vlaar (2002), The influence of rheological weakening and yield stress on the interaction of slabs with the 670 km discontinuity, *Earth Planet. Sci. Lett.*, *199*, 447–457.

Dalguer, L. A., and S. M. Day (2009), Asymmetric rupture of large aspect ratio faults at bimaterial interface in 3D, *Geophys. Res. Lett.*, *36*, L23307, doi:10.1029/2009GL040303.

Deichmann, N., J. Ansorge, F. Scherbaum, A. Aschwanden, F. Bernardi, and G. H. Gudmundsson (2000), Evidence for deep icequakes in an Alpine glacier, *Ann. Glac.*, *31*, 85–90, doi: 10.3189/172756400781820462.

Dor, O., T. K. Rockwell, and Y. Ben-Zion (2006), Geological observations of damage asymmetry in the structure of the San Jacinto, San Andreas and Punchbowl faults in Southern California: A possible indicator for preferred rupture propagation direction, *Pure Appl. Geophys.*, *163*, 301–349.

Eberhart-Phillips, D., and, A. J. Michael (1998), Seismotectonics of the Loma Prieta, California region determined from three-dimensional *Vp*, *Vp/Vs*, and seismicity, *J. Geophys. Res.*, *103*, 21099–21120.

Emmerson, B., and D. McKenzie (2007), Thermal structure and seismicity of subducting lithosphere, *Phys. Earth Planet. Inter.*, *163*, 191–208.

Frohlich, C. (2006), *Deep Earthquakes*, Cambridge Univ. Press, Cambridge, UK.

Fukao, Y., S. Widiyantoro, and M. Obayashi (2001), Stagnant slabs in the upper and lower mantle transition region, *Rev. Geophys.*, *39*(3), 291–323.

Gerya, T. V., D. A. Yuen, and W. V. Maresch (2004), Thermo-mechanical modelling of slab detachment, *Earth Planet. Sci. Lett.*, *226*(1–2), 101–106.

Green, H., and P. C. Brunley (1989), A new self-organizing mechanism for deep-focus earthquakes, *Nature*, *341*(6244), 733–737, doi:10.1038/341733a0.

Green, H. W., W.-P. Chen, and M. R. Brudzinski (2010), Seismic evidence of negligible water carried below 400-km depth in subducting lithosphere, *Nature*, *467*, 828–831.

Hibbitt, Karlsson & Sorensen, Inc. (2009), *Abaqus/Standard User's Manual Version 6.9*, Pawtucket, RI.

Hobbs, B., K. Regenauer-Lieb, and A. Ord (2007), Thermodynamics of folding in the middle to lower crust, *Geology*, *35*, 175–178, doi:10.1130/G23188A.

Houston, H. (2007), *Deep Earthquakes*, in G. Schubert (ed.), *Treatise on Geophysics, 4. Earthquake Seismology*, Elsevier, Amsterdam, 321–350.

Ide, S., G. C. Beroza, D. R. Shelly, and T. Uchide (2007), A scaling law for slow earthquakes, *Nature*, *447*, 76–79, doi:10.1038/nature05780.

Iidaka, T., and D. Suetsugu (1992), Seismological evidence for metastable olivine inside a subducting slab, *Nature*, *356*, 593–595.

Isacks, B., and P. Molnar (1971), Distribution of stress in descending lithosphere from a global survey of focal-mechanism solutions of mantle, earthquakes, *Rev. Geophys.*, *9*, 103–174.

Isacks, B., J. Oliver, and L. R. Sykes (1968), Seismology and the new global tectonics, *J. geophys. Res.*, *73*, 5855–5899.

Ito, E., and E. Takahashi (1989), Postspinel transformations in the system $Mg_2SiO_4–Fe_2SiO_4$ and some geophysical implications, *J. Geophys. Res.*, *91*, 10637–10646.

Jackson, D. D., and D. L. Anderson (1970), Physical mechanisms of seismic-wave attenuation. *Rev. Geophys. Space Phys.*, *8*, 1–63.

Ji, S., and M. H. Salisbury (1993), Shear-wave velocities, anisotropy and splitting in high-grade mylonites, *Tectonophysics*, *221*, 453–473.

John, T., S. Medvedev, L. H. Rüpke, T. B. Andersen, Y. Y. Podladchikov, and H. Austrheim (2009), Generation of intermediate-depth earthquakes by self-localizing thermal runaway, *Nat. Geosci.*, *2*, 137–140, doi:10.1038/ngeo419.

Jung, H., H. W. Green, and L. F. Dobrzhinetskaya (2004), Intermediate-depth earthquake faulting by dehydration embrittlement with negative volume change, *Nature*, *428*(6982), 545–549.

Karato, S., and M. Ogawa (1982), High-pressure recovery of olivine: implications for creep mechanisms and creep activation volume, *Phys. Earth Planet. Inter.*, *28*(2), 102–117.

Kaus, B. J. P., and T. W. Becker (2007), Effects of elasticity on the Rayleigh–Taylor instability: Implications for large-scale geodynamics, *Geophys. J. Int.*, *168*(2), 843–862, doi:10.1111/j.1365-246X.2006.03201.x.

Kirby, S. H., S. Stein, E. A. Okal, and D. C. Rubie (1996), Metastable mantle transformations and deep earthquakes in subducting oceanic lithosphere, *Rev. Geophys.*, *34*, 261–306.

Langer, S., L. M. Olsen-Kettle, D. K. Weatherley, L. Gross, and H.-B. Mühlhaus (2010), Numerical studies of quasi-static tectonic loading and dynamic rupture of bi-material interfaces, *Concurrency Comput. Pract. Exper.*, *22*(12), 1684–1702, doi: 10.1002/cpe.1540.

Li, L., and D. J. Weidner (2008), Effect of phase transitions on compressional-wave velocities in the Earth's mantle, *Nature*, *454*(7207), 984–986, doi:10.1038/nature07230.

Liu, L., and J. S. Zhang (2015), Differential contraction of subducted lithosphere layers generates deep earthquakes. *Earth and Planet. Sci. Lett.*, *421*, 98–106, doi:10.1016/j.epsl.2015.03.053.

Lutter, W. J., G. S. Fuis, T. Ryberg, D. A. Okaya, R. W. Clayton, P. M. Davis, C. Prodehl, J. M. Murphy, V. E. Langenheim, M. L. Benthien, N. J. Godfrey, N. I. Christensen, K. Thygesen, C. H. Thurber, G. Simila, and G. R. Keller (2004), Upper crustal structure from the Santa Monica mountains to the Sierra Nevada, southern California: tomographic results from the Los Angeles regional seismic experiment, Phase II (LARSE II), *Bull. Seism. Soc. Am.*, *94*(2), 619–632.

Maierová, P., T. Chust, G. Steinle-Neumann, O. Čadek, and H. Čízková (2012), The effect of variable thermal diffusivity on kinematic models of subduction, *J. Geophys. Res.*, *117*, B07202.

Meade, C., and R. Jeanloz (1991), Deep-focus earthquakes and recycling of water into the Earth's mantle, *Science*, *252*(5002), 68–72.

Myhill, R. (2013), Slab buckling and its effect on the distributions and focal mechanisms of deep-focus earthquakes, *Geophys. J. Int.*, *192*, 837–853.

Ogawa, M. (1987), Shear instability in a viscoelastic material as the cause of deep focus earthquakes, *J. Geophys. Res.*, *92*, 13801–13810, doi:10.1029/JB092iB13p13801.

Peacock, S. M., T. Rushmer, and A. B. Thompson (1994), Partial melting of subducting oceanic crust, *Earth Planet. Sci. Lett.*, *121*(1–2), 227–244.

Prieto, G. A., M. Florez, S. A. Barrett, G. C. Beroza, P. Pedraza, J. F. Blanco, and E. Poveda (2013), Seismic evidence for thermal runaway during intermediate-depth earthquake rupture, *Geophys. Res. Lett.*, *40*(23), 6064–6068.

Regenauer-Lieb, K., D. A. Yuen, and J. Branlund (2001), The initiation of subduction: Criticality by addition of water? *Science*, *294*, 578–581.

Regenauer-Lieb, K., F. Roberto, G. R. Weinberg (2012), The role of elastic stored energy in controlling the long term rheological behaviour of the lithosphere, *J. Geodyn.*, *55*, 66–75.

Ricard, Y., J. Matas, and F. Chambat (2009), Seismic attenuation in a phase change coexistence loop, *Phys. Earth Planet. Inter.*, *176*(1–2), 124–131, doi: 10.1016/j.pepi.2009.04.007.

Ringwood, A. E. (1991), Phase transformations and their bearing on the constitution and dynamics of the mantle, *Geochim. Cosmochim. Acta*, *55*, 2083–2110.

Rubie, D. C., and C. R. Ross II (1994), Kinetics of the olivine-spinel transformation in subducting lithosphere: experimental constraints and implications for deep slab processes, *Phys. Earth Planet. Inter.*, *86*, 223–241.

Rubin, A. M., and D. Gillard (2000), Aftershock asymmetry/rupture directivity among central San Andreas fault microearthquakes, *J. Geophys. Res.*, *105*, 19,095–19,109.

Schmalholz, S. M., and Y. Y. Podladchikov (2013), Tectonic overpressure in weak crustal-scale shear zones and implications for the exhumation of high-pressure rocks, *Geophys. Res. Lett.*, *40*(10), 1984–1988.

Schubert, G., D. A. Yuen, and D. L. Turcotte (1975), Role of phase transitions in a dynamic mantle, *Geophys. J. R. Astron. Soc.*, *42*, 705–735.

Schubnel, A., F. Brunet, N. Hilairet, J. Gasc, Y. Wang, and H. W. Green (2013), Deep-focus earthquake analogs recorded at high pressure and temperature in the laboratory, *Science*, *341*(6152), 1377–1380.

So, B.-D., and D. A. Yuen (2014), Stationary points in activation energy for heat dissipated with a power law temperature-dependent viscoelastoplastic rheology, *Geophys. Res. Lett.*, *41*, 4953–4960, doi:10.1002/2014GL060713.

So, B. D. and D. A. Yuen (2015). Generation of tectonic overpressure inside subducting oceanic lithosphere involving phase-loop of olivine–wadsleyite transition, *Earth Planet. Sci. Lett.*, *413*, 59–69, doi:10.1016/j.epsl.2014.12.048.

So, B.-D., and D. A. Yuen (2013), Influences of temperature-dependent thermal conductivity on surface heat flow near major faults, *Geophys. Res. Lett.*, *40*(15), 3868–3872, doi:10.1002/grl.50780.

So, B.-D., D. A. Yuen, K. Regenauer-Lieb, and S.-M. Lee (2012), Asymmetric lithospheric instability facilitated by shear modulus contrast: Implications for shear zones, *Geophys. J. Int.*, *190*(1), 23–36, doi:10.1111/j.1365-246X.2012.05473.x.

Stadler, G., M. Gurnis, C. Burstedde, L. C. Wilcox, L. Alisic, and O. Ghattas (2010), The dynamics of plate tectonics and mantle flow: From local to global scales, *Science*, *329*(5995), 1033–1038, doi:10.1126/science.1191223.

Sung, C. M., and R. G. Burns (1976), Kinetics of high-pressure phase transformations: implications to the evolution of the olivine-spinel transition in the downgoing lithosphere and its consequences on the dynamics of the mantle, *Tectonophysics*, *31*, 1–32.

Sykes, L. R. (1966), The seismicity and deep structure of island arcs, *J. Geophys. Res.*, *71*, 2981–3006.

Tackley, P. J. (1998), Self-consistent generation of tectonic plates in three-dimensional mantle convection, *Earth Planet. Sci. Lett.*, *157*(1–2), 9–22.

van den Berg, A. P., D. A. Yuen, and E. S. G. Rainey (2004), The influence of variable viscosity on delayed cooling due to variable thermal conductivity, *Phys. Earth Planet. Inter.*, *142*, 283–295.

Veveakis, E., S. Alevizos, and I. Vardoulakis (2010), Chemical reaction capping of thermal instabilities during shear of frictional faults, *J. Mech. Phys. Solids*, *58*(9), 1175–1194.

Weertman, J. (1980), Unstable slippage across a fault that separates elastic media of different elastic constant, *J. Geophys. Res.*, *85*(B3), 1455–1461.

Yoshino, T., T. Matsuzaki, S. Yamashita, and T. Katsura (2006), Hydrous olivine unable to account for conductivity anomaly at the top of the asthenosphere, *Nature*, *443*, 973–976, doi:10.1038/nature05223.

9

The Seismology of the Planet Mongo: The 2015 Ionospheric Seismology Review

Giovanni Occhipinti

ABSTRACT

The catastrophic seismic events of the last decades push forward the necessity to explore new techniques for source estimation, oceanic tsunami tracking, as well as tsunami warning systems. Early observations of the Rayleigh wave signature in the ionosphere by Doppler sounder were able to measure lithospheric properties, sounding the atmosphere at 200 km of altitude. After the Sumatra event (26 December 2004), the successful tsunami detection by altimeters validates the possibility of tsunami detection by ionospheric sounding. Today, the catastrophic tsunamigenic earthquake in Tohoku (11 March 2011) strongly affirms the potential of ionospheric sounding to visualize the vertical displacement of the ground and ocean: the Japanese GPS network, GEONET, imaged the source extent 8 min after the rupture; it also visualizes the radiation pattern and Rayleigh waves over the entire Japan, including the oceanic region overlooking the rupture; in the far field, the airglow camera located in Hawaii showed the internal gravity wave forced by the tsunami propagating in a zone of 180×180 km^2 around the island. The ionospheric sounding potential of 2D visualization could extend the present vision of seismology. This work highlights the actual capability and the potential improvement suggested by ionospheric seismology.

9.1. ORIGINS

In early history, *Aristoteles* [524] described his pneumatic theory suggesting that the cause of earthquake is the *pneuma*, and indirectly the atmosphere: indeed, the wind, heated by the Sun, gets inside the Earth-cavities where it produces strong pressure variations resulting in the earthquake. The simplistic (and unrealistic) vision of *Aristoteles* [524] contains the visionary idea that the solid Earth and the fluid parts of the planet, nominally the ocean and the atmosphere/ionosphere, continuously exchange energy. Today, the continuous excitation of normal modes, observed in the solid Earth by seismometers and usually called the "hum," is the clear proof of this hypothesis [*Nawa et al.*, 1998; *Suda et al.*, 1998; *Tanimoto et al.*, 1998; *Roult and Crawford*, 2000;

Kobayashi et al., 2001]. Several observational works show that the source of the hum is located mainly in the ocean [*Rhie and Romanowicz*, 2004; *Webb*, 2007] and partially in the atmosphere [*Nishida et al.*, 2000]. In particular, the atmospheric component of the hum seams to excite normal modes $_0S_{29}$ and $_0S_{37}$.

An additional proof of the coupling between the solid Earth and the atmosphere is the volcanic explosion of Mount Pinatubo in 1991. The atmospheric explosion was detected worldwide by the global network of seismometers in the form of a bichromatic signal with energy mainly located at 3.68 mHz and 4.40–4.65 mHz, corresponding to normal modes $_0S_{27-29}$ and $_0S_{34-37}$ [*Kanamori and Mori*, 1992; *Zurn and Widmer*, 1996; *Watada and Kanamori*, 2010].

In essence, the atmospheric energy excited by the volcanic explosion or by the atmospheric dynamics (hum-source) is transferred inside the solid Earth where it propagates at the surface as Rayleigh waves, detectable by seismometers.

Institut de Physique du Globe de Paris, Sorbonne Paris Cité, Université Paris Diderot, France

Subduction Dynamics: From Mantle Flow to Mega Disasters, Geophysical Monograph 211, First Edition.
Edited by Gabriele Morra, David A. Yuen, Scott D. King, Sang-Mook Lee, and Seth Stein.
© 2016 American Geophysical Union. Published 2016 by John Wiley & Sons, Inc.

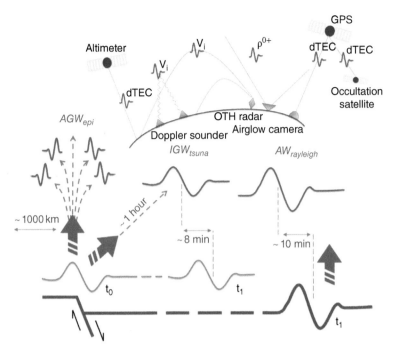

Figure 9.1 Schematic view of the coupling mechanism (bottom) and the ionospheric sounding techniques (top). Ground and oceanic displacement at the source produces AGW_{epi} that are observable in the ionosphere ~8 min after the rupture and observable until ~1000 km from the epicenter. The oceanic displacement initiates the tsunami that, during its propagation, creates IGW_{tsuna} that reach ionosphere in ~1* hr and keeps a delay of ~8* min compared to the tsunami at the sea surface. At teleseismic distance, the Rayleigh wave induces $AW_{Rayleigh}$ propagating vertically to the ionosphere in ~10 min. Times marked (*) are computed for a tsunami with a main period of 10 min, see *Occhipinti et al.* [2013] for different periods. Ionospheric sounding techniques (observable): Doppler sounder and OTH radar (vertical ion velocity); Altimeter and GPS (perturbation of the TEC), airglow (O^+ density perturbation).

Successive theoretical works [*Watada*, 1995; *Lognonné et al.*, 1998] strongly support this coupling observations and explain that the energy can be transferred in two ways: from the solid Earth to the atmosphere and *vice versa*. Normal modes computed for the Earth with ocean and atmosphere [*Lognonné et al.*, 1998] clearly show that Rayleigh waves produce acoustic waves in the atmosphere (see section 2); in the same way, tsunamis produce internal gravity waves in the overlying atmosphere (see section 4). Here we indicate the first with $AW_{Rayleigh}$ and the second IGW_{tsuna} (Figure 9.1).

The first observational evidence of the coupling between solid Earth and the fluid envelopes of the planet is an indirect consequence of the Cold War: the continuous monitoring to detect nuclear explosions via their signature in the atmosphere and in the solid Earth pushed scientists and engineers to compare atmospheric/ionospheric observations (barometers, Doppler sounders, and backscattered radars) with data from seismometers. This data matching from different instruments revealed that acoustic-gravity waves are generated not only after nuclear explosions but also after seismic events [*Row*, 1967]. As a consequence of the rich spectrum of energy characterizing the seismic rupture, the ground displacement at the epicenter (Figure 9.1) generates, simultaneously, acoustic and gravity waves in the atmosphere above the epicenter (AGW_{epi}).

After the Alaska earthquake in 1964 ($M\omega$ 9.2), the Berkeley barometer detected two unexpected signals: the first was correlated with the arrival time of Rayleigh wave ($AW_{Rayleigh}$), the second was essentially an acoustic-gravity wave propagation from the epicenter (AGW_{epi}) [*Bolt*, 1964]. The atmospheric waves propagate through the low neutral atmosphere [*Donn and Posmentier*, 1964] and up to the ionosphere where they are detected by ionospheric sounding [*Davies and Baker*, 1965; *Leonard and Barnes*, 1965; *Row*, 1966]. The 1964 Alaska earthquake opened the era of ionospheric seismology.

9.2. IONOSPHERIC SEISMOMETERS

Many observations followed the 1964 Alaska event and clearly showed that the signature of Rayleigh waves was detectable by ionospheric monitoring, mainly using Doppler sounders.

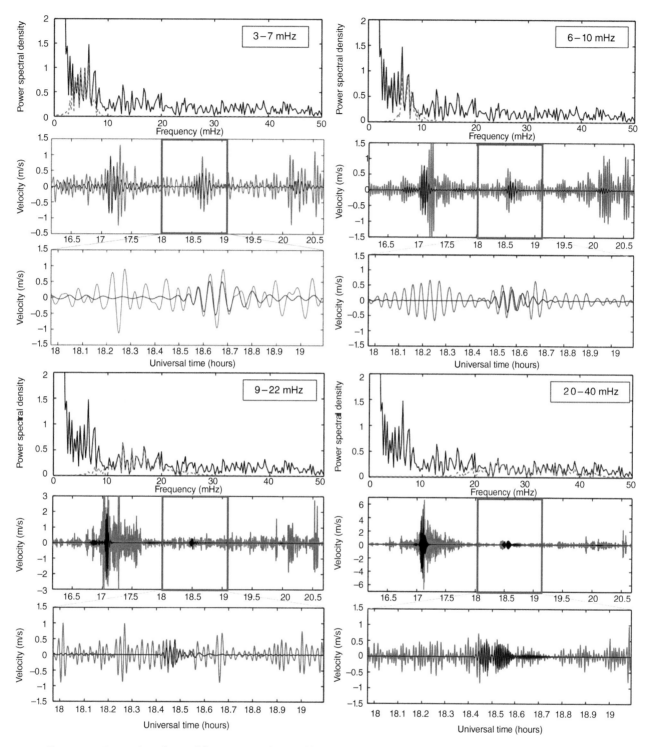

Figure 9.2 Spectral analysis of the $AW_{Rayleigh}$ detected by Doppler sounder and OTH radar after the Sumatra event (28 April 2005, M_ω 8.1). In each triptych (three plots for each highlighted band-pass range): top, spectrum of the Doppler sounder raw-data (black) and band pass filtered data (red dotted) in the frequency range showed in the right corner. Middle, Doppler sounder filtered data (red) and modeling (black), showing clearly R1 and R2, and sometimes a clear signature of R3. Bottom, OTH radar filtered data (blue) and modeling (black) showing only R2 (the timescale corresponds to the blue square in the middle plot). The R1 $AW_{Rayleigh}$ in the Doppler sounder is still recognizable independent of the frequency range.

In essence, the vertical ground displacement at the epicenter or at teleseismic distance (induced by Rayleigh waves) generates, by dynamic coupling, an acoustic-gravity wave that propagates in the neutral atmosphere. During the wave upward propagation, the wave is strongly amplified by the coupled effect of the conservation of kinetic energy $E_k = \frac{1}{2}\rho v^2$ and the exponential decrease of the air density ρ. The perturbation v of the particle velocity induced by wave propagation grows exponentially with the altitude. Reaching the ionosphere, the generated acoustic-gravity wave strongly affects the plasma density and plasma velocity becoming easily detectable by ionospheric monitoring.

The remarkable work of *Tanaka et al.* [1984] focused on the seismic event of Urakawa-Oki (21 March 1982, $M\omega$ 7.1): authors compared the signal of seismometers and Doppler sounders, both located in Japan. The spectral analysis of the two kinds of instruments clearly showed that the seismic signal with frequencies higher that the Brunt Väisääilaä frequency is transferred in the atmosphere/ionosphere and detected by Doppler sounders.

Similar Doppler sounder observations of the 1968 Tokachi-Oki event ($M\omega$ 7.5) and the 1969 Kurile Island event ($M\omega$ 7.9) were presented by *Najita and Yuen* [1979]. Authors supported theoretically the hypothesis that Rayleigh waves produce only acoustic waves in the overlying atmosphere/ionosphere, and they reproduced the dispersion curve of the Rayleigh wave [*Olivier*, 1962] using Doppler sounder observations. This is the first time that lithospheric properties are estimated observing ionosphere.

Today, $AW_{Rayleigh}$ are routinely detected by Doppler sounders for events with a magnitude larger than 6.5 [*Artru et al.*, 2004]. Additionally, *Occhipinti et al.* [2010] clearly proved that over-the-horizon (OTH) radars also are able to detect the ionospheric signature of Rayleigh waves with the same sensitivity as Doppler sounders (Figure 9.2).

9.3. THE GPS REVOLUTION: FROM POINT MEASUREMENTS TO IMAGES

The advent of global positioning system (GPS) and its ability to measure the total electron content (TEC) [*Manucci et al.*, 1993, 1998] introduced a new and revolutionary tool to sound the ionosphere: the TEC is the ionospheric electron density integrated along the ray-path between receivers and satellites. As the density of the ionospheric plasma is strongly peaked, the TEC measurements are usually located at the altitude of maximum electron density, that is, at around 300 km. TEC is expressed in TEC units (TECU); 1 TECU = $10^{16} e^-/m^2$.

A series of exploratory works found that the TEC measured by GPS receivers was able to reveal ionospheric perturbations induced by blast explosions [*Calais et al.*, 1998], shuttle launches [*Calais and Minster*, 1996], and, of particular interest here, earthquakes [*Calais and Minster*, 1995].

The following TEC measurements, performed with dense GPS arrays, allow one to visualize the signature of Rayleigh waves in the ionosphere. In essence, the plasma perturbation detected previously in a single point by Doppler sounders is now imaged in two dimensions. Using the dense californian GPS array, *Ducic et al.* [2003] imaged for the first time the propagation of $AW_{Rayleigh}$ generated by the Alaska earthquake (3 November 2002, M_ω 7.9). The estimation of the propagation speed of Rayleigh waves from the ionospheric measurement revealed a group velocity of 3.48 km/s, in accord with seismological studies [*Larson and Ekstrom*, 2001].

Similar observations were performed in Japan with the world's densest GPS array: GEONET [*Heki and Ping*, 2005]. Authors analyzed the TEC perturbations appearing close to the epicenter after the Tokachi-Oki earthquake (25 September 2003, M_ω 8.0) and after the Southeast Off-Kii-Peninsula earthquake (5 September 2004, M_ω 6.9). This last work highlighted for the first time the difference between AGW_{epi} and $AW_{Rayleigh}$, but authors interpreted both as acoustic waves, neglecting the gravity component of the AGW_{epi}. Additionally, *Heki and Ping* [2005] observed for the first time the north/south heterogeneity induced by the magnetic field inclination. This effect plays an important role in the coupling between the neutral atmosphere and the ionospheric plasma. In essence, the postseismic TEC perturbations are amplified more when they propagate southward than when they propagate northward. Later, the effect of the magnetic field was clearly explained, matching together TEC observations and normal-mode modeling, by *Rolland et al.* [2011a] in the case of $AW_{Rayleigh}$ generated by the 12 May 2008 Wenchuan earthquake (M_ω 7.9) and by the 25 September 2003 Tokachi-Oki earthquake (M_ω 8.3).

At the present time, the postseismic TEC perturbations are routinely detected. Nevertheless, the integrated nature of TEC limits the detection to events with magnitude larger than 7.

9.4. THE GREAT SUMATRA TSUNAMI IN THE IONOSPHERE

The indirect tsunami observation by ionospheric sounding slowly followed the ionospheric detection of Rayleigh waves, mainly because of observational difficulties. The idea that tsunamis produce internal gravity waves (IGW_{tsuna}) in the atmosphere, that are detectable by ionospheric sounding, was theoretically anticipated by *Hines* [1972] and *Peltier and Hines* [1976]. During the upward propagation, the IGW_{tsuna} is strongly amplified by the effect of the exponential decrease of the air density. The interaction of the IGW_{tsuna} with the ionospheric plasma environment produces strong variations in the plasma velocity and plasma density observable by ionospheric sounding, exactly as $AW_{Rayleigh}$ and AGW_{epi} described above (Figure 9.1).

The first observational work trying to detect the IGW_{tsuna} by ionospheric sounding was presented by *Artru et al.* [2005]. Authors measured TEC perturbations with the Japanese dense GPS network GEONET. The observed IGW_{tsuna} was supposed to be related to the Peruvian tsunamigenic quake on 23 June 2001 ($M\omega$ 8.4).

In essence, *Artru et al.* [2005] showed ionospheric traveling waves reaching the Japanese coast 22 hours after the tsunami generation, with an azimuth and arrival time consistent with tsunami propagation. Moreover, a period between 22 and 33 min, consistent with the tsunami, was identified in the observed TEC signals. The IGW_{tsuna} was, however, superimposed by other signals associated to traveling ionospheric disturbances (TIDs) [*Aframovich et al.*, 2003; *Balthazor and Moffett*, 1997]. The ionospheric noise is large in the gravity domain [*Garcia et al.*, 2005], consequently, the identification of the tsunami signature in the TEC was ambiguous.

The giant tsunami following the Sumatra-Andaman seismic event (26 December 2004, M_ω 9.3 [*Lay et al.*, 2005]), one order of magnitude larger than the Peruvian tsunami, provided worldwide remote sensing observations in the ionosphere, and provided the opportunity to explore ionospheric tsunami detection with a vast dataset. In addition to seismic waves detected by global seismic networks [*Park et al.*, 2005], coseismic displacement measured by GPS [*Vigny et al.*, 2005], oceanic sea surface variations measured by altimetry [*Smith et al.*, 2005], detection of magnetic anomaly [*Iyemori et al.*, 2005; *Balasis and Mandea*, 2007] and acoustic-gravity waves [*Le Pichon et al.*, 2005], a series of ionospheric disturbances, observed with different techniques, have been reported in the literature [*Liu et al.*, 2006a, b; *DasGupta et al.*, 2006; *Occhipinti et al.*, 2006, 2008b, 2013].

Two ionospheric anomalies in the plasma velocities were detected north of the epicenter by a Doppler sounding network in Taiwan [*Liu et al.*, 2006a]. The first was triggered by the vertical displacement induced by Rayleigh waves ($AW_{Rayleigh}$). The second, arriving one hour later with a longer period, is interpreted by *Liu et al.* [2006a] as the response of ionospheric plasma to the atmospheric gravity waves generated at the epicenter (AGW_{epi}).

A similar long-period perturbation, with an amplitude of 4 TECU peak to peak, was observed by GPS stations located on the coast of India [*DasGupta et al.*, 2006].

Authors didn't discriminate the origin of the observed TEC perturbation, highlighting both possibilities: the IGW_{tsuna} and the AGW_{epi}. Comparable TEC observations were done for 5 GPS stations (12 station-satellite pairs) scattered in the Indian Ocean [*Liu et al.*, 2006b]. The observed amplitude was comparable to the TEC perturbations observed by *DasGupta et al.* [2006], but the propagation speed observed in the middle of the Indian Ocean by *Liu et al.* [2006b] was clearly matched by the Deep-ocean Assessment and Reporting of Tsunamis (DART) measurements of the tsunami arrival time, showing that the ionospheric perturbation and the tsunami were following each other. This result strongly supported the IGW_{tsuna} hypothesis.

Close to those observations, the Topex/Poseidon and Jason-1 satellites acquired the key observations of the Sumatra tsunami. The measured sea level displacement observed by the two altimeters was well explained by tsunami propagation models with realistic bathymetry, and provided useful constraints on source mechanism inversions [e.g., *Song et al.*, 2005]. In addition, the inferred TEC data, required to remove the ionospheric effects from the altimetric measurements [*Bilitza et al.*, 1996], showed strong ionospheric anomalies [*Occhipinti et al.*, 2006].

In essence altimetric data from Topex/Poseidon and Jason-1 showed at the same time the tsunami signature on the sea surface and the supposed tsunami signature in the ionosphere. Using a three-dimensional numerical modeling, *Occhipinti et al.* [2006] computed the atmospheric IGW_{tsuna} generated by the Sumatra tsunami as well as the interaction of the IGW_{tsuna} with the ionospheric plasma. The quantitative approach reproduced the TEC observed by Topex/Poseidon and Jason-1 in the Indian Ocean on 26 December 2004. Those observations, supported by modeling, clearly explained the nature and the existence of the tsunami signature in the ionosphere. Later, the results obtained by *Occhipinti et al.* [2006] were reproduced by *Mai and Kiang* [2009]. Other theoretical works followed *Occhipinti et al.* [2006] to calculate the effect of dissipation, nominally viscosity and thermal conduction, on the IGW_{tsuna} [*Hickey et al.*, 2009]. Additional works theoretically explored the detection capability of IGW_{tsuna} by airglow monitoring [*Hickey et al.*, 2010] and over-the-horizon radar [*Coïsson et al.*, 2011].

The method developed by *Occhipinti et al.* [2006] was also used to estimate the role of the geomagnetic field in the tsunami signature at the E-region and F-region [*Occhipinti et al.*, 2008a]. Nominally, the authors showed that the amplification of the electron density perturbation in the ionospheric plasma at the F-region is strongly dependent on the geomagnetic field inclination. This effect is explained by the Lorenz force term in the momentum equation, characterizing the neutral plasma coupling (equation 8 in *Occhipinti et al.* [2008a]). Consequently, the detection of tsunamigenic perturbation in the F-region plasma is more easily observed at equatorial and midlatitude than at the high latitude. The heterogeneous amplification driven by the magnetic field is not observable in the E- region, consequently, detection at low altitude by HF sounding (i.e., Doppler sounding and OTH radar) is not affected by the geographical location.

Recent studies have shown the ionospheric detection of several tsunamis (Kuril, 2006, $M\omega$ 8.3; Samoa, 2009, $M\omega$ 8.1; Chile, 2010, $M\omega$ 8.8) in far field by GPS-derived TEC [*Rolland et al.*, 2010], generalizing the ionospheric detection of IGW_{tsuna} for events with lower magnitude ($M\omega \approx 8$) compared to the Sumatra event. The observed tsunami-related ionospheric perturbations, detected by the Hawaiian GPS network, appeared in the ionosphere overlying the ocean, and followed the propagation of the tsunamis at the sea level. Comparison with oceanic DART data showed similarity in the waveform as well as in the spectral signature of the ionospheric and oceanic data. This result proved again that ionosphere is a sensitive medium for tsunami propagation.

9.5. THE TOHOKU EARTHQUAKE AND TSUNAMI

9.5.1. In the Near Field

Particular attention has been paid recently to the Tohoku-Oki event (11 March 2011, $M\omega$ 9.0 [*Wei et al.*, 2012]). Thanks to the really dense GPS network in Japan (GEONET), the coseismic TEC perturbations observed at the source gave a clear image of the ionospheric perturbation in the near field [*Tsugawa et al.*, 2011; *Saito et al.*, 2011; *Rolland et al.*, 2011b], including the acoustic-gravity wave (AGW_{epi}) generated by the vertical displacement of the source [*Astafyeva et al.*, 2011, 2013], the acoustic waves coupled with Rayleigh waves ($AW_{Rayleigh}$), as well as the gravity wave induced by the tsunami propagation (IGW_{tsuna}) [*Liu et al.*, 2011; *Galvan et al.*, 2012]. The analysis of the first arrival in the TEC data in the epicentral area also allowed the localization of the epicenter with a discrepancy of less than 100 km from the official USGS location [*Tsugawa et al.*, 2011; *Tsai et al.*, 2011; *Astafyeva et al.*, 2011, 2013].

The remarkable result of *Astafyeva et al.* [2011, 2013] clearly showed that the ionospheric perturbation at the epicenter (AGW_{epi}) appears only 8 min after the rupture and shows the horizontal extension of the source. This result clearly opens the potential application of ionospheric seismology for a tsunami warning system. Indeed, the horizontal extent of the source is key information for early estimation of the tsunami amplitude.

Additionally, qualitative perturbation was also observed by four ionosondes [*Liu and Sun*, 2011] and by the Japanese SuperDARN Hokkaido radar [*Nishitami et al.*, 2011], which showed detection of the ionospheric signature of the Rayleigh waves, already observed in the past by the French OTH radar Nostradamus [*Occhipinti et al.*, 2010].

Notwithstanding the huge amount of GPS data and the clear image of the TEC perturbations at the epicentral area, discrimination between acoustic-gravity waves (AGW_{epi}) generated at the epicenter by the direct vertical displacement of the source-rupture and the internal gravity wave coupled with the tsunami (IGW_{tsuna}) was still difficult. The specific horizontal high speed of Rayleigh waves (≈ 3.5 km/s) separates and makes really recognizable the $AW_{Rayleigh}$ in the ionosphere.

The recent work of *Occhipinti et al.* [2013] collects ionospheric TEC data from several ground networks in order to visualize and analyze the ionospheric perturbations at the epicenter of the following events: Sumatra, 26 December 2004, $M\omega$ 9.1, and 12 September 2007, $M\omega$ 8.5; Chile, 14 November 2007, $M\omega$ 7.7; Samoa, 29 September 2009, $M\omega$ 8.1; and the catastrophic Tohoku-Oki event, 11 March 2011, $M\omega$ 9.0 (Figure 9.3). The work introduces some theoretical bases to interpret the data and discriminate between AGW_{epi} and IGW_{tsuna}. Section 6 summarizes the physical properties of the different waves described.

9.5.2. In the Far Field

Far away from the epicentral area, the AGW_{epi} generated by the direct vertical displacement of the ground and directly linked to the rupture are strongly attenuated and not detectable anymore by ionospheric sounding. In the far field, the propagation of $AW_{Rayleigh}$ and IGW_{tsuna} is completely separated as a consequence of the tremendous difference of horizontal speed of Rayleigh waves (≈ 3.5 km/s) and tsunamis (≈ 200 m/s). This is the main reason why, as described above, the $AW_{Rayleigh}$ and IGW_{tsuna} have been already detected by ionospheric sounding by different techniques. Notwithstanding that the $AW_{Rayleigh}$ and IGW_{tsuna} are today routinely detected in the far field, the Tohoku event observations opened new perspectives in ionospheric seismology.

First of all, the $AW_{Rayleigh}$ and IGW_{tsuna} were detected initially in the neutral atmosphere instead of the ionosphere [*Garcia et al.*, 2013, 2014]. The gravity mission Gravity field and Ocean Circulation Explorer (GOCE) crossed the wave front of $AW_{Rayleigh}$ [*Garcia et al.*, 2013] and IGW_{tsuna} [*Garcia et al.*, 2014] measuring the air-speed variation and the air-density variation. Those measurements represent a useful benchmark to validate the propagation modeling of $AW_{Rayleigh}$ and IGW_{tsuna} in the neutral atmosphere, and, combined with ionospheric observations, additionally allow one to explore the physics of the neutral-plasma coupling.

Second, the detection of IGW_{tsuna} by airglow, mentioned before, has been recently validated by observation [*Makela et al.*, 2011] and modeling [*Occhipinti et al.*, 2011] in the case of the Tohoku event (11 March 2011, M_ω 9.0). The airglow is measuring the photon emission at 630 nm, indirectly linked to the plasma density of O^+ [*Link and Cogger*, 1988] and it is commonly used to detect transient events in the ionosphere [*Kelley et al.*, 2002; *Makela et al.*, 2009; *Miller et al.*, 2009]. The modeling of the IGW_{tsuna} clearly reproduced the pattern of the airglow measurement observed over Hawaii (Figure 9.4). The comparison between the observation and the modeling

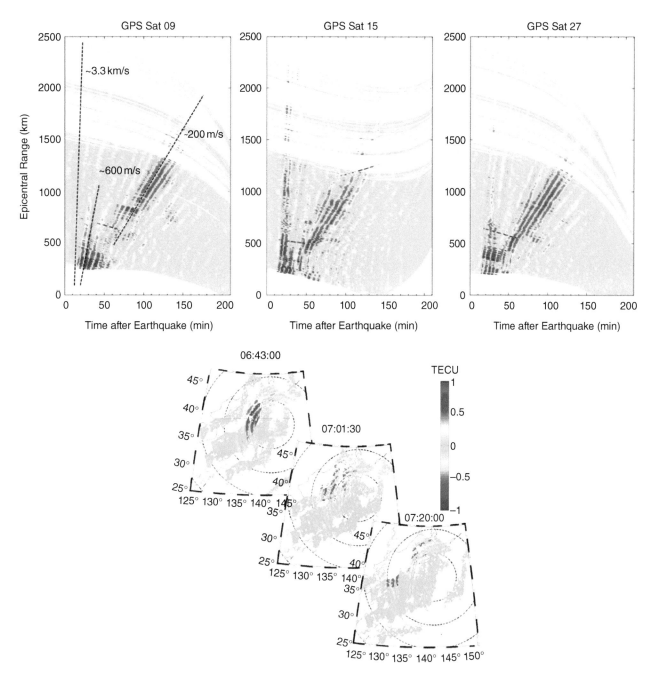

Figure 9.3 Hodochrones (top) of the TEC perturbation observed by the Japanese GEONET network and by three satellites (indicated in the top) following the recent Tohoku-Oki tsunamigenic earthquake (2011, M: 9.0). We observe $AW_{Rayleigh}$ with a speed of around 3.3 km/s, AGW_{epi} moving with a speed of around 650 m/s. The AGW_{epi} disappears after 500–1000 km, and it is replaced by IGW_{tsuna} with a speed of around 200 m/s. (Adapted from *Occhipinti et al.* [2013].) On the bottom, the data are represented over Japan using all satellites. (Adapted from *Rolland et al.* [2011b].)

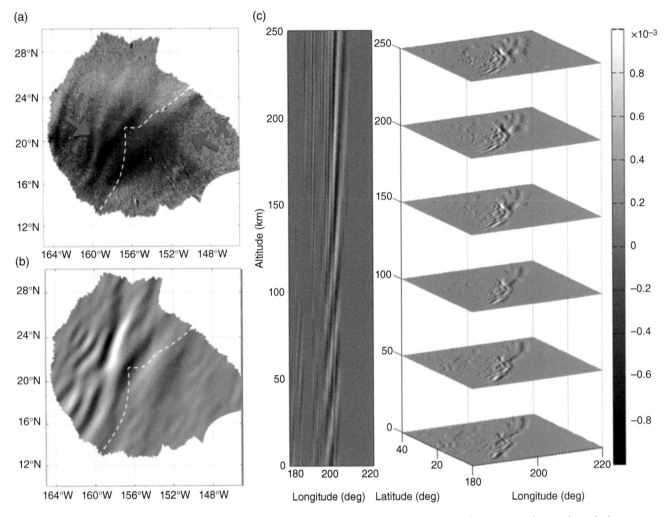

Figure 9.4 Airglow perturbation of the IGW_{tsuna} (a) observed by the camera located in Hawaii during the Tohoku tsunami (2011), and (b) the modeling of the IGW_{tsuna} for the layer corresponding to the observation, and (c) the entire structure in 3D. (Adapted from *Occhipinti et al.* [2011].)

allows one to recognize not only the evident Y shape, but also the longer wavelength perturbation (indicated with X in Figure 9.4) that arrives before the tsunami wavefront by the effect of bathymetry [*Occhipinti et al.*, 2011]. Approaching the Hawaiian archipelago, the tsunami propagation is slowed down (reduction of the sea depth), but, the IGW_{tsuna}, propagating in the atmosphere/ionosphere, conserves its speed.

An additional measurement following the Queen Charlotte event (27 October 2012, M_ω 7.8) has been recently detected, proving that the technique can be generalized for smaller events [*Makela et al.*, 2012]. In order to deeply explore the detection by airglow, four cameras were recently installed in Chile, Hawaii, and Tahiti in order to detect the signature of future tsunamis.

The potential echo of the detection of IGW_{tsuna} by airglow, using ground-based or on-board cameras, could change the future of tsunami detection and warning systems.

9.6. PHYSICAL PROPERTIES OF $AW_{Rayleigh}$, IGW_{tsuna}, AND AGW_{epi}

The physical properties of the three components can be summarized as follow: The $AW_{Rayleigh}$, clearly observed in the far field, has a horizontal speed of ≈ 3.5 km/s imposed by the forcing source, nominally the Rayleigh wave. The $AW_{Rayleigh}$ has two main frequencies of 3.68 mHz and 4.44 mHz, corresponding to the coupled normal modes $_0S_{29}$ and $_0S_{37}$, respectively. Consequently, as the propagating wave is an acoustic wave, the vertical velocity of $AW_{Rayleigh}$ is close to the acoustic wave speed, and the corresponding time to reach the ionospheric layers is on the order of 8–15 min (Figure 9.5). Additional frequencies around those major peaks, with small amplitudes, have been observed in the past by *Najita and Yuen* [1979] using Doppler sounders and recently by *Bourdillon et al.* [2014] using OTH radar. Additional spectral analysis of the

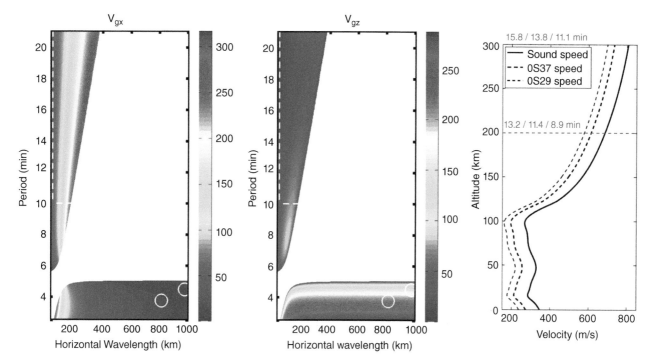

Figure 9.5 (Left) Horizontal and (center) vertical group velocity at the atmospheric bottom boundary, computed following *Watada* [2009]. The white dotted line limits the λ/T region of IGW_{tsuna}; the two white circles highlight the position of Rayleigh waves linked to $_0S_{29}$ and $_0S_{37}$. (Right)Supposing that the vertical evolution of the speed follows the sound speed, the vertical group velocity of $_0S_{29}$ and $_0S_{37}$ is compared with the sound speed, and the propagation time to reach 200mkm and 300 km of altitude is highlighted in red.

signal observed by *Occhipinti et al.* [2010] by Doppler sounder and OTH radar is computed here (Figure 9.2), and clearly shows that the Rayleigh wave signature is still visible at higher frequency (until 50 mHz). Comparison between data and modeling by normal mode summation [*Artru et al.*, 2004; *Occhipinti et al.*, 2010] for a complete stratified 1D Earth with solid and fluid (ocean and atmosphere) parts, clearly proved that our description of $AW_{Rayleigh}$ is satisfying enough (Figure 9.2) for using ionospheric data for magnitude estimation, source localization, as well as moment tensor inversion. Indeed, as showed by Figure 9.2, the amplitude of the main phase is well described by the modeling. At the higher frequencies, the effect of the lithospheric heterogeneity is not reproduced by the modeling showing the limit of the normal mode summation for 1D Earth.

The IGW_{tsuna}, clearly observable in the far field, has horizontal and vertical speeds that are clearly defined by *Occhipinti et al.* [2013] and depending of the physical properties of the tsunami:

$$v_g^h = \frac{k_h N^2 \left(D - k_h^2\right)}{\omega D^2} \qquad v_g^z = \frac{k_z k_h^2 N^2}{\omega D^2}$$

where k_z and k_h are the vertical and horizontal k−vectors, ω the frequency of the tsunami, N is the Brunt Väisäilää

frequency, and $D = k_z^2 + k_h^2 + \left(\frac{N^2}{2g}\right)^2$. Please, note that the vertical group velocity v_g^z has an opposite sign compared to the vertical phase velocity (ω/k_z). This is a typical propagation characteristic of internal gravity waves: the atmosphere falls down by the effect of the gravity, and consequently the vertical phase velocity is negative while the vertical group velocity is positive, generating the upward propagation of the IGW_{tsuna}. Using vertical group velocity v_g^z it is possible to compute the propagation time delay to reach the ionospheric layers (Figure 9.6).

The horizontal phase velocity of the IGW_{tsuna} is the same as the tsunami $\left(\omega/k_h = v_{tsuna} = \sqrt{gH}\right)$. However, this is no longer the case for the horizontal group velocity v_g^z, which is furthermore dispersive. In the case of tsunami, the group and phase velocities are the same. The horizontal group velocity of the generated IGW_{tsuna} is always smaller than the horizontal phase velocity (Figure 9.6).

The horizontal group velocity v_g^h does not play a role in the vertical propagation delay but it is useful to estimate the epicentral distance where the IGW_{tsuna} starts to interact with the ionosphere, and it is also useful to estimate the delay δt between the tsunami propagating at the sea surface and the IGW_{tsuna} propagating in the atmosphere at the altitude z_{iono} (Figure 9.6). The period of a tsunami and

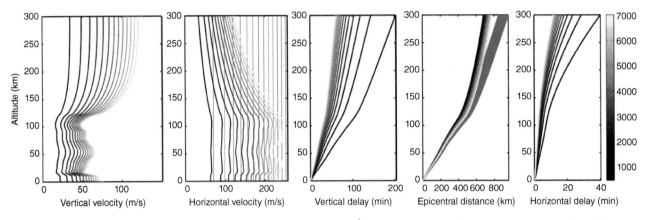

Figure 9.6 From left to right: vertical (v_g^z) and the horizontal (v_g^h) group velocity of the IGW_{tsuna}, generated at different oceanic deep h (m), see gray scale, and a characteristic period T of 10 min. Red lines in v_g^h show the tsunami speed ($v_{tsuna} = \sqrt{hg}$, where g is the gravity acceleration). Consequent vertical propagation delay (third from left); epicentral distances covered by the tsunami (red) and the coupled IGW (gray scale) during the delay spent by the IGW_{tsuna} to reach the altitude shown in the y-axis (fourth from left); last on the right: horizontal delay ∂t between the tsunami at sea level and the coupled IGW_{tsuna} at fixed altitudes (time to cross the same zenith). (Adapted from *Occhipinti et al.* [2013].)

the consequent IGW_{tsuna} is generally between 10 min and 40 min. As the group velocity depends on ω, the period variation (10–40 min) introduces strong variation in the vertical propagation delay (60–240 min to reach, e.g., 200 km of altitude), the epicentral distance (500–2500 km), as well as the delay δt (8–2 min) [*Occhipinti et al.*, 2013].

The AGW_{epi} is the more complex part of the signal observed in the ionosphere. The rupture and the consequent ground displacement at the source area has a rich spectral signature: consequently, both acoustic and gravity waves are simultaneously generated. The preliminary work of *Matsumura et al.* [2011] partially reproduce, by a 2D numerical modeling, some of the observed AGW_{epi}. Anyway, until now, no exhaustive modeling of the ionospheric perturbations at the epicentral area has been performed. Based on the interesting theoretical work of *Watada* [2009], about the acoustic-gravity wave properties at the bottom boundary of the atmosphere, it is possible to generalize the horizontal and vertical speed of the AGW_{epi} at the surface/atmosphere boundary:

$$_{boundary}\, v_g^h = \frac{c_s^2 \omega}{\omega^4 - k_h^2 N^2 c_s^2} k_h \left(\omega^2 - N^2\right)$$

$$_{boundary}\, v_g^z = \frac{c_s^2 \omega}{\omega^4 - k_h^2 N^2 c_s^2} k_z \omega^2$$

where c_s, in addition to the notation described above, is speed of the sound. Again, to avoid misunderstanding, *Watada* [2009] describes a formalism in an isothermal atmosphere to compute group speed of AGW_{epi}. Consequently, the $_{boundary}\, v_g$ is valid only at the surface/

atmosphere boundary. In the λ/T region describing tsunamis, the values of $_{boundary}\, v_g$ are consistent with the values of v_g computed at the surface/atmosphere boundary (Figure 9.5). Figure 9.5 clearly shows the inversion of vertical and horizontal group velocity for acoustic and gravity waves. The $AW_{Rayleigh}$ has a vertical speed slightly slower than the sound speed, consequently it takes around 10–15 min to reach the ionosphere. Consistent with the formalism described above [*Occhipinti et al.*, 2013], IGW_{tsuna} needs 1–2 hours to reach the ionosphere. The AGW_{epi}, generated by the vertical ground displacement at the epicenter, contains a more high-frequency component (0.5–1 Hz) that is transferred to the atmosphere with the sound speed c_s. Consequently, the high-frequency component of AGW_{epi} appears in the ionosphere in around 8 min, as observed by *Astafyeva et al.* [2011, 2013].

I wish to highlight again that today no numerical methods are able to fully describe the AGW_{epi}. Indeed, the work of *Matsumura et al.* [2011] is limited to the 2D neutral atmosphere and partially reproduces the observed waves. The normal mode summation for a complete planet [*Lognonné et al.*, 1998] has been used only to explain the $AW_{Rayleigh}$, and emphasized the limit of a 1D model. As shown in Figure 9.2, the synthetics strictly satisfy the data only for frequency under 10 mHz; higher frequencies are more sensitive to the heterogeneities of the upper crust and need 3D modeling. An appropriate method could be the spectral element method (SEM) usually applied for seismology with excellent results [e.g., *Komatitsch and Tromp*, 2002]. One quick test with SEM clearly shows that the methodology can easily take into account the propagation of acoustic waves (Figure 9.7),

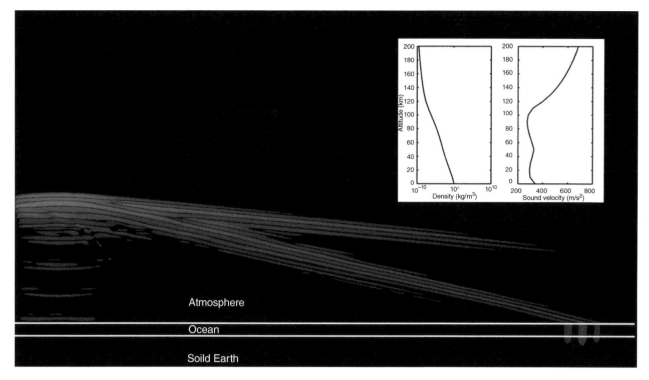

Figure 9.7 Propagation of seismic and acoustic waves in the solid Earth crust (20 km, ρ = 2.5 kg/m^3, V_p = 3.4 km/s, V_s = 1.96 km/s), ocean (10 km, ρ = 1.02 kg/m^3, V_p = 1.45 km/s, V_s = 0.0 km/s), and atmosphere (200 km, ρ, and sound speed in the inset) modeled by SPECFEM2D [e.g., *Komatitsch and Tromp*, 2002]. The propagation of Rayleigh wave in the solid part is transferred quickly to the ocean, then into the atmosphere. We also see the signature in the atmosphere of the P-waves, their amplitude becomes smaller when we are far away from the source, compared to Rayleigh wave signature.

but the propagation of gravity waves and tsunamis is not implemented yet in the SEM. In the other side, the method developed by *Occhipinti et al.* [2006, 2008a, 2013], even if it is totally 3D, propagates only gravity waves and is able to reproduce only the IGW_{tsuna} plus the gravity component of AGW_{epi}. One of the objectives of this work is to push forward the theoretical development of the propagation of acoustic-gravity waves coupled with the vertical displacement of the ground and ocean, in order to completely reproduce the AGW_{epi}. The full understanding and the dense measurement of AGW_{epi} is a key step for future tsunami warning systems.

9.7. CONCLUSION

The fact that humans don't fly is the main reason why seismology studied only solid Earth. If humans were Hawk Men of the gravitating city of Mongo [*Raymond*, 1934], they would know from the beginning that earthquakes and tsunamis disturb the upper fluid envelope, mainly atmosphere/ionosphere.

This work traces the evolution of the idea of the unique planet, where the solid Earth, the ocean, as well as the atmosphere/ionosphere continuously exchange energy,

with particular emphasis during natural hazards like earthquakes; volcanic or, more generally, atmospheric explosions; and tsunamis.

On the basis of theoretical considerations and observational proofs, the natural hazard signatures in the ionosphere are classified here following the source and physical characteristic of the generated waves: the AGW_{epi} is the acoustic-gravity wave generated by rupture and observable at the epicentral area within the first 1000 km; the $AW_{Rayleigh}$ is the acoustic wave generated by propagation of Rayleigh wave; and the IGW_{tsuna} is the internal gravity wave coupled and forced by the tsunami propagation.

Starting from the early atmospheric/ionospheric detection related to seismic events and nuclear explosions during the Cold War, this work highlights the role of Doppler sounders and OTH radar as ionospheric seismometers able to measure the $AW_{Rayleigh}$ (and consequently the lithospheric properties intrinsic in the Rayleigh wave propagation) sounding ionosphere at 200 km of altitude. This work also highlights the capability of GPS, measuring the total electron content (TEC), to image the propagation of the atmospheric/ionospheric signature of the AGW_{epi}, the $AW_{Rayleigh}$, and the IGW_{tsuna}.

The recent Tohoku event in Japan generated a tremendous earthquake and a consequent tsunami. The event generated catastrophic consequences, but the huge amount of collected data strongly helps the earthquake science evolution. The ionosphere sounded by the world's densest GPS network GEONET showed a clear image of the rupture extent observing the AGW_{epi} propagating in the atmosphere/ionosphere overlying Japan (including the epicentral area and the oceanic regions). In the far field, the airglow camera located in Hawaii showed without doubt the IGW_{tsuna} propagation over a region of $180 \times 180 km^2$ in the ionosphere overlying the islands. Never before was a tsunami measured off shore with such high spatial resolution.

Unfortunately, the estimation of the ground and sea surface displacement via ionospheric sounding is still not a direct measurement. Indeed, the ionospheric perturbation is modulated by the magnetic field inclination, and, even if this effect could be controlled by some a priori information, the integrated nature of TEC measured by GPS imposes the solution of an inverse problem to estimate the tsunami amplitude at sea level. Consequently, the measurement, the full understanding, and the accurate quantitative modeling of AGW_{epi}, $AW_{Rayleigh}$, and IGW_{tsuna} are necessary steps to fruitfully use ionospheric monitoring to prevent disasters.

The monitoring by dense GPS networks of the ionosphere overlying the subduction zones could greatly help the estimation of the source extent. Future satellites with on-board airglow cameras could track tsunami propagation over entire oceans before they hit the coast.

With this work I wish to encourage the use of ionospheric seismology as a potential support for future earthquake and tsunami warning systems.

ACKNOWLEDGMENT

This project is supported by the Programme National de Télédétection Spatiale (PNTS), grant n PNTS-2014-07, by the CNES grant SI- EuroTOMO and by the ONR project TWIST. I thanks J.-P. Avouac (Caltech Tectonic Observatory), Jeroen Tromp (Princeton), Claudio Satriano (Institut de Physique du Globe de Paris), and Shingo Watada (Earthquake Research Institute) for friendly and constructive suggestions. I thank A. De Santis and two anonymous reviewers for their constructive remarks. This is IPGP contribution 3573.

REFERENCES

Aframovich, E. L., N. P. Perevalova, and S. V. Voyeikov (2003), Traveling wave packets of total elctron content disturbances as deduced from GPS network data, *J. Atmo. Solar-Terr. Phys.*, 65, 1245–1262.

Aristoteles (524), *Meteorologia Aristotelis*, Book II, 300–380.

Artru, J., T. Farges, and P. Lognonné (2004), Acoustic waves generated from seismic surface waves: Propagation properties determined from Doppler sounding and normal-mode modeling, *Geophys. J. Int.*, 158(3), 1067–1077.

Artru, J., V. Ducic, H. Kanamori, P. Lognonné, and M. Murakami (2005), Ionospheric detection of gravity waves induced by tsunamis, *J. Geophys. Res.*, 160, 840.

Astafyeva, E., L. Rolland, P. Lognonne, K. Khelfi, and T. Yahagi (2013), Parameters of seismic source as deduced from 1 Hz ionospheric GPS data: Case study of the 2011 Tohokuoki event, *J. Geophys. Res. Space Physics*, 118, 59425950, doi:10.1002/jgra.50556.

Astafyeva, E., P. Lognonné, and L. Rolland (2011), First ionosphere images for the seismic slip of the Tohokuoki earthquake, *Geophys. Res. Lett.*, doi:10.1029/2011GL049623.

Balasis G., and M. Mandea (2007), Can electromagnetic disturbances related to the recent great earthquakes be detected by satellite magnetometers? *Special issue Mechanical and Electromagnetic Phenomena Accompanying Preseismic Deformation: From Laboratory to Geophysical Scale*, ed. by K. Eftaxias, T. Chelidze, and V. Sgrigna, *Tectonophysics*, 431, doi:10.1016/j.tecto.2006.05.038.

Balthazor, R. L., and R. J. Moffett (1997), A study of atmospheric gravity waves and travelling ionospheric disturbances at equatorial latitudes, *Ann. Geophysicae*, 15, 1048–1056.

Bilitza, D., C. Koblinsky, S. Zia, R. Williamson, and B. Beckley (1996), The equator anomaly region as seen by the TOPEX/Poseidon satellite, *Adv. Space Res.*, 18, 6, 23–32.

Bolt, B. A. (1964), Seismic air waves from the great 1964 Alaskan earthquake, *Nature, 202*, 1095–1096.

Bourdillon, A., G. Occhipinti, J.-P. Molinié, and V. Rannou (2014), HF radar detection of infrasonic waves generated in the ionosphere by the 28 March 2005 Sumatra earthquake, *J. Atmo. Solar-Terr. Phys.*, 109 (March), 75–79, doi:10.1016/j.jastp.2014.01.008.

Calais, E., and J. B. Minster (1995), GPS detection of ionospheric perturbations following the January, 1994, Northridge earthquake, *Geophys. Res. Lett.*, 22(9), 1045–48.

Calais, E., and J. B. Minster(1996), GPS detection of ionospheric perturbations following a Space Shuttle ascent, *Geophys. Res. Lett.*, 23, 15, 1897–1900.

Calais, E., J. B. Minster, M. A. Hofton, and M. A. H. Hedlin (1998), Ionospheric signature of surface mine blasts from Global Positioning System measurement, *Geophys. J. Int.*, 132,191–202.

Coïsson, P., G. Occhipinti, P. Lognonné, J.-P. Molinié, and L. M. Rolland (2011), Tsunami signature in the ionosphere: A simulation of OTH radar observations, *Radio Sci.*, 46, RS0D20, doi:10.1029/2010RS004603.

DasGupta, A., A. Das, D. Hui, K. K. Bandyopadhyay, and M. R. Sivaraman (2006), Ionospheric perturbation observed by the GPS following the December 26th, 2004 Sumatra-Andaman earthquake, *Earth Planet. Space*, 35, 929–959.

Davies, K., and D. M. Baker, Ionospheric effects observed around the time of the Alaskan earthquake of March 28, 1964, *J. Geophys. Res.*, 70, 2251–2253, 1965.

Donn, W., and E. S. Posmentier, Ground-coupled air waves from the great Alaskan earthquake, *J. Geophys. Res.*, 69, 5357–5361, 1964.

Ducic, V., J. Artru, and P. Lognonné (2003), Ionospheric remote sensing of the Denali Earthquake Rayleigh surface waves, *Geophys. Res. Lett.*, 30, 18, 1951, doi:10.1029/2003 GL017812.

Galvan, D. A., A. Komjathy, M. P. Hickey, P. Stephens, J. Snively, Y. T. Song, M. D. Butala, and A. J. Mannucci (2012), Ionospheric signatures of Tohoku-Oki tsunami of March 11, 2011: Model comparisons near the epicenter, *Radio Science*, 47, 4003, doi 10.1029/2012RS005023.

Garcia, R., F. Crespon, V. Ducic, and P. Lognonné (2005), 3D ionospheric tomography of post-seismic perturbations produced by Denali earthquake from GPS data, *Geophys. J. Int.*, *163*, 1049–1064, doi: 10.1111/j.1365-246X.2005.02775.x.

Garcia, R. F., S. Bruinsma, P. Lognonné, E. Doornbos, and F. Cachoux (2013), GOCE: The first seismometer in orbit around the Earth, *Geophys. Res. Lett.*, *40*, doi:10.1002/grl.50205.

Garcia, R. F., E. Doornbos, S. Bruinsma, and H. Hebert (2014), Atmospheric gravity waves due to the Tohoku-Oki tsunami observed in the thermosphere by GOCE, *J. Geophys. Res. Atmos.*, *119*, doi:10.1002/2013JD021120.

Heki, K., and J. Ping (2005), Directivity and apparent velocity of the coseismic traveling ionospheric disturbances observed with a dense GPS network, *Earth Planet Sci. Lett.*, *236*, 3–4, 15, 845–855.

Hickey, M. P., G. Schubert, and R. L. Walterscheid (2009), The propagation of tsunami-driven gravity waves into the thermosphere and ionosphere, *J. Geophys. Res.*, *114*, A08304, doi:10.1029/2009JA014105.

Hickey, M. P., G. Schubert, and R. L. Walterscheid (2010), Atmospheric airglow fluctuations due to a tsunami-driven gravity wave disturbance, *J. Geophys. Res.*, *115*, A06308, doi:10.1029/2009JA014977.

Hines, C.O. (1972), Gravity waves in the atmosphere, *Nature*, *239*, 73–78.

Iyemori, T., M. Nose, D. S. Han, Y. F. Gao, M. Hashizume, N. Choosakul, H. Shinagawa, Y. Tanaka, M. Utsugi, A. Saito, H. McCreadie, Y. Odagi, and F. X. Yang (2005), Geomagnetic pulsations caused by the Sumatra earthquake on December 26, 2004, *Geophys. Res. Lett.*, *32*, L20807.

Kanamori, H., and J. Mori (1992), Harmonic exitation of mantle Rayleigh waves by the 1991 eruption of Mount Pinatubo, Philippines, *Geophys. Res. Lett.*, *19*, 721–724.

Kelley, M. C., J. J. Makela, B. M. Ledvina, and P. M. Kintner (2002), Observations of equatorial spread?F from Haleakala, Hawaii, *Geophys. Res. Lett.*, *29*(20), 2003, doi:10.1029/2002 GL015509.

Kobayashi, N., K. Nishida, and Y. Fukao (1998), Continuous exitation of Earth's free oscillations, *Nature*, *395*, 357–360.

Komatitsch, D., and J. Tromp (2002), Spectral-element simulations of global seismic wave propagation, I Validation, *Geophys. J. Int.*, *149*(2), 390–412.

Larson, E. W., and G. Ekstrom (2001), Global models of surface wave group velocity, *Pure Ap. Geophys.*, *158*, 13771399.

Lay, T., H. Kanamori, C. J. Ammon, M. Nettles, S. N. Ward, R. C. Aster, S. L. Beck, S. L. Bilek, M. R. Brudzinski, R. Butler, H. R. DeShon, G. Ekstrom, K. Satake, and S. Sipkin (2005), The great Sumatra-Andaman earthquake of 26 December 2004, *Science*, *308*, 1127–1133.

Leonard, R. S., and R. A. Barnes, Jr (1964), Observation of ionospheric disturbances following the Alaskan earthquake, *J. Geophys. Res.*, *70*, 1250–1253.

Le Pichon, A., P. Herry, P. Mialle, J. Vergoz, N. Brachet, M. Garces, D. Drob, and L. Ceranna (2005), Infrasound associated with 2004-2005 large Sumatra earthquakes and tsunami, *Geophys. Res. Lett.*, *32*, L19802.

Link, R., and L. L. Cogger (1988), A reexamination of the O I 6300 nightglow, *J. Geophys. Res.*, *93*(A9), 98839892.

Liu, J., Y. Tsai, K. Ma, Y. Chen, H. Tsai, C. Lin, M. Kamogawa, and C. Lee (2006a), Iono- spheric GPS total electron content (TEC) disturbances triggered by the 26 December 2004 Indian Ocean tsunami, *J. Geophys. Res.*, *111*, A05303.

Liu, J. Y, Y. B. Tsai, S. W. Chen, C. P. Lee, Y. C. Chen, H. Y. Yen, W. Y. Chang, and C. Liu (2006b), Giant ionospheric distrubances excited by the M9.3 Sumatra earthquake of 26 December 2004, *Geophys. Res. Lett.*, *33*, L02103.

Liu, J.-Y., and Y.-Y. Sun (2011), Seismo-traveling ionospheric disturbances of ionograms observed during the 2011 Mw 9.0 Tohoku Earthquake, *Earth Planets Space*, *63*, 897–902.

Liu, J.-Y., C.-H. Chen, C.-H. Lin, H.-F. Tsai, C.-H. Chen, and M. Kamogawa (2011), Ionospheric disturbances triggered by the 11 March 2011 M9.0 Tohoku earthquake, *J. Geophys. Res.*, *116*, doi10.1029/2011JA016761.

Lognonné, P., E. Clévédé, and H. Kanamori (1998), Computation of seismograms and atmospheric oscillations by normal-mode summation for a spherical Earth model with realistic atmosphere, *Geophys. J. Int.*, *135*, 388–406.

Mai, C.-L., and J.-F. Kiang (2009), Modeling of ionospheric perturbation by 2004 Sumatra tsunami, *Radio Sci.*, *44*, RS3011, doi:10.1029/2008RS004060.

Makela, J. J., M. C. Kelley, and R. T. Tsunoda (2009), Observations of midlatitude ionospheric instabilities generating meter scale waves at the magnetic equator, *J. Geophys. Res.*, *114*, A01307, doi:10.1029/ 2007JA012946.

Makela, J. J., P. Lognonné, H. Hébert, T. Gehrels, L. Rolland, S. Allgeyer, A. Kherani, G. Occhipinti, E. Astafyeva, P. Coïsson, A. Loevenbruck, E. Clévédé, M. C. Kelley, and J. Lamouroux (2011), Imaging and modelling the ionospheric response to the 11 March 2011 Sendai Tsunami over Hawaii, *Geophys. Res. Let.*, *38*, L20807, doi:10.1029/2011GL047860.

Makela, J., A. Bablet, and G. Occhipinti (2012), Internal gravity wave induced by the Queen Charlotte event (27 October 2012, Mw 7.8): Airglow observation and modeling, AGU, Fall Meeting 2015, abstract 64936.

Manucci, A. J., B. D. Wilson, and C. D. Edwards (1993), A new method for monitoring the Earth's ionospheric total electron content using GPS global network, Paper presented at *ION GPS-93*, Salt Lake City, September 22–24.

Manucci, A. J., B. D. Wilson, D. N. Yuan, C. H. Ho, U. J. Lindqwister, and T. F. Runge (1998), A global mapping technique for GPS-derived ionospheric electron content measurements, *Radio Sci.*, *33*, 565–582.

Matsumura, M., A. Saito, T. Iyemori, H. Shinagawa, T. Tsugawa, Y. Otsuka, M. Nishioka, and C. H. Chen (2011), Numerical simulations of atmospheric waves excited by the 2011 off the Pacific coast of Tohoku Earthquake, *Earth Planets Space*, *63*, 885889, doi:10.5047/eps.2011.07.015.

Miller, E. S., J. J. Makela, and M. C. Kelley (2009), Seeding of equatorial plasma depletions by polarization electric fields from middle latitudes: Experimental evidence, *Geophys. Res. Lett.*, *36*, L18105, doi:10.1029/ 2009GL039695.

Najita, K., and P. C. Yuen (1979), Long-period oceanic Rayleigh wave group velocity dispersion courve from HF doppler sounding of the ionosphere, *J. Geophys. Res.*, *84*, 1253–1260.

Nawa, K., N. Suda, Y. Fukao, T. Sato, Y. Aoyama, and K. Shibuya (1998), Incessant excitation of the earths free oscillations, *Earth Planet. Space*, *50* (38).

Nishida, K., N. Kobayashi, and Y. Fukao (2000), Resonant oscillations between the solid Earth and the atmosphere, *Science*, *287*, 2244–2246.

Nishitani, N., T. Ogawa, Y. Otsuka, K. Hosokawa, and T. Hori (2011), Propagation of large amplitude ionospheric disturbances with velocity dispersion observed by the SuperDARN

Hokkaido radar after the 2011 off the Pacific coast of Tohoku Earthquake, *Earth Planets Space*, *63*, 891896.

Occhipinti, G., A. Komjathy, and P. Lognonné (2008b), Tsunami detection by GPS: How ionospheric observation might improve the Global Warning System, *GPS World, Feb.*, 50–56,

Occhipinti, G., E. Alam Kherani, and P. Lognonné (2008a), Geomagnetic dependence of ionospheric disturbances induced by tsunamigenic internal gravity waves, *Geophys. J. Int.*, doi: 10.1111/j.1365-246X.2008.03760.x.

Occhipinti, G., L. Rolland, P. Lognonné, and S. Watada (2013), From Sumatra 2004 to Tohoku-Oki 2011: The systematic GPS detection of the ionospheric signature induced by tsunamigenic earthquakes, *J. Geophys. Res. Space Physics*, *118*, doi:10.1002/jgra.50322.

Occhipinti, G., P. Coïsson, J. J. Makela, S. Allgeyer, A. Kherani, H. Hébert, and P. Lognonné (2011), Three-dimensional numerical modeling of tsunami-related internal gravity waves in the Hawaiian atmosphere, *Earth Planet, Science*, *63* (7), 847–851, doi:10.5047/eps.2011.06.051.

Occhipinti, G., P. Dorey, T. Farges, and P. Lognonné (2010), Nostradamus: The radar that wanted to be a seismometer, *Geophys. Res. Lett.*, doi:10.1029/2010GL044009.

Occhipinti, G., P. Lognonné, E. Alam Kherani, and H. Hebert (2006), Three-dimensional wave-form modeling of ionospheric signature induced by the 2004 Sumatra tsunami, *Geophys. Res. Lett.*, *33*, L20104.

Olivier, J. (1962), A summary of observed seismic surface wave dispersion, *Bull. Seismol. Soc. Amer.*, *44*, 127.

Park, J., K. Anderson, R. Aster, R. Butler, T. Lay, and D. Simpson (2005), Global Seismographic Network records the Great Sumatra-Andaman earthquake, *EOS Trans. AGU*, *86*(6), 60.

Peltier, W. R., and C. O. Hines (1976), On the possible detection of tsunamis by a monitoring of the ionosphere, *J. Geophys. Res.*, *81*, 12.

Raymond, A. (1934), Flash Gordon, King Features Syndicate.

Rhie, J., and B. Romanowicz (2004), Excitation of Earths continuous free oscillations by atmosphere-ocean-seafloor coupling, *Nature*, *431*, 552–556.

Rolland, L., G. Occhipinti, P. Lognonné, and A. Loevenbruck (2010), The 29 September 2009 Samoan tsunami in the ionosphere detected offshore Hawaii, *Geophys. Res. Lett.*, *37*, L17191 doi:10.1029/2010GL044479.

Rolland, L. M., P. Lognonné, and H. Munekane (2011a), Detection and modeling of Rayleigh waves induced patterns in the ionosphere, *J. Geophys. Res.*, *116*:A05320. DOI: 10.1029/2010JA016060.

Rolland, L. M., P. Lognonné, E. Astafyeva, E. A Kherani, N. Kobayashi, M. Mann, and H. Munekane (2011b), The resonant response of the ionosphere imaged after the 2011 off the Pacific coast of Tohoku Earthquake, *Earth Planets Space*, *63*, 853–857.

Roult, G., and W. Crawford (2000), Analysis of background free oscillations and how to improve resolution by subtracting the atmospheric pressure signal, *Phys. Earth Planet. Int.*, *121*, 325–338.

Row, R. V. (1966), Evidence of long-period acoustic-gravity waves launched into the *F* region by the Alaskan earthquake of March 28, 1964, *J. Geophys. Res.*, *71*, 343–345.

Row, R. V. (1967), Acoustic-gravity waves in the upper atmosphere due to a nuclear detonation and an earthquake, *J. Geophys. Res.*, *72*, 1599–1610.

Saito, A., T. Tsugawa, Y. Otsuka, M. Nishioka, T. Iyemori, M. Matsumura, S. Saito, C. H. Chen, Y. Goi, and N. Choosakul (2011), Acoustic resonance and plasma depletion detected by GPS total electron content observation after the 2011 off the Pacific coast of Tohoku Earthquake, *Earth Planets Space*, *63*, 863867.

Smith, W., R. Scharroo, V. Titov, D. Arcas, and B. Arbic (2005), Satellite altimeters measure Tsunami, *Oceanography*, *18* (2), 102.

Song, Y. Tony, Chen Ji, L.-L. Fu, Victor Zlotnicki, C. K. Shum, Yuchan Yi, and Vala Hjorleifsdottir (2005), The 26 December 2004 tsunami source estimated from satellite radar altimetry and seismic waves, *Geophys. Res. Lett.*, *32*, L20601.

Suda, N., K. Nawa, and Y. Fukao (1998), Earth's background free oscillations, *Science*, *279*, 2085–2091.

Tanaka, T., T. Ichinose, T. Okuzawa, and T. Shibata (1984), HF-Doppler observations of acoustic waves excited by the Urakawa-Oki earthquake on 21 March 1982, *J. Atmospheric Terrest. Phys.*, *46*, 233–245.

Tanimoto, T., J. Um, K. Nishida, and N. Kobayashi (1998), Earth's continuous oscillations observed on seismically quiet days, *Geophys. Res. Lett.*, *25* (10), 1553–1556.

Tsai, H.F., J.-Y. Liu, C.-H. Lin, and C.-H. Chen (2011), Tracking the epicenter and the tsunami origin with GPS ionosphere observation, *Earth Planets Space*, *63*, 859–862.

Tsugawa, T., A. Saito, Y. Otsuka, M. Nishioka, T. Maruyama, H. Kato, T. Nagatsuma, and K. T. Murata (2011), Ionospheric disturbances detected by GPS total electron content observation after the 2011 off the Pacific coast of Tohoku Earthquake, *Earth Planets Space*, *63*, 875–879.

Vigny, C., W. J. F. Simons, S. Abu, R. Bamphenyu, C. Satirapod, N. Choosakul, C. Subarya, A. Socquet, K. Omar, H. Z. Abidin, and B. A. C. Ambrosius (2005), Insight into the 2004 Sumatra-Andaman earthquake from GPS measurement in southeast Asia, *Nature*, *436*, 201–206.

Watada, S. (1995), Part 1: Near-source acoustic coupling between the atmosphere and the solid Earth during volcanic eruptions, Ph. D. thesis, California Institute of Technology.

Watada, S. (2009), Radiation of acoustic and gravity waves and propagation of boundary waves in the stratified fluid from a time-varying bottom boundary, *J. Fluid Mech.*, *627*, 361–377.

Watada, S., and H. Kanamori (2010), Acoustic resonant oscillations between the atmosphere and the solid earth during the 1991 Mt. Pinatubo eruption, *J. Geophys. Res.*, *115*, B12319, doi:10.1029/2010JB007747.

Webb, S. C. (2007), The Earth's "hum" is driven by ocean waves over the continental shelves, *Nature* 445 (15 February), 754–756, doi:10.1038/nature05536.s.

Wei, S., R. Graves, D. Helmberger, J.-P. Avouac, and J. Jiang (2012), Sources of shaking and flooding during the Tohoku-Oki earthquake: A mixture of rupture styles, *Earth and Planetary Science Letters* 333334 (2012) 91100.

Zurn, W., and R. Widmer, Worldwide observation of bichromatic long-period Rayleigh waves exited during the June 15, 1991, eruption of Mount Pinatubo, *Fire and Nud, eruption of Mount Pinatubo, Philippines*, 615–624. C. Newhall, R. Punongbayan J., Philippine Institute of Volcanology and Seismology, Quezo City and University of Washington Press.

10

Why We Need a New Paradigm of Earthquake Occurrence

Robert J. Geller[1], Francesco Mulargia[2], and Philip B. Stark[3]

ABSTRACT

Like all theories in any branch of physics, theories of the seismic source should be testable (i.e., they should be formulated so that they can be objectively compared to observations and rejected if they disagree). Unfortunately, many widely held theories of the seismic source, such as the elastic rebound paradigm and characteristic earthquake model, and theories for applying them to make probabilistic statements about future seismicity, such as probabilistic seismic hazard analysis (PSHA), either disagree with data or are formulated in inherently or effectively untestable ways. Researchers should recognize that this field is in a state of crisis and search for a new paradigm.

Discussion of economic policy is monopolized by people who learned nothing after being wrong.

Paul Krugman, 2015

10.1. INTRODUCTION

Seismologists can perhaps learn from other branches of physics. At one time, it was widely believed that light traversed a hypothetical physical medium called the "ether." *Michelson and Morley* [1887] set out to measure the velocity of the Earth with respect to the ether, but instead found that the velocity of light was constant. This discrepancy was resolved by Einstein, who discarded the notion of the ether altogether, and instead proposed the special theory of relativity. The point of this episode is that an intuitively appealing concept, such

as the existence of the ether, must be ruthlessly discarded when it is shown to conflict with observed data. In this paper we suggest that the time has come to discard some intuitively appealing concepts regarding the seismic source.

In spite of the Tbytes of data recorded daily by tens of thousands of seismographic and geodetic observatories, there is no satisfactory model of the earthquake source process. We just have semiqualitative models dominated by the elastic rebound paradigm. This paradigm identifies stick-slip on preexisting fault planes as the predominant mechanism for generating large earthquakes, while failing to provide any mechanism for producing the vastly larger number of smaller earthquakes [*Bak*, 1996; *Corral*, 2004]. If this paradigm was correct, it would support the characteristic earthquake (hereafter abbreviated as CE) model, which would provide enough regularity to make it possible to produce reliable hazard maps and might also facilitate reliable predictions of individual large events. On the other hand, the surprise inevitably induced by the occurrence of "unexpected" destructive earthquakes (Figure 10.1) reminds us how wrong this picture is [*Geller*, 2011; *Stein et al.*, 2012; *Kagan and Jackson*, 2013; *Mulargia*, 2013]; also see critical comment by *Frankel* [2013] and reply by *Stein et al.* [2013]. Yet the CE model

[1]*Department of Earth and Planetary Science, Graduate School of Science, University of Tokyo, Hongo, Bunkyo-ku, Tokyo, Japan*

[2]*Department of Physics and Astronomy, University of Bologna, Bologna, Italy*

[3]*Department of Statistics, University of California, Berkeley; Berkeley, California, USA*

Subduction Dynamics: From Mantle Flow to Mega Disasters, Geophysical Monograph 211, First Edition.
Edited by Gabriele Morra, David A. Yuen, Scott D. King, Sang-Mook Lee, and Seth Stein.
© 2016 American Geophysical Union. Published 2016 by John Wiley & Sons, Inc.

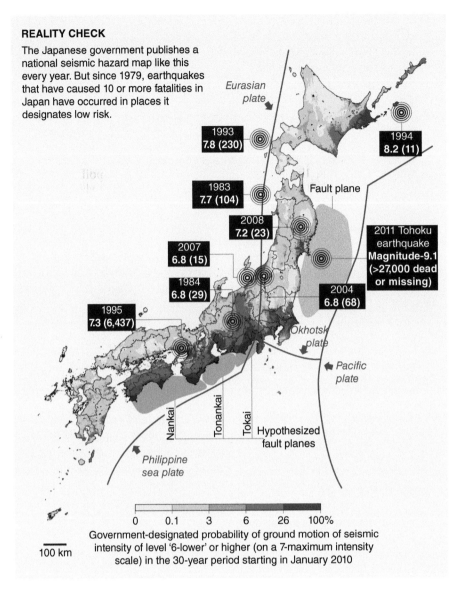

REALITY CHECK

The Japanese government publishes a national seismic hazard map like this every year. But since 1979, earthquakes that have caused 10 or more fatalities in Japan have occurred in places it designates low risk.

Eurasian plate

1993
7.8 (230)

1994
8.2 (11)

1983
7.7 (104)

Fault plane

2008
7.2 (23)

2011 Tohoku earthquake
**Magnitude-9.1
(>27,000 dead or missing)**

2007
6.8 (15)

1984
6.8 (29)

2004
6.8 (68)

1995
7.3 (6,437)

Okhotsk plate

Pacific plate

Nankai

Tonankai

Tokai

Hypothesized fault planes

Philippine sea plate

100 km

0   0.1   3   6   26   100%

Government-designated probability of ground motion of seismic intensity of level '6-lower' or higher (on a 7-maximum intensity scale) in the 30-year period starting in January 2010

Figure 10.1 Comparison of Japanese government hazard map to the locations of earthquakes since 1979 (the year in which it became the official position of the Japanese government that the Tokai region was at high risk of an imminent large earthquake) that caused 10 or more fatalities [*Geller*, 2011].

and other variants based on classical continuum mechanics remain the prevalent framework for modeling earthquakes and earthquake hazards.

The phenomenological laws describing the statistics of earthquake occurrence (i.e., general trends over a large number of events) are well known. These include the law of *Gutenberg and Richter* [1956] (hereafter abbreviated as GR) for the relationship between magnitude and frequency of occurrence, and the Omori law, for the time evolution of the rate of aftershock occurrence [*Utsu et al.*, 1995; *Rundle et al.*, 1995; *Kagan*, 1999; *Scholz*, 2002; *Turcotte et al.*, 2009]. Both prescribe a power-law behavior, which suggests scale-invariance in size and time. Note, however, that above some corner magnitude (typically on the order of $M_w \approx 8$ for shallow events), the

scale invariance breaks down, as the power-law behavior is tapered by a gamma function for large magnitudes [*Kagan*, 1999].

From the point of view of recent developments in nonlinear physics, earthquakes are an example of self-organized criticality [*Bak et al.*, 1987; *Bak and Tang*, 1989]. According to this view, supported by the fact that tiny perturbations such as the dynamic stress induced by earthquakes at distances of more than 100 km [*Felzer and Brodsky*, 2006] and the injection of fluids in the crust at modest pressures [*Mulargia and Bizzarri*, 2014] are capable of inducing earthquakes, the Earth's crust is governed by strongly nonlinear processes and is always on the verge of instability [*Turcotte*, 1997; *Hergarten*, 2002].

In contrast to this nonlinear picture, the classical deterministic model based on continuum mechanics starts from the CE model: a large earthquake is assumed to be the result of the elastic strain released when some limit is exceeded. Under this view, earthquakes are assumed to possess both a characteristic energy and time scale [*Scholz*, 2002; *Turcotte et al.*, 2009].

The elastic rebound paradigm [*Gilbert*, 1884; *Reid*, 1911], which is based on geological intuition, predates the formulation of plate tectonics. This paradigm requires earthquakes to be stick-slip instabilities ruled by the physical laws of rock friction on preexisting planes [*Marone*, 1998; *Scholz*, 2002; *Dieterich*, 2009; *Tullis*, 2009]. Modeling efforts based on this paradigm assume both that frictional behavior at low pressures and velocities can be extrapolated to the higher pressure and velocity regime in actual faulting *in situ*, and that the simple geometries and small sizes that can be handled in laboratory experiments are representative of conditions on actual faults in the Earth. The continuum approach is commonly applied to CE models through the rate- and state-dependent friction (RSF) law, even though experiments suggest that the RSF law no longer holds at high sliding velocities due to various mechanochemical reactions that are induced by the frictional heat [*Tsutsumi and Shimamoto*, 1997; *Goldsby and Tullis*, 2002; *Di Toro et al.*, 2004; *Hirose and Shimamoto*, 2005; *Mizoguchi et al.*, 2006].

10.2. MODELS OF EARTHQUAKES

Earthquake models can be classified into two categories [cf. *Mulargia and Geller*, 2003; *Bizzarri*, 2011]. The first uses as-detailed-as-possible representations of the source and attempts to reproduce the mechanical aspects through the constitutive laws of linear elasticity; the introduction of body force equivalents to dislocations simulates sudden sliding and produces a transient perturbation in the strain field, which—if fast enough—generates seismic waves in the frequency range detectable by seismographs. All continuum models belong to this category [*Ben-Menahem*, 1961; *Ben-Menahem and Toksöz*, 1962; *Kostrov*, 1964, 1966; *Aki*, 1972; *Tse and Rice*, 1986; *Rice*, 1993]. Such models depict earthquakes as fracture plus a slip transient on a preexisting plane of weakness.

The benchmark for such models is their capability to reproduce the observed wavefield, at least in some (usually very limited) temporal and spectral domain. Discriminating among models is difficult, since essentially any model capable of producing a transient slip in the form of a ramp of finite duration will produce acceptable results [*Bizzarri*, 2011]. Research on this category of models is focused on the minutiae, typically one specific earthquake (selected, perhaps, because it caused considerable damage or is of tectonic interest), with a reasonable fit ensured by having a large number of adjustable parameters, and with

considerable variation in estimates of the source parameters depending on the choice of data set and the details of the inversion method [e.g., *Mai and Thingbaijam*, 2014]. While these models can fit waveform data for individual earthquakes, they cannot reproduce the basic empirical laws of earthquake recurrence, that is, the GR and Omori laws, unless further parameters are added to the model. For example, *Hillers et al.* [2007] incorporated *ad hoc* spatial heterogeneity in the characteristic slip distance of the model and obtained a relation between earthquake magnitude and frequency similar to the GR law.

The second category uses simpler models, typically in form of mechanical analogs or cellular automata. Their constitutive laws are simplified to the point that they are often called "toy" models. An example of this category is the Olami-Feder-Christensen model [*Olami et al.*, 1992]. The rationale for such simplifications has its roots in Occam's razor (the principle of choosing the simplest possible model that can explain the data) and aims at reproducing—with a small number of parameters—the dynamical phenomenology that continuum models fail to reproduce (i.e., the GR and Omori laws). On the other hand, this approach is incapable of reproducing the mechanics of the system, that is, the wavefield.

The Burridge-Knopoff (BK) model, a discrete assembly of blocks coupled via springs [*Burridge and Knopoff*, 1967], takes an intermediate approach between these two extremes. It is much simpler than continuum models, but has more parameters than cellular automata models. BK can account for the classical low-velocity friction equations and is generally capable of reproducing the dynamical behavior, but not the radiated wavefield. As such, according to Occam's razor, BK has little advantage over either end-member category of model.

10.3. THE CHARACTERISTIC EARTHQUAKE MODEL

Continuum mechanics models, which are still largely favored in the seismological literature, are basically all variations on the CE model [*Schwartz and Coppersmith*, 1984], which is still widely accepted and used in estimating seismic risk. A typical example is an application to the San Francisco Bay area [*WGCEP*, 1999]. *Schwartz and Coppersmith* [1984] has been cited 750 times (as of 22 Feb. 2015, according to the Web of Science Core Collection database), which demonstrates its high impact on research in this field. These authors (p. 5681) summarize the case for the CE model as follows.

Paleoseismological data for the Wasatch and San Andreas fault zones have led to the formulation of the characteristic earthquake model, which postulates that individual faults and fault segments tend to generate essentially same size or characteristic earthquakes having a relatively narrow range of magnitudes near the maximum…. . Comparisons of earthquake recurrence relationships on both the Wasatch and San Andreas faults based on historical seismicity data and geologic data show that a linear (constant *b* value) extrapolation of the cumulative recurrence curve from the smaller magnitudes leads to gross underestimates of the frequency of occurrence of the large or characteristic earthquakes. Only by assuming a low *b* value in the moderate magnitude range can the seismicity data on small earthquakes be reconciled with geologic data on large earthquakes. The characteristic earthquake appears to be a fundamental aspect of the behavior of the Wasatch and San Andreas faults and may apply to many other faults as well.

Debate over the CE model has been ongoing since it was first proposed. For some early contributions to the controversy see, for example, *Kagan* [1993, 1996] and *Wesnousky* [1994, 1996]. *Kagan* [1993, p. 7] said that evidence cited as supporting CE could be explained by statistical biases or artifacts:

Statistical methods are used to test the characteristic earthquake hypothesis. Several distributions of earthquake size (seismic moment-frequency relations) are described. Based on the results of other researchers as well as my own tests, evidence of the characteristic earthquake hypothesis can be explained either by statistical bias or statistical artifact. Since other distributions of earthquake size provide a simpler explanation for available information, the hypothesis cannot be regarded as proven.

On the other hand, *Wesnousky* [1994, p. 1940] said that the seismicity on some particular faults was in accord with CE:

Paleoearthquake and fault slip-rate data are combined with the CIT-USGS catalog for the period 1944 to 1992 to examine the shape of the magnitude-frequency distribution along the major strike-slip faults of southern California. The resulting distributions for the Newport-Inglewood, Elsinore, Garlock, and San Andreas faults are in accord with the characteristic earthquake model of fault behavior. The distribution observed along the San Jacinto fault satisfies the Gutenberg-Richter relationship. If attention is limited to segments of the San Jacinto that are marked by the rupture zones of large historical earthquakes or distinct steps in fault trace, the observed distribution along each segment is consistent with the characteristic earthquake model. The Gutenberg-Richter distribution observed for the entirety of the San Jacinto may reflect the sum of seismicity along a number of distinct fault segments, each of which displays a characteristic earthquake distribution. The limited period of instrumental recording is insufficient to disprove the hypothesis that all faults will display a Gutenberg-Richter distribution when averaged over the course of a complete earthquake cycle. But, given that (1) the last 5 decades of seismicity are the best indicators of the expected level of small to moderate-size earthquakes in the next 50 years, and (2) it is generally about this period of time that is of interest in seismic hazard and engineering analysis, the answer to the question posed in the title of the article, at least when concerned with practical implementation of seismic hazard analysis at sites along these major faults, appears to be the "characteristic earthquake distribution."

In a recent contribution to the debate, *Kagan et al.* [2012, p. 952] pointed out that CE had failed many statistical tests:

The seismic gap model has been used to forecast large earthquakes around the Pacific Rim. However, testing of these forecasts in the 1990s and later revealed that they performed worse than did random Poisson forecasts [see *Rong et al.*, 2003 and its references]. Similarly, the characteristic earthquake model has not survived statistical testing [see *Jackson and Kagan*, 2011 and its references]. Yet, despite these clear negative results, the characteristic earthquake and seismic gap models continue to be invoked.

On the other hand, *Ishibe and Shimazaki* [2012, p. 1041] argued that CE more appropriately explained the seismicity on some particular faults:

A total of 172 late Quaternary active fault zones in Japan are examined to determine whether the Gutenberg-Richter relationship or the characteristic earthquake model more adequately describes the magnitude-frequency distribution during one seismic cycle. By combining seismicity data for more than 100 active fault zones at various stages in their seismic cycles, we reduced the short instrumental observation period compared to the average recurrence interval. In only 5% of the active fault zones were the number of observed events equal to or larger than the number of events expected by the Gutenberg-Richter relationship. The average and median frequency ratios of the number of observed events to the number of expected events from the Gutenberg-Richter relationship are only 0.33 and 0.06, respectively, suggesting that the characteristic earthquake model more appropriately describes the magnitude-frequency distribution along the late Quaternary active faults during one seismic cycle.

The failure to reach a consensus about the appropriateness, or lack thereof, of the CE model after 30 years of continuing debate reflects poorly on the earthquake science community. It is comparable to what might have happened in a parallel universe in which the physics research community was still arguing about the existence or nonexistence of the "ether" in 1917, 30 years after the Michelson-Morley experiment.

Why is the controversy over CE (needlessly in our opinion) ongoing? Before getting into the details, let us recall the basic problem afflicting research on earthquake occurrence. In most fields of physics a hypothesis is formulated and is then tested against experimental data. On the other hand, in some areas of physics, including seismology (and astrophysics, etc.), realistic experiments are impossible and the data must come from nature, over a long period of time. This can lead to a situation in which data from past events are retrospectively sifted to find patterns that match the investigator's preconceptions, while much larger volumes of data that do not match are ignored [cf. *Mulargia*, 2001].

In many cases, including the works of *Schwartz and Coppersmith* [1984], *Wesnousky* [1994], and *Ishibe and Shimazaki* [2012], the study region is divided into many subregions on the basis of what is already known (or believed) about its geology and seismicity. Thus, effectively, a huge number of parameters have been used to choose the subregions. Even if analyses of the datasets for the various subregions produce results that nominally support the CE model, the retrospective parameter adjustment means that the statistical force of such arguments is virtually nil. Furthermore, it is well known [e.g., *Howell*, 1985] that dividing a large region into many small subregions and then studying the frequency-magnitude distribution in the various subregions is essentially guaranteed to produce artifacts.

The arguments in favor of CE also have other serious flaws. *Ishibe and Shimazaki* [2012], for example, assume that the CE model is correct, and then use the CE model for each subregion to compute the expected number of earthquakes. They then compare this to the GR curve and argue that an insufficient number of small earthquakes is evidence for the CE model. However, this argument is essentially circular. The apparent discrepancy could also easily be the result of occasional larger-than-characteristic earthquakes that break multiple fault segments at one time. Furthermore, GR is a statistical law that should not necessarily be expected to apply strictly to small subsets. In summary, convincing arguments have not been advanced by the CE proponents.

Two popular variants of the CE model—the time- and slip-predictable models—have also been proposed. These models implicitly are based on the intuitive arguments that there must exist definite and fixed limits to either (1) the breaking strength or (2) the ground level of strain [*Shimazaki and Nakata*, 1980]. However, these models were also shown to fail statistical tests [*Mulargia and Gasperini*, 1995].

10.4. UNCHARACTERISTIC EARTHQUAKES

The real test of a model in physics is its ability to predict future data, not its ability to explain past data. The CE model has had no notable successes, but we cite three failures below.

The first and most serious failure is the 2011 Tohoku earthquake. As summarized by *Kanamori et al.* [2006], a Japanese government agency claimed that characteristic M7.5 earthquakes occurred off the Pacific coast of Miyagi Prefecture at about 37 ± 7-year intervals and could be expected to continue to do so. *Kanamori et al.* [2006] pointed out that the previous off-Miyagi events were much less similar to one another than would be expected if the CE model were correct. What actually happened, however [see *Geller*, 2011; *Kagan and Jackson*, 2013], is that, rather than the "next characteristic M7.5 earthquake," the M9 Tohoku earthquake occurred in 2011. This earthquake simultaneously ruptured many segments that government experts had stated would rupture as separate characteristic earthquakes.

The second failure is Parkfield [for details see *Geller*, 1997; *Jackson and Kagan*, 2006]. To make a long story short, *Bakun and Lindh* [1985] said that repeated M6 characteristic earthquakes had occurred at Parkfield, California, and that there was a 95% chance of the next characteristic earthquake occurring by 1993. Their prediction was endorsed by the US government. *Savage* [1993] pointed out that even if an M6 earthquake did occur at Parkfield within the specified time window, it was more reasonable to explain it as a random event than as the predicted characteristic event. An M6 earthquake near Parkfield in 2004 was 11 years "late" and failed to match other parameters of the original prediction.

Another problematic instance involves a cluster of small earthquakes off the coast of northern Honshu, Japan, near the city of Kamaishi. In an abstract for the Fall AGU Meeting (held in December 1999) *Matsuzawa et al.* [1999] made the following prediction:

> ... we found that one of the clusters is dominated by nearly identical and regularly occurring small earthquakes (characteristic events). This cluster is located about 10 km away from the seashore and its depth is around 50 km. By relocating the hypocenters using JMA (Japan Meteorological Agency) catalogue, we confirmed that small earthquakes with JMA magnitude (Mj) of 4.8 ± 0.1 have repeatedly occurred with a recurrence interval of 5.35 ± 0.53 years since 1957; eight characteristic events in total are identified. ... If this cluster really has characteristic nature, the next event with Mj 4.8 will occur there in July, 2000 ± 6 months.

An earthquake similar to the earlier events in the cluster off the coast near Kamaishi occurred on 13 November 2001, and was claimed by *Matsuzawa et al.* [2002, abstract and paragraph 20] as a successful prediction:

> The next event was expected to occur by the end of November 2001 with 99% probability and actually M4.7 event occurred on November 13, 2001.
>
> ...
>
> Since the recurrence interval was so stable, *Matsuzawa et al.* [1999] predicted that the next event would occur by the end of January 2001 with 68% probability and by the end of November 2001 with 99% probability assuming that the recurrence interval would follow the normal distribution.

We now examine the statistical arguments in the final paragraph of the above quotation from *Matsuzawa et al.* [2002]. (Note that *Matsuzawa et al.* [1999] did not explicitly state the probability values of 68% by January 2001 and 99% by the end of November 2001.) The above probability values rest on unrealistic premises, including "that the recurrence interval would follow the normal distribution" and the implicit assumption that interarrival times are random, independent, and identically distributed (IID). The IID assumption implies, in particular, that interarrival times do not depend on magnitudes, which seems to contradict physical models that support CE.

Because a normal random variable can have arbitrarily large negative values, modeling interarrival times as IID normal variables implies that the ninth event in a sequence could occur before the eighth event in the sequence, and, indeed, even before the first event in the sequence: this is not a plausible statistical model. Moreover, even if interarrival times were IID normal variables, the number *Matsuzawa et al.* [2002] computed is not the chance of the next event occurring by November 2001, but something else, as we now explain.

The observed mean interarrival time for the eight events on which the prediction was based [see Figure 3a of *Matsuzawa et al.*, 2002] is 5.35 yr (64.20 mo), with an observed standard deviation of 0.53 yr (6.36 mo). The last event before the announcement of the prediction took place on 11 March 1995. The date 31 January 2001 is 70.74 mo after 11 March 1995, that is, the observed mean interarrival time plus 1.028 times the observed standard deviation. The date 30 November 2001 is 80.69 mo after 11 March 1995, that is, the observed mean interarrival time plus 2.593 times the observed standard deviation. The area under the standard normal curve from $-\infty$ to 1.028 is 64.8%, and the area from $-\infty$ to 2.593 is 99.5%. Hence, we infer that *Matsuzawa et al.* [2002] arrived at the 68% and 99% figures by calculating the area under a normal curve with known mean 5.35 yr and known standard deviation 0.53 yr.

This calculation ignores the difference between *parameters* (the true mean interarrival time and the true standard deviation of interarrival times, assuming that the interarrival times are IID normal random variables) and *estimates* (the observed sample mean of the seven interarrival times between the eight events and the observed sample standard deviation of those seven interarrival times).

The problem addressed by *Matsuzawa et al.* [1999] was to predict the next observation in a sequence of observations of IID normal random variables. The goal was not to estimate the (true) mean recurrence interval, which would involve calculating a standard Student's t confidence interval for the mean from the seven observations of interarrival times. Rather, the goal was to predict when the "next" (ninth) event would occur. The appropriate statistical device for addressing this problem is called a prediction interval. A prediction interval takes into account both the variance of the estimated mean interarrival time and the uncertainty of the time of the single event being predicted. Hence, a prediction interval is significantly longer than a confidence interval for the mean arrival time.

A proper one-sided $1-\alpha$% prediction interval based on seven IID normal observations with sample mean M and sample standard deviation S would go from $-\infty$ to $M + t_{1-\alpha,6} S\sqrt{1+1/7}$, where $t_{1-\alpha,6}$ is the $1-\alpha$ percentile of Student's t distribution with 6 degrees of freedom. Now $t_{.68,6} = 0.492$, $t_{.95,6} = 1.943$, and $t_{.99,6} = 3.143$. Hence, if the interarrival times were IID normal random variables, there would have been a 68% chance that the ninth event would occur within $5.35 + 0.492 \times 0.53 \times 1.069 = 5.629$ yr of the eighth event, that is, by 25 October 2000; and there would have been a 95% chance that the ninth event would occur within $5.35 + 1.943 \times 0.53 \times 1.069 = 6.451$ yr, that is, by 22 August 2001; and there would have been a 99% chance that the ninth event would occur within $5.35 + 3.143 \times 0.53 \times 1.069 = 7.131$ yr, that is, by 27 April 2002. Thus, under the assumption that interarrival times are IID normal random variables, the probability of an

event occurring by the end of January 2001 was 81.3% (not 68%) and the probability of an event by the end of November 2001 was about 97.4% (not 99%).

The ninth event (the event on 13 November 2001) occurred outside the 95% prediction interval; it was in fact on the boundary of the 97.1% prediction interval. The data thus allow us to reject the hypothesis that interarrival times are IID normal random variables at significance level 100% − 97.1% = 2.9%. In summary, this case should not be regarded as supporting the CE model.

Finally, we note in passing that the above discussion has been deliberately oversimplified in one important respect. A more rigorous statistical treatment would consider the conditional probability of a future earthquake as of the time of the announcement of the prediction—December 1999—given that the eighth and most recent event in the cluster occurred in March 1995 and had already been followed by over 4 years with no subsequent event.

10.5. THE PSHA APPROACH TO EARTHQUAKE-HAZARD MODELING

If the CE model were correct, it could be used to reliably and accurately estimate earthquake hazards, but, as discussed above, CE fails to explain observed seismicity data. Nonetheless, CE is used routinely to generate nominally authoritative, albeit effectively meaningless, estimates of seismic hazards through probabilistic seismic hazard analysis (PSHA). PSHA relies on two unvalidated premises: (1) earthquake occurrence is random and follows a parametric model, and (2) the parameters of that model can be estimated well from available data. Neither of these premises appears realistic.

No standard definition of probability allows one to make meaningful statements about "the probability of an earthquake" [cf. *Stark and Freedman*, 2003]. The frequentist approach fails because the "experiment" is not repeatable. The Bayesian approach separates probability from the underlying physical processes, becoming a pronouncement of an individual's degree of belief; moreover, most attempts to quantify "the probability of an earthquake" do not use Bayes' rule properly. The only viable interpretation of PSHA equates probability to a number in a mathematical model intended to describe the process [cf. *Stark and Freedman*, 2003; *Luen and Stark*, 2012]. There are two potential justifications for a probability model and the forecasts derived therefrom: (1) intrinsic physical validity (e.g., as in quantum mechanics or thermodynamics), or (2) predictive validity. Thus, justification for PSHA implicitly hinges either on the physical validity of the CE model or the ability of PSHA to make reliable and accurate predictions. As discussed above, the CE model has not been validated and conflicts with observed data. Furthermore, hazard maps based on PSHA have failed as a predictive tool [*Stein et al.*, 2012].

Present discussion of testing PSHA is phenomenological and ignores the physics. Given the formidable complexity of the problem, progress may require the collaboration of many groups, which the Collaboratory Study for Earthquake Predictability (CSEP) of the Southern California Earthquake Center (SCEC) is attempting to facilitate [*Zechar et al.*, 2010]. But there is a more fundamental question than collaboration and consensus building: namely, does the time interval covered by available data allow accurate parameter estimation and model testing? We suggest that the answer is no, as data from seismicity catalogs, macroseismic data [*Mulargia*, 2013], and geological studies of faults all fall far short of what is required.

In summary, the basic premises of PSHA lack foundation, and it has not been validated empirically or theoretically. PSHA therefore should not be used as the basis for public policy.

10.6. PROBABILISTIC FORECASTS OF INDIVIDUAL EARTHQUAKES

Finally, we consider studies that argue for increased probability of individual earthquakes in the aftermath of a large earthquake. *Parsons et al.* [2000, p. 661] calculated:

> We calculate the probability of strong shaking in Istanbul, an urban center of 10 million people, from the description of earthquakes on the North Anatolian fault system in the Marmara Sea during the past 500 years and test the resulting catalog against the frequency of damage in Istanbul during the preceding millennium. Departing from current practice, we include the time-dependent effect of stress transferred by the 1999 moment magnitude M = 7.4 Izmit earthquake to faults nearer to Istanbul. We find a 62 ± 15% probability (one standard deviation) of strong shaking during the next 30 years and 32 ± 12% during the next decade.

And *Toda and Stein* [2013, p. 2562] forecasted:

> We therefore fit the seismicity observations to a rate/state Coulomb model, which we use to forecast the time-dependent probability of large earthquakes in the Kanto seismic corridor. We estimate a 17% probability of a M ≥ 7.0 shock over the 5 year prospective period 11 March 2013 to 10 March 2018, two-and-a-half times the probability had the Tohoku earthquake not struck.

There is apparently no definition of "probability" for which this statement is meaningful: it is statistical gibberish. The probability is assumed, not inferred and estimated.

Statements of the above type should be abjured, because they cannot be tested objectively. If a large earthquake occurs in the specified space-time window it might be claimed as a success, but as *Savage* [1993] pointed out it could also be an event that just happened anyway. Also, note that the above statements do not precisely specify the bounds of the spatial area of the prediction, the magnitude scale being used, and so on. This invites potential controversy over whether or not the predictions were successful, of the type which has often accompanied controversies over the evaluation of earthquake prediction claims [e.g., *Geller*, 1996].

10.7. CONCLUSION

The CE paradigm is clearly at odds with reality, but it inexplicably continues to be widely accepted. As CE serves as the justification for PSHA, PSHA should not be used operationally in public policy. The "inconvenient truth" is that there is no satisfactory physical model for earthquakes and none is in sight. However, this is scarcely acknowledged, with a few notable exceptions [e.g., *Ben-Menahem*, 1995]. A new paradigm of earthquake occurrence is clearly required. However, the fact that we don't yet have a new paradigm in no way justifies continued reliance on the old and discredited elastic rebound paradigm and CE model to estimate seismic hazards.

REFERENCES

Aki, K. (1972), Earthquake mechanism, *Tectonophysics*, *13*, 423–446.

Bak, P. (1996), *How Nature Works: The Science of Self-Organized Criticality*, Copernicus, New York, pp. 85–88.

Bak, P., and C. Tang (1989), Earthquakes as a self-organized critical phenomenon, *J. Geophys. Res.*, *94*, 15,635–15,637.

Bak, P., C. Tang, and K. Wiesenfeld (1987), Self-organized criticality—an explanation of 1/f noise, *Phys. Rev. Lett.*, *59*, 381–384.

Bakun, W. H., and A. G. Lindh (1985), The Parkfield, California, earthquake prediction experiment, *Science*, *229*, 619–624.

Ben-Menahem, A. (1961), Radiation of seismic surface-waves from finite moving sources, *Bull. Seismol. Soc. Am.*, *51*, 401–435.

Ben-Menahem, A. (1995), A concise history of mainstream seismology: Origins, legacy, and perspectives, *Bull. Seismol. Soc. Am.*, *85*, 1202–1225.

Ben-Menahem, A., and M. N. Toksöz (1962), Source mechanism from spectra of long-period seismic surface-waves, 1. The Mongolian earthquake of December 4, 1957, *J. Geophys. Res.*, *67*, 1943–1955.

Bizzarri, A. (2011), On the deterministic description of earthquakes, *Rev. Geophys.*, *49*, RG3002, doi: 10.1029/2011RG000356.

Burridge, R., and L. Knopoff (1967), Model and theoretical seismicity, *Bull. Seismol. Soc. Am.*, *57*, 341–371.

Corral, A. (2004), Universal local versus unified global scaling laws in the statistics of seismicity, *Physica A*, *340*, 590–597.

Dieterich, J. H. (2009), Applications of rate- and state-dependent friction to models of fault slip and earthquake occurrence, in H. Kanamori (ed.), *Earthquake Seismology, Treatise on Geophysics*, *4*, 107–129, Elsevier, Amsterdam.

Di Toro, G., D. L. Goldsby, and T. E. Tullis (2004), Natural and experimental evidence of melt lubrication of faults during earthquakes, *Nature*, *427*, 436–439.

Felzer, K. R., and E. E. Brodsky (2006), Decay of aftershock density with distance indicates triggering by dynamic stress, *Nature*, *441*, 735–738.

Frankel, A. (2013), Comment on "Why earthquake hazard maps often fail and what to do about it," *Tectonophysics*, *592*, 200–206.

Geller, R. J. (1996), VAN: A Critical evaluation, in J. H. Lighthill (ed.), *A Critical Review of VAN*, 155–238, World Scientific, Singapore.

Geller, R. J. (1997), Earthquake prediction: A critical review, *Geophys. J. Int.*, *131*, 425–450.

Geller, R. J. (2011), Shake-up time for Japanese seismology, *Nature*, *472*, 407–409.

Gilbert, G. K. (1884), A theory of the earthquakes of the Great Basin, with a practical application, *Am. J. Sci., ser. 3*, *27*, 49–54.

Goldsby, D. L., and T. E. Tullis (2002), Low frictional strength of quartz rocks at subseismic slip rates, *Geophys. Res. Lett.*, *29*, doi:10.1029/ 2002GL015240.

Gutenberg, B., and C. F. Richter (1956), Magnitude and energy of earthquakes, *Annali di Geofisica*, *9*, 1–15 (republished in *Annals of Geophysics*, *53*, 7–12, 2010).

Hergarten, S. (2002), *Self-Organized Criticality in Earth Systems*, Springer, Berlin.

Hillers, G., P. M. Mai, Y. Ben-Zion, and J. P. Ampuero (2007), Statistical properties of seismicity of fault zones at different evolutionary stages, *Geophys. J. Int.*, *169*, 515–533.

Hirose, T., and T. Shimamoto (2005), Growth of molten zone as a mechanism of slip weakening of simulated faults in gabbro during frictional melting, *J. Geophys. Res.*, *110*, doi:10.1029/2004JB003207.

Howell, B. F. Jr. (1985), On the effect of too small a data base on earthquake frequency diagrams, *Bull. Seismol. Soc. Am.*, *75*, 1205–1207.

Ishibe, T., and K. Shimazaki (2012), Characteristic earthquake model and seismicity around late Quaternary active faults in Japan, *Bull. Seismol. Soc. Am.*, *102*, 1041–1058.

Jackson, D. D., and Y. Y. Kagan (2006), The 2004 Parkfield earthquake, the 1985 prediction, and characteristic earthquakes: Lessons for the future, *Bull. Seismol. Soc. Am.*, *96*, S397–S409.

Jackson, D. D., and Y. Y. Kagan (2011), Characteristic earthquakes and seismic gaps, in H. K. Gupta (ed.), *Encyclopedia of Solid Earth Geophysics*, 37–40, Springer, doi: 10.1007/978-90-481-8702-7.

Kagan, Y. Y. (1993), Statistics of characteristic earthquakes, *Bull. Seismol. Soc. Am.*, *83*, 17–24.

Kagan, Y. Y. (1996), Comment on "The Gutenberg-Richter or characteristic earthquake distribution, which is it?" by S. G. Wesnousky, *Bull. Seismol. Soc. Am.*, *86*, 275–284.

Kagan, Y. Y. (1999), Universality of the seismic moment-frequency relation, *Pure Appl. Geophys.*, *55*, 537–573.

Kagan, Y. Y., and D. D. Jackson (2013), Tohoku earthquake: A surprise? *Bull. Seismol. Soc. Am.*, *103*, 1181–1194.

Kagan, Y. Y., D. D. Jackson, and R. J. Geller (2012), Characteristic earthquake model, 1884–2011, R.I.P., *Seismol. Res. Lett.*, *83*, 951–953.

Kanamori, H., M. Miyazawa, and J. Mori (2006), Investigation of the earthquake sequence off Miyagi prefecture with historical seismograms, *Earth Planets and Space*, *58*, 1533–1541.

Kostrov, B. V. (1964), Self-similar problems of propagation of shear cracks, *J. Appl. Math. Mech.*, *28*, 1077–1087.

Kostrov, B. V. (1966), Unsteady propagation of longitudinal shear cracks, *J. Appl. Math. Mech.*, *30*, 1241–1248.

Krugman, P. (2015), Cranking up for 2016, *The New York Times*, Feb. 21.

Luen, B., and P. B. Stark (2012), Poisson tests of declustered catalogs, *Geophys. J. Int.*, *189*, 691–700.

Mai, P. M., and K. K. S. Thingbaijam (2014), SRCMOD: An online database of finite-fault rupture models, *Seismol. Res. Lett.*, *85*, 1348–1357.

Marone, C. (1998), Laboratory-derived friction laws and their application to seismic faulting, *Ann. Rev. Earth Planet. Sci.*, *26*, 643–696.

Matsuzawa, T., T. Igarashi, and A. Hasegawa (1999), Characteristic small-earthquake sequence off Sanriku, Japan, *Eos Trans. AGU*, *80*(46), Fall Meet. Suppl., Abstract S41B-07, F724.

Matsuzawa, T., T. Igarashi, and A. Hasegawa (2002), Characteristic small-earthquake sequence off Sanriku, northeastern Honshu, Japan, *Geophys. Res. Lett.*, *29*, 1543, doi:10.1029/2001GL014632.

Michelson, A. A., and E. W. Morley (1887), On the relative motion of the Earth and the luminiferous ether, *Am. J. Sci., ser. 3*, *34*, 333–345.

Mizoguchi, K., T. Hirose, T. Shimamoto, and E. Fukuyama (2006), Moisture-related weakening and strengthening of a fault activated at seismic slip rates, *Geophys. Res. Lett.*, *33*, doi:10.1029/2006GL026980.

Mulargia, F. (2001), Retrospective selection bias (or the benefit of hindsight), *Geophys. J. Int.*, *145*, 1–16.

Mulargia, F. (2013), Why the next large earthquake is likely to be a big surprise, *Bull. Seismol. Soc. Am.*, *103*, 2946–2952.

Mulargia, F., and A. Bizzarri (2014), Anthropogenic triggering of large earthquakes, *Nature Sci. Rep.*, *4*, 6100, 1–7, doi: 10.1038/srep06100.

Mulargia, F., and P. Gasperini (1995), Evaluation of the applicability of the time-and slip-predictable earthquake recurrence models to Italian seismicity, *Geophys. J. Int.*, *120*, 453–473.

Mulargia, F., and R. J. Geller (eds.) (2003), *Earthquake Science and Seismic Risk Reduction*, Kluwer, Dordrecht, The Netherlands.

Olami, Z., H. J. S. Feder, and K. Christensen (1992), Self-organized criticality in a continuous, nonconservative cellular automaton modeling earthquakes, *Phys. Rev. Lett.*, *68*, 1244–1247.

Parsons, T., S. Toda, R. S. Stein, A. Barka, and J. H. Dieterich (2000), Heightened odds of large earthquakes near Istanbul: an interaction-based probability calculation, *Science*, *288*, 661–665.

Reid, H. F. (1911), The elastic-rebound theory of earthquakes, *Univ. Calif. Publ. Bull. Dept. Geol. Sci.*, *6*, 413–444.

Rice, J. R. (1993), Spatio-temporal complexity of slip on a fault, *J. Geophys. Res.*, *98*, 9885–9907.

Rong, Y.-F., D. D. Jackson, and Y. Y. Kagan (2003), Seismic gaps and earthquakes, *J. Geophys. Res.*, *108*, doi:10.1029/ 2002JB002334.

Rundle, J. B., W. Klein, S. Gross, and D. L. Turcotte (1995), Boltzmann fluctuations in numerical simulations of nonequilibrium threshold systems, *Phys. Rev. Lett.*, *75*, 1658–1661.

Savage, J. C. (1993), The Parkfield prediction fallacy, *Bull. Seismol. Soc. Am.*, *83*, 1–6.

Scholz, C. H. (2002), *The Mechanics of Earthquakes and Faulting*, 2nd ed., Cambridge Univ. Press, New York.

Schwartz, D. P., and K. J. Coppersmith (1984), Fault behavior and characteristic earthquakes — examples from the Wasatch and San Andreas fault zones, *J. Geophys. Res.*, *89*, 5681–5698.

Shimazaki, K., and T. Nakata (1980), Time-predictable recurrence model for large earthquakes, *Geophys. Res. Lett.*, *7*, 279–282.

Stark, P. B., and D. Freedman (2003), What is the chance of an earthquake?, in F. Mulargia and R. J. Geller (eds.), *Earthquake Science and Seismic Risk Reduction*, NATO Science Series IV: Earth and Environmental Sciences, *32*, 201–213, Kluwer, Dordrecht, The Netherlands.

Stein, S., R. J. Geller, and M. Liu (2012), Why earthquake hazard maps often fail and what to do about it, *Tectonophysics*, *562–563*, 1–25.

Stein, S., R. J. Geller, and M. Liu (2013), Reply to comment by Arthur Frankel on "Why earthquake hazard maps often fail and what to do about it," *Tectonophysics*, *592*, 207–209.

Toda S., and R. S. Stein (2013), The 2011 M = 9.0 Tohoku oki earthquake more than doubled the probability of large shocks beneath Tokyo, *Geophys. Res. Lett.*, *40*, 2562–2566, doi:10.1002/grl.50524.

Tse, S. T., and J. R. Rice (1986), Crustal earthquake instability in relation to the depth variation of frictional slip properties, *J. Geophys. Res.*, *91*, 9452–9472, doi:10.1029/JB091iB09p09452.

Tsutsumi, A., and T. Shimamoto (1997), High-velocity frictional properties of gabbro, *Geophys. Res. Lett.*, *24*, 699–702.

Tullis, T. E. (2009), Friction of rock at earthquake slip rates, in H. Kanamori (ed.), *Earthquake Seismology, Treatise on Geophysics*, *4*, 131–152, Elsevier, Amsterdam.

Turcotte, D. L. (1997), *Fractals and Chaos in Geology and Geophysics*, 2nd ed., Cambridge Univ. Press, New York.

Turcotte, D. L., R. Scherbakov, and J. B. Rundle (2009), Complexity and earthquakes, in H. Kanamori (ed.), *Earthquake Seismology, Treatise on Geophysics*, *4*, 675–700, Elsevier, Amsterdam.

Utsu T., Y. Ogata, and R. S. Matsu'ura (1995), The centenary of the Omori formula for a decay law of aftershock activity, *J. Phys. Earth*, *43*, 1–33.

Wesnousky, S. G. (1994), The Gutenberg-Richter or characteristic earthquake distribution, which is it? *Bull. Seismol. Soc. Am.*, *84*, 1940–1959.

Wesnousky, S. G. (1996), Reply to Yan Kagan's comment on "The Gutenberg-Richter or characteristic earthquake distribution, which is it?" *Bull. Seismol. Soc. Am.*, *86*, 286–291.

WGCEP (Working Group on California Earthquake Probabilities) (1999), *Earthquake Probabilities in the San Francisco Bay Region: 2000 to 2030—A Summary of Findings*, U.S. Geological Survey Open-File Report 99-517. http://pubs.usgs.gov/of/1999/0517/.

Zechar, J. D., D. Schorlemmer, M. Liukis, J. Yu, F. Euchner, P. J. Maechling, and T. H. Jordan (2010), The collaboratory for the study of earthquake predictability perspective on computational earthquake science, *Concurrency and Computation: Practice and Experience*, *22*, 1836–1847, doi:10.1002/cpe.1519.

INDEX

Accretionary prisms, 81, 83, 84f, 86, 86f, 87, 93
Accretionary wedge, 1, 3
    deformation pattern of, 81, 86–88
Acoustic-gravity wave from epicenter (AGW_{epi})
    coupling mechanism with, 170, 170f
    physical properties of, 176–79, 177f, 178f
    Tohoku earthquake and tsunami with, 174, 175f, 176f
Adakite, 69
Adakitic arc magmatism
    calculated temperatures in model for, 74f
    global plate reconstruction model with, 70
    late Jurassic–Cenozoic, 70, 70f
    mantle plume with, 69, 75–77, 75f
    mechanisms for, 73, 77
    migration from South China to Russia, 69
    numerical models for, 71–73, 72f
    partial melting of subducted oceanic crust with, 71
    plate reconstruction model with, 70f, 71, 73, 76, 77
    southwest-to-northeast migration of, 70, 70f, 75f
    tectonic model for genesis of, 69–77, 70f, 72f, 74f, 75f
    three-dimensional subduction model with, 71
    trench-normal convergence rates with, 71, 72f, 73
    two-dimensional subduction model with, 71
Adakitic rocks, 54f
    migration of intracontinental, 76
AGW_{epi}. See Acoustic-gravity wave from epicenter
Akiyoshi belt, geologic and tectonic subdivisions and, 39f
Alaska earthquake, $AW_{Rayleigh}$ generated by, 172
Aseismic zone (ASZ), 157–58, 161
    deviatoric stress and, 157, 163
    location between two seismically active zones, 165
Asia. See also Cretaceous East Asia
    Cenozoic volcanism in central, 97–110, 98f, 100f, 101f, 103t, 104f, 105f, 106f, 107f, 109f
    distribution of volcanism in, 101f
    Eurasia plate, 8f
    slab tomography lying below, 100f
    volcanism, implications in central and northeastern, 108
Asian tectonics
    map of northeastern, 22f
    overview, 3–4, 3f
    seismic stations with, 8, 8f
ASZ. See Aseismic zone
A-type granitoids, migration of intracontinental, 69, 75–76
Augen gneisses, 45f
$AW_{Rayleigh}$. See Rayleigh wave

Baikal lake, 99
Bangong-Nujiang suture, 54f, 59f
Basement topography, 81, 84, 85f, 93
Bibong eclogite, 24f
    garnet and omphacite in, 26f
Big Mantle Wedge hypothesis, 97, 99, 110
Bimaterial shear heating
    earthquake distributions in slabs from, 157–65, 158f
    elastic energy density with, 161–62, 163f
    elastic shear modulus with, 158, 158f, 159, 160
    equations for model of, 159
    input parameters for model of, 159t
    model settings for, 158f, 159–60, 159t
    numerical method for model of, 158f, 159–60, 159t
    phase loop in, 157–65, 158f, 160f, 161f, 162f, 163f, 164f
BK model. See Burridge-Knopoff model
Bureya, 51f
Burridge-Knopoff (BK) model, 185
Byerlee-type plastic yielding, 119

Carboneferous-Permian subduction complex, 29f
Cathaysia block, 29t, 51t
Caustic waveform modeling
    data sources, 6–9, 7f, 8f, 9t
    diagram of, 7f
    dynamic implication for, 12–15, 14f, 15f
    dynamic simulation of slab thickening in, 11–12, 12f
    earthquakes used in, 9t
    essential features of, 11–12
    for long slab thickening, 5–15
    mantle transition zone in, 5, 6, 8–9, 10f, 13, 14f, 15
    P wave in, 4, 6, 8f, 9, 9t, 14f
    seismic triplication as cause in, 6–8, 7f
    SH wave in, 8f, 9, 9t
    slab thickness estimate uncertainty in, 11, 11f
    S wave in, 4, 6, 7f, 9
    thermal expansivity in, 2f, 4, 11, 13
    velocity models obtained from, 10f
CE. See Characteristic earthquake model
Cenozoic era, missing Pacific slab from, 5
Cenozoic volcanism
    Big Mantle Wedge hypothesis in explaining, 97, 99, 110
    buoyancy driven viscous instabilities with, 100–101
    in central Asia, 97–110, 98f, 100f, 101f, 103t, 104f, 105f, 106f, 107f, 109f
    cluster models, results for, 102–6, 103t, 104f, 105f
    continental rifting with, 99
    cylinder models, results for, 106–7, 106f, 107f

Subduction Dynamics: From Mantle Flow to Mega Disasters, Geophysical Monograph 211, First Edition.
Edited by Gabriele Morra, David A. Yuen, Scott D. King, Sang-Mook Lee, and Seth Stein.
© 2016 American Geophysical Union. Published 2016 by John Wiley & Sons, Inc.